FUNDAMENTALS OF
AQUATIC ECOLOGY

FUNDAMENTALS OF
AQUATIC ECOLOGY

Edited by R.S.K. Barnes
St Catharine's College and Department of Zoology,
University of Cambridge, UK

and

K.H. Mann
Department of Fisheries and Oceans,
Bedford Institute of Oceanography,
Nova Scotia, Canada

Second Edition of
Fundamentals of Aquatic Ecosystems

OXFORD

BLACKWELL SCIENTIFIC PUBLICATIONS

LONDON EDINBURGH BOSTON

MELBOURNE PARIS BERLIN VIENNA

© 1980, 1991 by
Blackwell Scientific Publications
Editorial offices:
Osney Mead, Oxford OX2 0EL
25 John Street, London WC1N 2BL
23 Ainslie Place, Edinburgh EH3 6AJ
238 Main Street, Cambridge
    Massachusetts 02142, USA
54 University Street, Carlton
    Victoria 3053, Australia

Other Editorial Offices:
Librairie Arnette SA
2, rue Casimir-Delavigne
75006 Paris
France

Blackwell Wissenschafts-Verlag
Meinekestrasse 4
D-1000 Berlin 15
Germany

Blackwell MZV
Feldgasse 13
A-1238 Wien
Austria

First published 1980
Reprinted 1982, 1987, 1988
Second edition 1991
Reprinted 1993

Set by Excel Typesetters Company, Hong Kong
Printed and bound in Great Britain at
The University Press, Cambridge

DISTRIBUTORS
Marston Book Services Ltd
PO Box 87
Oxford OX2 0DT
(*Orders*: Tel: 0865 791155
            Fax: 0865 791927
            Telex: 837515)

USA
Blackwell Scientific Publications, Inc.
238 Main Street
Cambridge, MA 02142
(*Orders*: Tel: 800 759-6102
            617 876-7000)

Canada
Oxford University Press
70 Wynford Drive
Don Mills
Ontario M3C 1J9
(*Orders*: Tel: 416 441-2941)

Australia
Blackwell Scientific Publications Pty Ltd
54 University Street
Carlton, Victoria 3053
(*Orders*: Tel: 03 347-5552)

British Library
Cataloguing in Publication Data

Fundamentals of aquatic ecology. – 2nd ed.
    1. Aquatic environments. Natural history
    I. Barnes, R.S.K. *Fundamentals of aquatic ecosystems*
    II. Barnes, R.S.K. (Richard Stephen Kent) *1944–*
    III. Mann, K.H. (Kenneth Henry) *1923–*
    574.5263
    ISBN 0–632–02983–8

Library of Congress
Cataloging-in-Publication Data

Fundamentals of aquatic ecology/edited by
    R.S.K. Barnes and K.H. Mann. – 2nd ed.
        p.   cm.
    Rev. ed. of: Fundamentals of
    aquatic ecosystems, 1980.
    Includes index.
    ISBN 0–632–02983–8
    1. Aquatic ecology. I. Barnes, R.S.K.
    (Richard Stephen Kent)
    II. Mann, K.H. (Kenneth Henry), *1923–*
    III. Title: Fundamentals of aquatic ecosystems.
    QH541.5.W3F86   1991
    574.5′263—dc20

Contents

Contributors

R.S.K. BARNES *Department of Zoology, University of Cambridge, Cambridge, UK*

J.H.R. GEE *Department of Zoology, University College of Wales, Aberystwyth, UK*

B.T. HARGRAVE *Department of Fisheries and Oceans, Bedford Institute of Oceanography, Dartmouth, Nova Scotia, Canada*

R.N. HUGHES *School of Biological Sciences, University College of North Wales, Bangor, UK*

K.H. MANN *Department of Fisheries and Oceans, Bedford Institute of Oceanography, Dartmouth, Nova Scotia, Canada*

D.W. SCHINDLER *Departments of Zoology and Botany, University of Alberta, Edmonton, Alberta, Canada*

C.R. TOWNSEND *Department of Zoology, University of Otago, Dunedin, New Zealand*

I. VALIELA *Boston University Marine Program, Marine Biological Laboratory, Woods Hole, Massachusetts, USA*

M.J. WINTERBOURN *Department of Zoology, University of Canterbury, Christchurch, New Zealand*

Preface

In origin, this book is the second edition of our earlier *Fundamentals of Aquatic Ecosystems* (1980). We decided, however, not simply to update the first edition, but to modify this earlier book in two major respects — the effect of this was to cause the second edition to depart so far from the first as to become a completely different and new book: hence the new title. Firstly, we felt that integration of the various chapters into a single book — always a problem in multi-authored works — would be improved if fewer authors were involved in the project; and, secondly, we endeavoured to make each chapter more inclusive of all types of aquatic habitat and more truly 'aquatic' by having it drafted by two authors, one a marine biologist and the other concerned with continental waters. If a chapter was written in the first instance by someone mainly with marine interests (as half were), then it was ghosted/refereed by a member of the writing team specializing in freshwater systems, and vice versa. We hope that both these innovations have led to a more balanced approach and to a book that describes exactly what its title indicates.

We are most grateful to our seven colleagues for sticking closely to the briefs that we supplied, in respect of both writing and ghosting their various chapters, and to Simon Rallison and Susan Sternberg of Blackwell Scientific Publications for support and encouragement.

R.S.K.B.

K.H.M.

PART 1
INTRODUCTION

1
Organisms and Ecosystems

K.H. MANN

1.1 INTRODUCTION

The term aquatic habitat covers a whole spectrum from the world's oceans to the bays and estuaries around their fringes, from major lakes (including inland salt seas) to small ponds and to the marshes and swamps that are often found associated with them. It also includes rivers characterized by a one-way flow from the uplands, where they were fed by rainfall and springs, to their junctions with the sea at estuaries. At first sight these habitats may seem so diverse that it is not sensible to try to discuss their ecology in one volume. However, we hope to show that there are many processes that occur in all these types of environment, and that there is a fundamental unity between them.

1.2 THE SPECIAL PROPERTIES OF WATER

All living organisms contain a large proportion of water, and life as we know it would not be possible if it were not for the special properties of that water. For example, its specific heat is very high; that is to say, for a given input of heat, its temperature changes relatively little. Pure water is taken as the standard, so that 1 calorie (4.17 joules) raises the temperature of 1 gram of water by 1 degree Celsius (i.e. the specific heat is $4.17\,\mathrm{J\,g^{-1}\,°C^{-1}}$). Most other substances in the biosphere, such as the common rocks, have a temperature rise of about 5°C for an input of 4.17 J (i.e. a specific heat of $0.83\,\mathrm{J\,g^{-1}\,°C^{-1}}$). Hence, water forms a valuable buffer against changing environmental temperature, both for the water within organisms and for the aquatic environment.

Density relationships are also important. Again, pure water is the standard, with a maximum density of $1000\,\mathrm{kg\,m^{-3}}$. It reaches this density at a temperature close to 4°C. As it is warmed above this temperature it becomes lighter, but it also becomes lighter as it cools between 4°C and its freezing point, at 0°C. This is of critical importance for preserving an ice-free environment in a lake or pond. Suppose the weather is getting colder and the surface of a lake is cooling from about 10°C to 4°C. The density of the surface water is increasing so it sinks through the layers below and convective mixing occurs. The lake may eventually have a uniform temperature and density from top to bottom. If the surface cooling process continues, the surface water may drop to 3°C, but instead of becoming more dense, the water now becomes less dense, and floats at the surface. Convective mixing no longer occurs and freezing of the lower layers is delayed. Once the surface temperature reaches 0°C, ice forms, with a density about 8% lower than that of the water. It remains at the surface and still further delays freezing of the water below. In this way, lakes of moderate depth retain a lower layer of unfrozen water in which plants and animals can survive the coldest winters.

Salt content depresses the freezing point of water. For sea water with a salt content of 35%, the freezing point is −1.91°C. However, the temperature of maximum density is also changed, and as salt water cools towards its freezing point it becomes progressively more dense, so that convective sinking occurs continuously. The oceans are prevented from freezing by their sheer volume, ceaseless movement (driven by wind and tides) and convective currents, not by the special density properties found in fresh water.

Small organisms often have a specific gravity close to that of water. They are thus close to neutral buoyancy and compared with terrestrial organisms, expend very little energy in counteracting the forces of gravity. Even larger organisms, with dense skeletal material, obtain some buoyancy support from being immersed in the water, and are able to save energy otherwise needed for counteracting gravity.

On the other hand, the high viscosity of water compared with air means that there is increased frictional resistance to the movement of organisms through it. Viscosity decreases as temperature increases, but at 10°C for example, the frictional resistance to an organism moving through water is about 100 times what it would be for that organism in air. Furthermore, viscosity assumes greater importance for smaller organisms. If a human being makes a few swimming strokes, it seems natural that the person will glide through the water for many seconds before viscosity eventually stops the motion. But if a small flagellate cell makes a swimming movement and then rests, it will come to a stop in milliseconds. This is because it has very little momentum. The ratio of momentum to viscosity is expressed in the Reynolds number. For small organisms the Reynolds number is low, and simple activities become unbelievably complicated. For example, the techniques used for feeding by copepods must allow for the small momentum of a food particle in relation to the viscosity of the water. To appreciate what it might be like, try using two forks to pick a rice grain out of a jar of molasses! These problems are discussed in Purcell (1977) and in Strickler (1984).

1.3 LIVING ORGANISMS IN THE AQUATIC ENVIRONMENT

1.3.1 The biota of marine and continental habitats

In almost all kinds of aquatic habitat we can find three communities: the pelagic community of the open water, the benthic community living on or in the bottom deposits and the fringing community where water is shallow and there is usually an abundance of rooted aquatic plants. The pelagic community has two components; those close to

neutral buoyancy, suspended relatively passively in the water, which we call the plankton, and those larger, actively swimming animals known as the nekton. Plankton, nekton, benthic and fringing communities are found in almost all aquatic habitats, though their proportions may differ widely. A small pond may be dominated by its fringing community and have only a small area of open water, while a fast-flowing river may have a very poorly-developed plankton community.

In general, marine habitats have a wider diversity of plant and animal types than freshwater (though waters of intermediate salinity in estuaries may have the lowest diversity of all). Microscopic algae are widely distributed in the plankton of both marine and fresh waters, but the sea has a greater proportion of large attached algae, the seaweeds. It is thought that the early evolution of all the major animal phyla took place in the sea. Freshwater fauna can be seen to have two major components; those that invaded directly from the sea and those that colonized from the land. Flatworms, oligochaete worms, prosobranch snails, bivalves, crustaceans and fish probably colonized primarily from the sea, but flowering plants and insects underwent major coevolution on land and subsequently invaded fresh water, along with the air-breathing snails. Freshwater communities contain a rich assortment of insects, many of which spend their larval life in water, then metamorphose to flying insects. They are perhaps best represented in the fringing communities, but the larvae of dipterous flies are abundant in other benthic habitats and have a few representatives in the plankton. The echinoderms, so abundant in marine habitats, appear to have failed to make the transition to fresh waters.

Of the kinds of animals that colonized freshwater habitats from the sea, what proportion travelled up estuaries into rivers and what proportion crossed the intertidal zone onto the land, before invading freshwater habitats? Clues to the riddle can be discovered by studying the way in which each type of animal adapts to salinity changes. Most marine organisms have a salt concentration in the blood approximately equivalent to the concentration in sea water. If they are suddenly transferred to fresh water they absorb large quantities of fresh water by osmosis, usually with fatal results. Adaptations to survival in fresh water include reduction of the salt content of the body fluids and evolution of an efficient excretory mechanism for ridding the body of excess water. Some groups, such as the echinoderms, seem never to have made this adaptation, but free-living freshwater flatworms (Platyhelminthes) are found to have a complex set of organs for osmoregulation and excretion not present in their marine relatives. It is reasonable to suppose that these were evolved as an adaptation to salinity change as they moved up estuaries into rivers and lakes.

The annelids are sharply divided into: (i) the polychaetes, which have a poor system of osmoregulation and shed their eggs and sperm freely into the water for fertilization; and (ii) the oligochaetes and leeches, which have an effective system of osmoregulation and enclose their eggs in a cocoon after internal fertilization. The polychaetes have very few freshwater representatives, while the oligochaetes have freely invaded both freshwater and the land: their cocoons would have protected the eggs from being swept away as they adjusted to the salinity changes and moved up the rivers, so we may conclude that this was the route that they took.

Freshwater snails are of two types: the prosobranchs with a gill and the pulmonates with a primitive kind of lung. The former probably moved up estuaries to rivers, but the pulmonates are thought to have first colonized the land by way of salt marshes, and radiated to become the very successful land snails. Their soft tissue is vulnerable to desiccation, but they are able to survive dry periods by retreating into their shells. A small proportion of these snails secondarily returned to fresh water. Most of them still breathe air and return periodically to the water surface in order to do so.

About the bivalve molluscs — the mussels, clams, etc. — there is really no doubt. They could not possibly colonize the land without giving up their filter-feeding habit and the characteristic structures that go with it. They must, therefore, have moved up estuaries to rivers and lakes. The same argument applies to filter-feeding soft-bodied animals, such as the sponges and the bryozoans.

Evidence concerning the benthic crustaceans in fresh water is not clear. Amphipods and isopods are abundant along the sea shore and those that invaded fresh water may well have moved up estuaries and rivers, along with members of the crayfish family. However, amphipods and isopods have also successfully colonized damp habitats on land by way of the littoral zone and the possibility that some terrestrial forms successfully invaded freshwater habitats cannot be ruled out.

The origin of terrestrial insects is speculative, but some have postulated a marine origin, with colonization of the land by way of damp habitats in the littoral zone and in low-lying land. By developing a waterproof cuticle, insects became terrestrial animals *par excellence*. Undoubtedly they had their main adaptive radiation on land at the same time as flowering plants, but a significant proportion of present-day insects now spend part of their life history in water. It is of interest to ask why there has not been an equivalent reinvasion of marine habitats. Two explanations have been offered: one is that many insects make extensive use of the relatively calm surface of rivers, lakes and ponds for locomotion, respiration, feeding or mating — such calm surfaces are not common in marine habitats; another is that the recolonizing insects were unable to compete with the abundant crustaceans found in the fringing marine habitats. In fresh water the insects may be filter feeders (e.g. larvae of chironomids, simuliids, net-spinning caddises, etc.), they may browse on the surfaces of plants and stones (e.g. larvae of mayflies, stoneflies) or they may be predators (e.g. dragonfly larvae). In marine fringing habitats, many of these niches are filled with crustaceans of various kinds. Barnacles are filter feeders, isopods and amphipods are browsers and crabs and lobsters are predators. The fact that relatively few representatives of these groups invaded fresh water probably made it easier for insects to carve out their niches.

Analysis of the blood of most marine bony fish shows that the salt content is one-half to one-third that of sea water. The most common explanation is that bony fish spent a large proportion of their evolutionary history in fresh water and only secondarily invaded the sea. Some have challenged this view, but there is no convincing mechanism to explain the lowering of the salt content of blood of fish that lived and evolved entirely within the marine environment. The fish, then, are the outstanding example of colonization of the sea from fresh water.

An obvious difference between freshwater lakes and the sea is the absence of significant tides and strong tidal currents in the lakes. This may explain why filter feeders are much less prominent in the benthic communities of lakes, compared with the sea. The tidal currents of continental shelves transport large quantities of organic detritus in suspension just above the bottom, and this is filtered out and used to advantage by many kinds of large mussels and clams, by sponges and by a wide variety of coelenterates (e.g. sea anemones), polychaete worms and sea cucumbers (holothurians). None of these groups are as abundant in the benthos of lakes as they are on continental shelves.

On studying the benthos of the River Thames, in England, the author found that in contrast to the situation in lakes, filter-feeding mussels, sponges and bryozoans comprised the greater part of the benthic biomass (Mann, 1972). They were feeding on the large quantities of organic matter in suspension, much of it derived from treated human sewage! Thus, in rivers, the one-way flow may be the analogue of the tidal currents in marine habitats, and may transport suspended organic matter, if available, to filter feeders.

The scarcity of animals with tentacles (polychaetes, coelenterates and holothurians) may be in part related to the absence of tidal currents, but may also reflect the fact that very few members of these groups have successfully colonized freshwater habitats, presumably because they failed to adapt to the change in salinity.

The problem of adapting to life in saline lakes is the reverse of that confronting marine organisms colonizing fresh waters. If the concentration of salts in the environment is higher than the concentration in blood, there will be a loss of water by osmosis. The Dead Sea, with a salinity of 226‰, has no animal life, but the classification 'saline lake' covers the whole range from only a little more saline than freshwater lakes up to the range 200–300‰. Moreover, it is characteristic of saline lakes that their salinity tends to fluctuate greatly, for example between wet and dry seasons of the year. This does not necessarily limit their primary productivity. For example, in the saline lakes of the East African Rift Valley, blue-green algae commonly dominate the phytoplankton and these support huge flocks of flamingoes, which filter them directly from the water. In saline lakes in general, protozoans and rotifers are often abundant in the zooplankton and copepods may be present if the salinities are not excessive. It has been shown that while some adult copepods can withstand salinities in excess of 100‰, the eggs require substantially lower values for successful hatching and development. Perhaps the most characteristic planktonic animals are those that have come to be known as 'brine shrimps' (Entomostraca). Not only are they tolerant of high salinities, but their life history is adapted to periods when the lake bed may become totally dry. They produce resting eggs which can remain dormant for months or even years and yet hatch as soon as water becomes available. (The resting eggs of brine shrimps are sold commercially. They may be hatched in brine and later used as live food for pet fish.) In nature the adult brine shrimps are slow-moving and are probably restricted to habitats devoid of predatory fish.

At the lower end of the salinity range, various other invertebrates colonize saline lakes. Insects, being relatively impermeable, may be locally abundant. Hemipteran bugs, dipteran larvae and occasionally beetles and damsel fly larvae are present in varying proportions. In general however, the problems of osmoregulation result in the animal communities of saline lakes being species-poor in comparison with ordinary freshwater lakes.

1.3.2 Body size, metabolism, growth and production

More than 60 years ago, Charles Elton (1927) wrote a very perceptive and forward-looking book called, simply, *Animal Ecology*. In it, he drew attention to the relationship between size and abundance of animals. In general, the animals at the base of the food chain (feeding on plants) are small and very abundant. Those that prey on them are larger and less abundant, and there is a progressive decrease in abundance until the final stage in the food web, where the carnivores are large and few in number. This has come to be known as the 'pyramid of numbers'. Elton (1927) further observed that small herbivorous animals increase at a very high rate, thus providing a surplus to support the carnivores which prey on them. The process continues to the top of the pyramid where the animals are too few in number and increasing too slowly to support any further stage in the food web. Elton worked primarily in terrestrial ecosystems and plants did not fit his scheme, for the small herbivores to which he referred were feeding on plants much larger than themselves. In aquatic ecosystems, however, the pyramid of size and numbers can be extended to include the phytoplankton, since they are usually eaten by animals larger than themselves.

These early ideas have since been refined and expressed in quantitative form. It has been found that a whole range of properties of organisms decrease with increasing size, and do so according to the same basic rule. Respiration is a good example. If R_s is the rate of respiration per unit bodyweight (W), it follows the rule

$$R_s = aW^{-b} \qquad (1.1)$$

where the value of a varies with type of organism, temperature, etc. but b is always close to 0.25.

This is called the allometric rule, in which b is the other (allo-) measure of respiration.

The rate of increase of populations (r_m), to which Elton referred, was shown by Fenchel (1974) to be related to body size by the equation

$$r_m = aW^{-0.27}. \qquad (1.2)$$

Thus, the original insight — that progressing up the pyramid from small organisms to large the rate of increase slows down — has been supported in a very precise and predictable way (Fig. 1.1).

This gives us the possibility of envisaging the life process in an aquatic environment in a dynamic way. In the absence of nutritional or other constraints, bacteria, with a bodyweight of about 10^{-12} g are able to produce new tissue at a rate of about 50 times their own weight per day; protozoa at about 10^{-9} g and feeding on the bacteria could produce one to ten times their own weight per day; zooplankton of 10^{-3} g would produce their own weight about every 10 days, and so on to a fish of 100 g that would double its weight in about 100 days.

In practice, it is usually easier to determine the living weight, or biomass, of a particular size fraction of a community, rather than the number of organisms. The rate at which that biomass (B) increases is called its rate of production (P), and the ratio P/B is found to obey the same allometric

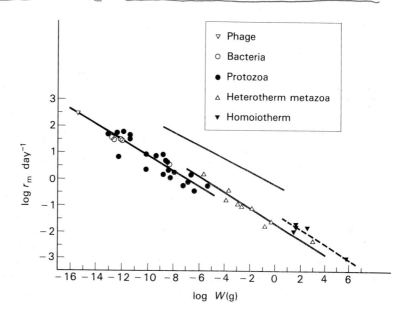

Fig. 1.1 Plot of the relationship between 'intrinsic rate of natural increase' of a population (r_m) and bodyweight (W). The line with no points has a slope of -0.249 and represents a generalized relationship between bodyweight and metabolic rate per unit weight. From Fenchel (1974).

laws. Taking the whole range of organisms from unicells to mammals the relationship is close to

$$P/B = aW^{-0.25} \qquad (1.3)$$

although within groups of organisms like microplankton, macrobenthos, fish, etc., allometric relationships with higher values of the exponent have been demonstrated.

We may now turn from consideration of the characteristics of organisms to descriptions of the communities of which they are part. This is most conveniently done under the headings of plankton, nekton, fringing communities and benthos.

1.4 PLANKTON

The open water of both lakes and sea is colonized by a rich assortment of algae and animals that drift passively, or, if they swim, are in general not able to move against the prevailing currents.

Many planktonic animals, zooplankton, make extensive vertical excursions through the water column. The significance of this is still being debated, but we may note that they exert some control over their horizontal movement by spending a certain amount of time in a surface current going in one direction and then change to a deeper current going in a different direction. It should not be assumed that planktonic animals are totally passive in their horizontal movements. Another important reason for vertical migration is that some zooplankton avoid visual predators by sinking to the dark depths by day and rising to the surface waters to feed at night (see Section 8.5).

For a planktonic alga to obtain its supplies of inorganic nutrients by diffusion from the water, it is necessary to have a large surface area relative to its volume. For this reason, phytoplankton are relatively small, mostly under 1 mm diameter. For experiments with natural mixtures of phyto-

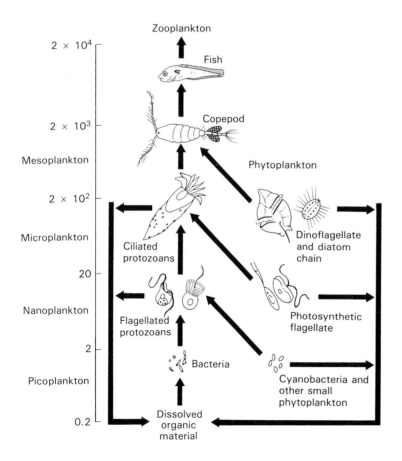

Fig. 1.2 The most common groups of phytoplankton and zooplankton, with size indicated by the scale of micrometres. Arrows indicate feeding transfers. From Fenchel (1987).

plankton cells it is convenient to sort them by size. For example, the size range 20–200 µm contains most of the diatoms and desmids and is called the microplankton. The range 2.0–20 µm contains most of the flagellates and is called the nanoplankton, while those in the range 0.2–2.0 µm are called the picoplankton (Fig. 1.2). To this could be added recently discovered marine bacteria and viruses that are smaller than 0.2 µm and are called the femtoplankton. The abundance and function of the microplankton and nanoplankton is quite well understood, but it is only with the advent of new electronic methods of sorting cells that we have begun to discover how abundant and potentially important are the picoplankton.

Diatoms, which tend to dominate the 20–200 µm class have internal protoplasm that has a density very close to that of water, but their cell walls often contain large amounts of silicon which has a density two to three times greater. As a result, diatoms tend to sink, typically at the rate of about 1 m day^{-1}. Phytoplankton cells, stationary in the water column and taking up inorganic nutrients, would tend to develop zones of nutrient depletion round them. Sinking is the way by which diatoms alleviate this problem. However, the disadvantage is that the cells sink into lower and lower light intensities. What is needed is large-scale mixing in the water column that will bring diatoms to the surface again. Diatoms are particularly characteristic of waters where wind-driven mixing is of frequent occurrence. The cells have a limited control over their own buoyancy, for example by producing fats and oils internally, but it is harder for freshwater plankton to attain neutral buoyancy than it is for marine forms, on account of the differing densities of the two media. In part, this may explain why freshwater plankton span fewer size classes than their marine counterparts, and do not attain the really large size categories.

The flagellates of the nanoplankton lack heavy cell walls and the accompanying tendency to sink. Instead they use flagella to propel themselves through water at speeds of up to 20 body lengths second^{-1}, say 5–10 m day^{-1}. This is often enough to get them out of a zone of low nutrient concentration and into one of higher concentration. Incidentally, dinoflagellates (those with two flagella) use the first for locomotion, and the second to move water over their surfaces to enhance nutrient uptake. Flagellate populations are characteristic of waters with low wind-driven turbulence, such as occur in calm summer weather in temperate zones, or near the lower part of the lighted (euphotic) zone in the tropical lakes and seas.

The importance of picoplankton in marine and freshwater ecosystems was only realized in the late 1970s. Two kinds of organism are involved: small algae, related to the flagellates in the larger size classes, and a group of photosynthetic bacteria known as the cyanobacteria. Under the microscope, an important diagnostic difference is that the bacteria lack a well-defined nucleus (are prokaryotic), while the algae have the nucleus (are eukaryotic). One reason that the importance of these organisms went unrecognized for so long is that ecologists did not routinely use filters with a fine enough pore size to retain them. It is now believed that in situations in which essential nutrients are in short supply, the picoplankton, with their large ratio of surface area to mass, have an advantage over larger forms and may be the dominant primary producers (for review see Stockner, 1988).

If we separate out from a plankton sample the organisms 200–2000 µm, the zooplankton dominate, and the most abundant organisms are usually crustaceans such as copepods or cladocerans. These are the primary consumers of diatoms and flagellates. In the smaller size classes, larval stages of the crustaceans are found, but also protozoa such as ciliates and non-photosynthetic flagellates. These are the consumers of the picoplankton and of the non-photosynthetic bacteria. It used to be thought that zooplankton removed phytoplankton from the water primarily by various filtering mechanisms. However, as mentioned earlier, recent work, recognizing the problems created by the low ratio of momentum to viscosity at this scale, reveals that feeding consists more in 'guiding' the food particles into the mouths of the organisms by a variety of indirect actions, often involving very little contact between the feeding appendages and the food (for review see Strickler, 1984).

Organisms over 2000 µm are conventionally

referred to as the macrozooplankton. Samples in this size range often include fish eggs and larvae. In the ocean, various 'gelatinous' animals, such as jellyfish or salps may clog the nets and make sorting difficult. In lakes, predaceous plankton such as larvae of the midge *Chaoborus* may be included in macroplankton.

Any natural sample of the plankton also contains dead organic material, detritus, the origins of which are various. Phytoplankton and zooplankton that die may break up in the water column, releasing plant and animal fragments. Living zooplankton release faecal pellets that sink slowly through the water column. Phytoplankton give off dissolved organic matter which condenses on the surfaces of bubbles and solid objects to form organic flakes. In the sea, mucus given off by large planktonic filter feeders sinks slowly, as flakes that are known as 'marine snow'. The surfaces of the detritus become colonized by bacteria, and these in turn attract flagellates and ciliates that feed upon them. In coastal waters it is common to find in the plankton, finely divided dead remains of the large rooted plants and algae that grow in the fringing zone. As these plants decompose, wave action breaks them into fine particles, brings them into suspension and allows currents to carry them out into open water.

Large zooplankton are eaten by pelagic fish, and for many years the main emphasis in planktonic food web studies was the growth and production of the diatoms and flagellates, the grazing upon them by mesozooplankton and macrozooplankton and the consumption of these by larval and adult fish. In recent years, a great deal of attention has been paid to alternative pathways: (i) growth and production of picoplankton and their consumption by protozoa; (ii) growth of a microbial food web on detritus; and (iii) feeding of the larger zooplankton on the detritus particles. These processes emphasize the importance of very small organisms and the 'recycling' of dead organic matter back into food webs leading to higher trophic levels. The latter process has come to be known as 'the microbial loop'.

Clearly, a characteristic of plankton is small size. In this respect, planktonic systems may be contrasted with terrestrial systems. The phyto-plankton population at any given time has a low biomass, but by growth and reproduction may produce annually 15–45 times that biomass. Terrestrial systems, on the other hand, tend to have a large standing biomass of plants, of which the annual production/biomass (*P/B*) ratio is less than 1. Many terrestrial plants have to devote a significant portion of their resources to manufacturing distasteful or harmful substances that will deter herbivores. Phytoplankton are protected by their small size, large numbers and rapid growth rate. It is very difficult for herbivores to cause local extinction of a phytoplankton population when the cells are so small, numerous and widely distributed. In the plankton, it is also common for the biomass of consumers to greatly exceed the biomass of primary producers; but in a forest, for example, the biomass of consumers may be 1000 times less than that of the plants.

1.5 NEKTON

The consumer organisms that inhabit the pelagic zone but are active swimmers are known as the nekton. Fish are the most frequently encountered, but, especially in the sea, invertebrates such as squid also constitute important nektonic predators. In lakes and coastal areas of the ocean, diving birds are locally important. Whales and seals tend to congregate in highly productive areas of the ocean and are then major components of the food web.

Compared with the plankton, the nekton are long-lived and slow-growing. Some invertebrates may complete their life histories in 1 year, but for most fish their lifespan is of the order of 5–10 years. The majority of fish species produce large numbers of eggs that are released into the plankton and experience heavy mortality (see Table 8.4). For example, a pair of cod may produce 7 million fertilized eggs, but in a state of zero growth of the population, only about two of those fertilized eggs survive and develop to enter the adult breeding population. A few fish species produce relatively small numbers of eggs with large yolks and some even provide parental care. For example, members of the shark family may produce small numbers of large-yolked eggs which they enclose in a protective shell. Some species even retain the developing eggs within their bodies until they hatch and the young are ready to swim independently. Sticklebacks lay relatively small numbers of eggs in nests in the sediment and guard them during development.

In the ocean, a pattern of migration is commonly associated with the breeding cycle. The fertilized eggs are released in an area where the prevailing currents will carry them, during development, to areas of high plankton and/or benthos production, known as the nursery areas. After some weeks or months in the nursery areas, the juveniles will migrate to adult feeding grounds, where there is good production of food organisms of a larger size range. Finally, the sexually mature adults will migrate back to the breeding grounds, where it all began. In many cases, very extensive migrations are undertaken. For example, the eels of western Europe breed in the Sargasso Sea, and their offspring are carried back to Europe in the Gulf Stream and North Atlantic Drift. Various species of tuna migrate thousands of kilometres between feeding grounds and breeding grounds. None of this would be possible without the large size and freedom of movement that characterize the nekton. In the case of eels, their mobility is so great that at one time or another in their life history, they occupy all types of aquatic habitat.

1.6 FRINGING COMMUNITIES

The margins of many aquatic habitats, salt water and fresh, running and standing water, are occupied by large plants which form marshes, swamps and submerged beds of various kinds. The type of development around the margins of the ocean depends very much on the type of substrate and on the climate. In some areas, where wave and current action is strong, the bottom consists of large boulders or exposures of bedrock. In these places, the dominant submerged primary producers are the algae known as seaweeds. They have no roots, but have a holdfast by which they attach themselves to the bottom. Uptake of nutrients takes place over the whole surface. By virtue of their attachment in one place while waves and currents pass over them, their

supply of nutrients is continually renewed and they are able to maintain very high rates of growth and production. They produce distasteful substances which deter most grazers, although a few animals, such as sea urchins and gastropod snails may browse on their tissues. For the most part, seaweeds continue growth until their tissues become senescent, break away and contribute to the pool of detritus in coastal waters. Small particles of algal detritus are extensively used by filter-feeding bivalve molluscs and by browsing crustaceans and snails. On rocky shores in temperate climates, the algae may also form dense growths in the zone between high and low tide.

Shores on which the action of water movement is less intense, so that fine particles accumulate on the bottom, are known as sedimenting shores. Here, rooted flowering plants, colonizers from the land, will form intertidal marshes and subtidal seagrass beds. They are not much utilized by grazing animals, but eventually decompose to form detritus. For the most part, that detritus has a high content of fibre and is relatively unpalatable to animals, so that most of it ends up being degraded by bacteria and fungi, though some of it is consumed by specialized detritivores. Ecologically speaking, the most important role for vascular marine plants is to provide a substrate for microscopic and macroscopic algae which grow as periphyton on their surfaces. Salt marshes and seagrass beds also act as traps for sediment, and provide habitat for the young stages of fish, which are able to forage in shallow water in places where their larger predators cannot follow.

While salt marshes are the characteristic intertidal development of temperate climates, the intertidal zones of tropical areas are often colonized by salt-tolerant trees known as mangroves. Mangrove forests stabilize shorelines and provide a complex network of moist habitats in which shrimps, crabs and a variety of molluscs are able to browse on the decomposing mangrove leaf litter and on the algae that thrive on the mud surface and on the roots and stems of the trees.

In analogous fashion, the margins of lakes and slowly-moving rivers may be colonized by a variety of emergent, floating-leaved and submerged macrophytes. Among the largest emergent plants are the bullrush (or cat-tail) *Typha* that is found in richer soils and the reed *Phragmites* that is common around the perimeter of lakes on poorer soils. In parts of Europe this reed is so abundant that it has traditionally been used for thatching roofs. Smaller reeds and rushes include *Scirpus* and *Sparganium*. Since these plants have their roots in anaerobic sediments, they characteristically have special tissue to conduct air from the leaves down to the roots. The same adaptation occurs in the salt marsh plant *Spartina*.

The best known floating-leaved plants, at least in the temperate parts of the northern hemisphere, are the water lilies *Nuphar* and *Nymphaea*. Both have a well-developed rhizome that is found in the sediment at a water depth up to about 1 m. Species such as *Potamogeton natans* grow in considerably greater depths of water. If a small body of water becomes heavily colonized with floating-leaved plants, the leaves restrict the gas exchange between water and atmosphere and the water may become anaerobic in summer, greatly restricting the diversity of organisms that can live there.

Submerged freshwater macrophytes include a member of the fern family, *Isoetes*, many kinds of mosses, some algae and a variety of flowering plants which, in spite of their origins, may grow and multiply for very long periods entirely by vegetative reproduction, without flowering. Some members of this group, e.g. the water buttercup *Ranunculus*, live in fast-flowing clear streams. When the water is high, they develop finely divided leaves and reproduce asexually, but when the water level falls they develop entire leaves and produce flowers above the surface of the water. Various species of *Potamogeton* behave in much the same way.

Finally we should note that there are free-floating plants in the fringing communities of certain lakes and rivers, particularly those with a rich supply of dissolved nutrients. The duckweed *Lemna* is a small but often very abundant plant, probably originally named due to its association with small bodies of water enriched with the droppings of ducks. Larger plants, like the water hyacinth *Eichornia*, may become so abundant in nutrient-rich lakes in the tropics that they impede the use of the water for navigation and fishing.

As with marine fringing plants, freshwater macrophytes are seldom grazed directly by herbivores. They grow up, die and decay, producing a zone of litter that provides food for detritivores and habitat for a variety of invertebrates and young fish. Various crustaceans, such as the isopod *Asellus* are able to browse on this litter in spite of its high content of fibre, and these in turn are preyed upon by invertebrate and vertebrate carnivores.

The surfaces of growing plants are colonized by algae and bacteria; it has recently been shown that dissolved organic matter, released into the water by both macroscopic and microscopic forms, condenses on the plant surfaces to form a rich complex of amorphous organic matter mixed with microorganisms and known as detrital aggregate. This material is prime food for fish, for example an economically important tropical freshwater fish, *Saratherodon* (previously known as *Tilapia*), and for a wide range of browsing invertebrates.

The biomass of plants in fringing environments is typically very much greater than that of the phytoplankton, but it turns over much more slowly, so that the production to biomass (*P/B*) ratio is in the range 0.5–5. Even so, the actual production of plant tissue may be very much greater than that of the phytoplankton in neighbouring waters. Careful study of the fate of this material indicates that for vascular plants, only a small proportion is incorporated into food webs involving invertebrates and fish. The majority appears to be utilized by bacteria and microscopic consumers. Algal production, whether in the form of epiphytes or as seaweeds, is utilized more directly in the food webs leading to top consumers. For a review of detritus food webs in fringing communities see Mann (1988).

1.7 BENTHOS

If we take the global view, by far the greatest area of sea floor and lake bottom is in places where the water column is far too deep for light to penetrate and for algae to grow. In these areas, the community is almost entirely dependent upon a rain of organic matter from above. This material may be moved about by horizontal currents, but its origin is from algal growth in surface waters. In the water column, immediately below the euphotic zone, there will be found sinking live phytoplankton, fragments of dead plankton of all kinds, faecal pellets and various forms of particulate organic matter described earlier. However, there exists in the water column a community of planktonic consumers that progressively removes sinking material. In the deepest parts of the ocean, perhaps 1 or 2% of the material leaving the euphotic zone reaches the bottom. However, as we move to shallower depths, the proportion reaching the bottom increases. Not surprisingly, the population density of the organisms depending on this food supply also increases. As we approach the shallow fringing environments, enough light reaches the bottom for benthic algae or flowering plants to colonize and grow. There is then a sharp increase in the biomass of consumers that the community is able to support.

A small proportion of benthic production is supported by bacteria that derive their energy from sources other than the rain of organic matter from above. They make use of various inorganic chemical reactions and are known as chemolithotrophic bacteria. They are particularly abundant round the hydrothermal vents — those places on the sea floor where heated, chemically-enriched water emerges through the earth's crust.

In the world's oceans, the biomass of benthos at depths greater than 3000 m may be less than $1\,g\,m^{-2}$ wet weight, at depths between 200 and 3000 m the world average is about $20\,g\,m^{-2}$, and for waters shallower than 200 m the average is about $200\,g\,m^{-2}$. These averages are, of course, subject to wide variation: some highly productive fringing areas have several kilograms of benthic organisms per square metre. It has been calculated that more than 80% of the marine benthic biomass of the world is on the continental shelves at depths of 0–200 m. In analogous fashion, the benthic biomass of lakes is a function of their depth. The world average benthic biomass for shallow lakes (up to 25 m deep) is in the range $100–200\,g\,m^{-2}$, whereas for deep lakes it is of the order of $20\,g\,m^{-2}$.

There are two ways for benthic animals to make a living: one is to burrow into the sediments and either feed on those sediments or collect food at the surface, at the mouth of a burrow; the other is to live on the surface, either in a sedentary mode or free-ranging. Hence it is usual to divide the benthos into infauna and epifauna. Infauna can be collected quantitatively by taking samples of the sediment using a grab or a corer. The usual method of sorting is to wash the bottom deposits through a series of sieves that retain different size fractions. From this has arisen the convention of recognizing three types of benthic infauna: those retained by a 1-mm mesh are called the macrobenthos; those passing through a 0.05-mm mesh the microbenthos; and those in between the meiobenthos. The macrobenthos from lake and river soft sediments is often dominated by larvae of chironomid midges, oligochaete worms and bivalve molluscs. In the ocean, polychaete worms and molluscs tend to be most abundant. The meiobenthos of both types of habitat tend to be dominated by nematode worms and small crustaceans known as harpacticoid copepods. A functional distinction between macrobenthos and meiobenthos is that the macrobenthic animals tend to form burrows that displace the bulk sediment, while meiobenthos tend to live between grains of the sediment. Finally, the microbenthos, dominated by bacteria, fungi and the protozoans that prey on them, can be regarded as living primarily on the surfaces of the sediment particles.

Flow of water over the bottom has a profound influence on the benthos, especially in shallow water. If water movement is slight, organic carbon in various forms may be transported horizontally until it reaches this quiet water where it sinks to the bottom. Such benthic deposits tend to be rich in organic matter relative to inorganic matter, and a rich infauna is associated with this. If water movement is moderate, much of the light organic matter may be scoured away, leaving sandy or rocky bottom deposits. Under these circumstances, animals that make their living by filtering the water above the sediment tend to do better than the infauna; bivalve molluscs, like clams and mussels, do well under these conditions, as do various tube worms that protrude their tentacles to filter the water above the bottom. In rivers, where the flow of water is unidirectional and relatively constant, we find that in addition to filter-feeding molluscs and sponges, several kinds of insect larvae, such as dipteran flies and caddises have developed methods of filtering the water. Recent work has also shown that dissolved organic matter in the water readily condenses on the surfaces of stones in rivers, to form a film of solid material. This becomes colonized by bacteria and mixed with microscopic algae on the rock surface, to form a rich source of food for browsing insect larvae, snails, etc.

Strong water movement tends to inhibit the development of benthic communities. A physicist has pointed out that in areas of coastal waters where the tidal currents are strong, benthic animals are subjected to forces equivalent to hurricane force winds, up to four times a day. In marine areas where this occurs, benthic communities tend to be impoverished. The exceptions are relatively shallow areas where large seaweeds (kelps) are able to establish a hold on the bottom. These have the effect of cutting down water movement within the kelp bed, providing conditions for good growth of a wide variety of epibenthic animals such as sea urchins, crabs, lobsters and mussels.

In rivers, the effect of the current has an analogous effect. Slowly-moving rivers may have highly organic bottom deposits with a rich infauna. In moderate currents, the filter-feeding organisms thrive, but in regions of very strong currents, the rapids, the fauna is liable to be restricted.

Benthic epifauna comprises those animals not resident in burrows. Marine examples of mobile organisms are shrimps, crabs and lobsters, starfish and sea urchins; while mussels, barnacles and anemones live more sedentary lives. Fish which live close to the bottom may also be regarded as part of the benthic epifauna. Large areas of estuaries often consist of mud-flats containing fine organic and inorganic particles that provide ideal habitat for burrowing worms, crustaceans and bivalve molluscs. Estuarine fish, such as flounders and eels, roam over the mud-flats when the tide is in, preying on the infauna. Shoals of shrimps take smaller organisms and organic detritus. In these shallow waters microscopic algal life flourishes on the mud surface, augmenting the supply of organic matter already present.

In lakes, water movement in deeper parts is usually slight, so that organic matter tends to accumulate there. Benthic communities of shallow lakes

are often dominated by tubificid worms and the larvae of chironomid midges, along with a selection of bivalve molluscs. On approaching the fringing community, wave action may remove organic matter from the sediments, leaving a gravelly bottom on which snails, leeches and a variety of insect larvae find their home. In the fringing areas themselves, there is usually extensive growth of submerged or emergent flowering plants, which provide protection from predators for a variety of animals, and a good substrate for the growth of epiphytic algae. It is here that we usually find the most abundant and diverse benthic fauna.

1.8 THE HYDRODYNAMIC BASIS OF PLANKTON PRODUCTIVITY

1.8.1 The paradox of stratification

The growth of phytoplankton populations, a process which provides fundamental support to all other biological processes in the plankton, is strongly controlled by hydrodynamic events that are operative in both freshwater and salty habitats. Imagine a body of water that is completely still. Light enters from above, and its intensity I diminishes exponentially with depth, z, according to the equation

$$I = I_o e^{-\alpha z}$$

where I_o is the light intensity at the surface and α is the extinction coefficient. Light intensity decreases with depth because the short wave radiation is absorbed in the water column, releasing heat. If the water column remained perfectly still, the distribution of temperature would be similar to that shown in Fig. 1.3a. Each layer is slightly warmer than the layer below, and is therefore less dense, or more buoyant. The situation is stable so long as the light input remains stable.

In the real world, there are winds at the surface and their friction drag at the surface sets up turbulent mixing, driving some of the surface water downwards and allowing cooler water to come to the surface and be warmed. The net result is that the distribution of temperature changes to that

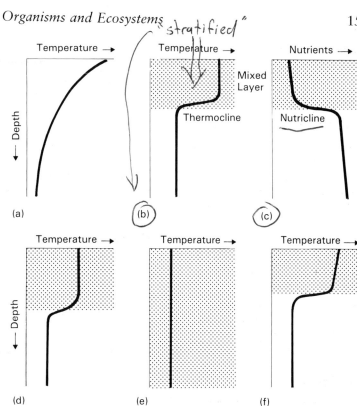

"stratified"

Fig. 1.3 Diagrams of temperature and nutrient profiles in a lake: (a) hypothetical temperature profile in a completely still body of water; (b) formation of mixed layer and thermocline as a result of wind at the surface; (c) formation of a nutricline after wind mixing; (d) reduction of temperature gradient during surface cooling; (e) distribution of temperature after mixing from top to bottom; (f) consequence of spring warming and formation of shallow mixed layer.

shown in Fig. 1.3b. There is an upper, mixed layer in which the temperature is relatively constant. Then comes a thermocline, in which temperature changes rapidly, and finally there is the deep layer of cool water. The body of water is said to be stratified. (See Chapter 5 for further discussion of stratification.)

Any phytoplankton cells in the upper, mixed layer, are liable to be circulated at random through that mixed layer. On average they will experience enough light to enable them to gain a photosynthetic surplus, so that they can grow and divide. Cells below the thermocline will circulate throughout the lower layer and, on average, the light will be insufficient for them to grow and multiply.

After a time, the active phytoplankton cells in the upper, mixed layer will have used most of the essential nutrients, such as nitrate, ammonium, phosphate, etc. Their only remaining supply is from the excretion of bacteria and animals in the mixed layer, and these recycled nutrients are usually quite limited in amount. Meanwhile, in the lower layer, nutrients have not been depleted, and are being augmented by the dead organisms, faecal pellets, etc. which sink into the lower layer and decompose. Nutrient concentrations are high in the lower layer but low in the upper layer. In addition to the thermocline, we can demonstrate the presence of a nutricline at the boundary between upper and lower layers (Fig. 1.3c). This condition is typical of lakes and oceans, in temperate latitudes, in summertime, and is a more or less permanent condition of tropical lakes and seas. It presents the phytoplankton with a paradoxical situation: the light needed for photosynthesis is available in the upper layer but not the lower layer, but the nutrients needed for growth are concentrated in the lower layer. Any process which breaks down the barrier and brings nutrient-rich water from the lower layer to the surface, mixed layer, will stimulate primary production. These processes are known as upwelling, or vertical mixing. Understanding the control of primary production in planktonic habitats requires a good understanding of the physical processes causing vertical mixing. We shall consider five: (i) seasonal temperature change; (ii) wind-driven mixing; (iii) river runoff; (iv) tidal mixing; and (v) large-scale ocean circulation.

1.8.2 Seasonal changes in temperature

The processes described here are seen most clearly in a moderately deep lake in a temperate climate, but similar events occur throughout the temperate oceans. Starting from the stratified condition typical of summer (Fig. 1.3b and c), we find that as autumn approaches the surface waters cool to a temperature a little below that of the water a few centimetres down. These surface waters are now more dense, and sink until they find water of the same density. As the cooling process continues, convective sinking becomes more intense. The density difference between upper and lower layers decreases (Fig. 1.3d), and a strong wind can deepen the mixed layer, drawing in some nutrient-rich water from below. Then the wind may slacken off, and the phytoplankton have a little burst of production as they use the new nutrients. Eventually, the density difference between the layers above and below the thermocline becomes so small that a strong wind is able to mix the whole water column from top to bottom (Fig. 1.3e). The surface waters that were nutrient-deficient are replaced by waters that have abundant nutrients and there is often an autumn bloom of rapid phytoplankton growth. It is short-lived, however, because the duration and intensity of sunlight are decreasing and the phytoplankton cells are being mixed down into the deepest parts of the lake, where the light is insufficient for photosynthesis. A state of low phytoplankton growth lasts throughout the winter. Nutrients are abundant in the surface waters, but light is limiting.

Next spring, surface warming coupled with wind mixing, leads to the formation of a relatively shallow, less dense, mixed layer with a thermocline below (Fig. 1.3f). Those phytoplankton cells caught in the mixed layer are now prevented from spending time in the dark depths. Their daily ration of light is enough to permit growth and reproduction. There follows a great increase in phytoplankton numbers — the spring bloom. The depth of the mixed layer which just makes this possible is called the critical depth. Zooplankton begins to feed on the bloom, and typically it reaches a peak in number a few weeks after the phytoplankton.

After a few weeks, the phytoplankton have utilized most of the nutrients in the upper, mixed layer, while nutrients are being released below the thermocline. This is the summer condition with which we began this account. Vertical transport of nutrients took place during the autumn cooling phase, but the effective use of those nutrients by the phytoplankton was delayed until spring warming caused a shallowing of the mixed layer above the critical depth. Only then were the phytoplankton cells restricted in their vertical excursions and exposed to enough light to permit new growth. A similar cycle of seasonal changes occurs in oceanic waters of temperate latitudes, except that winter mixing often does not reach to the sea floor, but stops at 100–200 m, below which is a permanent thermocline delineating the upper limits of 'the interior of the ocean'.

Remote sensing, as well as ship-borne measurements, have confirmed that the spring bloom of the phytoplankton occurs first at the southern limits of temperate waters, and gradually migrates northwards with the changing altitude of the sun. A few weeks after the peak biomass of phytoplankton, there occurs a peak in biomass of zooplankton, as it feeds on the spring bloom, grows and reproduces.

At the northern limits of the temperate regions, adjacent to arctic waters, the spring bloom arrives approximately in midsummer. There is not time for an extended period of stratification before the autumn cooling sets in. As a result, we see a single midsummer bloom in phytoplankton, accompanied by a pulse of zooplankton activity, followed by a return to winter resting conditions.

If we view lakes in their global context, the same kind of seasonal variation in patterns of primary production may be discerned, with the spring bloom arriving earlier in more southerly lakes. However, each lake is isolated in its drainage basin, and considerable lake-to-lake differences in open water primary production can be attributed to differences in the nutrient status of

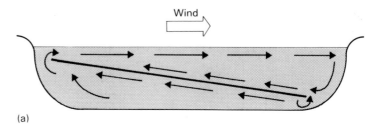

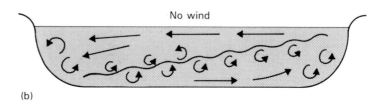

Fig. 1.4 Rocking of the thermocline and turbulence caused by wind over the surface of a lake.

the landscape, and to differences in the depth of each lake basin. These will be discussed later.

We should also note that there are lakes in which the mixing from top to bottom, illustrated in Fig. 1.3e, does not occur because the lower portion of the basin is filled with a more saline water that does not mix with the water column above. It is separated by a sharp salinity gradient known as a halocline — such lakes are known as meromictic lakes. The origin of the saline lower layer may be a marine intrusion, salt springs within the lake basin, or simply accumulation of nutrient salts from long periods of stagnation of bottom water.

1.8.3 Wind-driven vertical transport

During a period of strong wind in one direction over the surface of a lake, the wind stress will cause warm water to pile up at the downwind end (Fig. 1.4a). The surface waters will rise a few centimetres above normal, but more importantly, the thermocline at the downwind end becomes greatly depressed, with a corresponding rise at the upwind end. When the wind stops, this unstable situation leads to a rocking motion of the thermocline, with a periodicity that varies from a few hours to a few days, being longer in the larger basins. Under certain circumstances, internal waves are formed on the thermocline, and these may break

causing strong vertical transport across the thermocline. In any event, there is strong vertical mixing at each end of the lake, as the thermocline rises and falls in contact with the shore (Fig. 1.4b). From this we can see that even during the summer period of strong stratification, wind mixing may cause pulses of nutrient-rich water to move up into the surface waters and stimulate primary production.

In the ocean, the most spectacular examples of wind-driven vertical transport are known as coastal upwelling. The principle is that strong wind blowing parallel with the coast starts a surface current in the same direction as the wind. The Coriolis force, resulting from the rotation of the earth on its axis, causes currents to curve: to the right in the northern hemisphere, to the left in the south. If the wind direction is in the correct orientation in relation to the coast, the net result will be a massive transport of surface water away from the coast (Fig. 1.5). This necessitates the upwelling of nutrient-rich deep water to take its place. When the wind dies down, the upwelled water goes through a sequence in which the phytoplankton take up nutrients and multiply, and the zooplankton populations grow and multiply by feeding on phytoplankton. The process is widespread around the shores of the world's oceans, but there are four particular sites where really

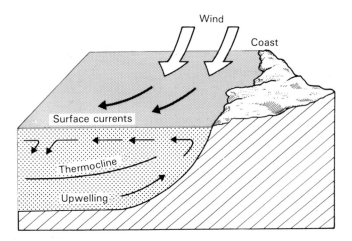

Fig. 1.5 Diagram illustrating coastal upwelling. For details see text.

massive upwelling occurs for much of the year: off the coasts of Peru, Oregon and California, north- and southwest Africa. Fish of the anchovy family are particularly well-adapted to make use of this biological production, and tens of millions of tonnes are harvested in good years.

1.8.4 Upwelling driven by river runoff

Rivers themselves are normally sufficiently turbulent for stratification into a nutrient-rich layer and a nutrient-depleted layer not to be a problem. There are situations where a river broadens out to form a true lake or a lake-like structure, and where seasonal stratification may occur, but in these cases, the mechanisms discussed in previous sections apply.

Quite different, however, is the situation at the mouth of a river, where fresh water meets salt. The fresh water, being lighter, floats as a discrete layer on top of the salt. Typically, this occurs in a semi-enclosed arm of the sea that we call an estuary (Fig. 1.6a). Between the two layers there is a density difference which we call a pycnocline. The movement of the fresh water towards the sea creates a shear, which in certain circumstances may cause internal waves on the pycnocline. Surface winds and waves may also cause instability at the pycnocline. The net result is that there is progressive mixing of fresh and salt water, and the freshwater flow has the effect of carrying out to sea several times its own volume of salt water

from below the pycnocline. This process is called entrainment. As Fig. 1.6a shows, there is a net inflow to the estuary of salt water near the bottom, and upward mixing of salt water to take the place of that which has been entrained. That bottom water has become nutrient-rich through its contact with the decomposing organisms on the sea floor, so there is a vertical transport of nutrient-rich water into the surface waters, especially at the head of the estuary.

During its journey down the estuary a regular food web develops, with phytoplankton taking up nutrients, zooplankton eating phytoplankton and in turn being eaten by young fish. All of these processes tend to produce sinking organic matter such as faeces and dead organisms, which enrich still more the water moving up the estuary near the sea floor. In this way the estuary becomes a trap for nutrients which recycle time and again, and give rise to very high levels of biological production. It is for this reason that estuaries are traditionally rich fishing grounds for shellfish and certain kinds of finfish, and why they are preferred sites for aquaculture.

When the estuarine water moves out onto the continental shelf, it forms a distinct plume, recognizable by its reduced salinity. The Coriolis force causes the plume to turn to the right (in the northern hemisphere) and such a plume is often recognizable for a great distance along the coast. One of its characteristics is that, because of the layer of less dense water at the surface, it takes

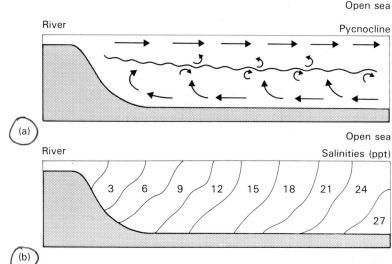

Fig. 1.6 (a) Diagram of estuarine circulation driven by freshwater runoff in the presence of weak to moderate tides. (b) Diagram of salinity distribution in a tidally mixed estuary. For details see text.

less spring warming to provide the degree of stratification needed to start the spring bloom. Hence, spring blooms usually start earlier in coastal waters than they do offshore. Those coastal waters also tend to be enriched with nutrients exported from the estuaries, and hence to be more productive.

The mechanisms just described operate in estuaries where the strength of tidal currents is low to moderate. The pattern changes drastically where tidal currents are very strong. This will be the subject of the next section.

1.8.5 Upwelling driven by tidal mixing

Tides are caused by the gravitational pull of the sun and the moon, coupled with the rotation of the earth. At any given place, the effect is of a rise and fall of sea level, most commonly about twice a day. The height of the rise and fall of the tide varies greatly from place to place. In estuaries in which there is a fairly large tidal range, tidal currents are generated that run strongly on the rising tide, slacken off, then run strongly on the falling tide. As these tidal currents move over the bottom, turbulence is generated in the water above. If they are strong enough, they can break down the stratification in the water column and cause the fresh water to be mixed completely with the underlying salt water. We then have a tidally mixed

estuary (Fig. 1.6b). In these situations the nutrient-rich water from the bottom is mixed throughout the water column, and nutrients are not limiting the growth of the phytoplankton. On the other hand, the phytoplankton is being mixed from top to bottom of the water column. Since estuaries are often turbid, the penetration of light is limited and phytoplankton production may be much less than in a stratified estuary.

1.8.6 Upwelling driven by large-scale circulation

Equatorial upwelling

In general the situation in tropical waters is that they are permanently stratified, there being no seasonal cooling that is strong enough to cause convective mixing of nutrient-rich water from depth, up into the surface layers. Hence, in general, tropical lakes and ocean waters are less productive than their counterparts in temperate climates. There are, however, some upwelling mechanisms that lead to enhanced primary productivity over very large areas of the tropics. Close to the equator in both the Atlantic and Pacific basins there are strong westward-flowing currents, driven by the trade winds. We saw earlier that the Coriolis force causes currents to diverge to the right in the northern hemisphere and to the left in the southern

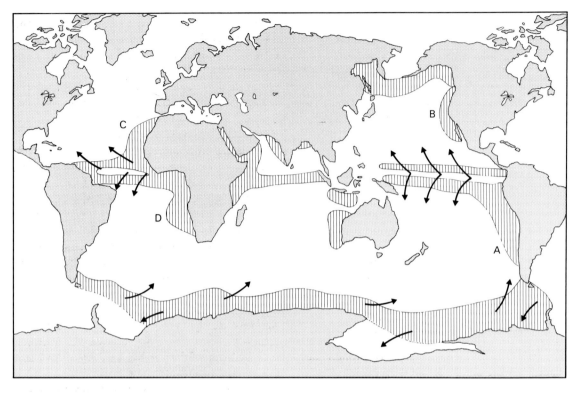

Fig. 1.7 Diagram of the main upwelling areas of the world (shaded). Arrows indicate tropical and Antarctic divergences. Coastal upwelling systems: A, Peru; B, California Current; C, northwest Africa; D, southeast Africa.

hemisphere. Hence, these westward-flowing waters have a strong tendency to diverge, those to the north of the equator turning northwards, those to the south of the equator turning southwards (Fig. 1.7). As a result of this divergence, there is an upwelling of nutrient-rich water at the equator, which stimulates phytoplankton production and has consequences at all levels of the food web. Satellite pictures of chlorophyll in tropical waters often show this line of enhanced phytoplankton biomass running along the equator (Fig. 1.7).

Antarctic Divergence

A similar kind of upwelling occurs along a line roughly parallel with the coast of Antarctica. In this case, the cause is the proximity of two currents moving in opposite directions. The one flowing in a westerly direction causes water to turn south towards Antarctica under the influence of the Coriolis force. Close to that current, on its northerly flank, is a current flowing to the east, which, under the influence of the Coriolis force, causes water to turn northward. The boundary of the two currents is therefore a site of divergence, where the surface waters are being replaced by upwelling from below (Fig. 1.7). Much of the high marine productivity associated with Antarctica, as shown by the great aggregations of sea mammals and sea birds, results from the upwelling at the Antarctic Divergence.

There are other mechanisms, too numerous to mention, whereby the productivity of surface waters of the oceans is enhanced by upwelling resulting from large-scale circulation patterns. The conclusion from this review is that a large proportion of the world's lakes and oceans are stratified, either seasonally or permanently, and during this stratification the supply of nutrients in surface waters is reduced to a low level by the uptake of the phytoplankton. Escape from this nutrient-limited condition is made possible by a variety of mechanisms for the vertical transport of a new supply of nutrients from the waters below the thermocline. Understanding these mechanisms is the key to understanding the broad-scale distribution of production in open water aquatic environments.

1.9 OTHER FACTORS AFFECTING AQUATIC PRODUCTIVITY

While the previous section has given us an insight into the factors influencing biological production in open waters, and especially in the oceans, there are many other factors that help to explain the local variations in primary and secondary production and in the yield of fish.

1.9.1 Dissolved salts and nutrient status

The chemistry of inland waters is strongly dependent on the nature of the rocks and soil from which their water is drained. If the basin is deficient in ions such as calcium and magnesium, there will be little to neutralize the humic acids derived from plant decomposition, and the water will tend to be acidic. Similarly, if the drainage area is deficient in nutrients such as nitrogen and phosphorus, this will be reflected in the chemistry of the river and lake waters. Even if conditions of solar radiation and circulation are ideal, rivers and lakes may have their productivity strongly limited by the availability of dissolved substances. The variability in productivity seen within relatively limited geographical areas is reflected in

the terminology developed to describe lakes, and only secondarily applied to marine habitats. Highly productive waters are termed eutrophic, unproductive waters oligotrophic, and those in the middle of the range mesotrophic. Table 1.1 shows the concentrations of nutrients, the biomass and productivity of phytoplankton commonly associated with those three types of water body.

In regions with appropriate surface topography and high temperature for at least part of the year, a high rate of evaporation may lead to progressive concentration of salts within the water body, and to the formation of saline lakes. Their chemical makeup varies according to the local geology, so that there are carbonate, chloride and sulphate lakes. In parallel with the accumulation of these salts, there is often an accumulation of plant nutrients such as nitrate and phosphate, so that algal productivity can be very high. Table 1.1 shows this very clearly. Although saline lakes are often fishless and have therefore received less attention than freshwater lakes, it has been calculated that the volume of inland saline lakes of the world is almost equal to the volume of freshwater lakes. A good proportion of this is accounted for by the Caspian Sea — the world's largest inland basin separated from the ocean.

The nutrient status of freshwater lakes changes with time. A great deal of limnology in Europe and North America has been carried out on lakes lying in basins that were glaciated, say 10–20 000 years ago. These basins were left with relatively little soil and in general a poor nutrient status, so the lakes were oligotrophic. As vegetation colonized the lake basins and facilitated the formation of soil, and as the underlying rocks gave up

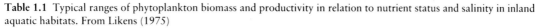

Table 1.1 Typical ranges of phytoplankton biomass and productivity in relation to nutrient status and salinity in inland aquatic habitats. From Likens (1975)

	Net primary productivity (g dry wt m^{-2} year^{-1})	Phytoplankton biomass (mg dry wt m^{-3})	Total phosphorus (ppb)	Inorganic nitrogen (ppb)	Total inorganic solutes (ppm)
Oligotrophic	15–50	20–200	1–5	1–200	2–20
Mesotrophic	50–150	200–600	5–10	200–400	10–200
Eutrophic	150–500	600–10 000	10–30	300–650	100–500
Saline	500–2500	1000–20 000	30–100	400–5000	1000–150 000

their salts and nutrients, the general direction of change was towards higher productivity. This has been detected, for example, in studies of the composition of cores from the lake sediments. For a time it was asserted that the natural 'succession' of characteristic organisms in the development of lakes was always towards eutrophication. However, in lakes with a long history free from glaciation, where soil and vegetation cover protects the underlying rocks from further erosion, the supply of new nutrients may be severely reduced. With continued loss of nutrients through, say, an outflowing stream, the overall direction of 'succession' may well be from eutrophic towards oligotrophic communities.

With the rapid growth of the human population in almost all parts of the world, there are now very few lakes that are free from man's influence. Since sewage is rich in plant nutrients, it is almost inevitable that lakes suffer to a greater or lesser degree from 'cultural eutrophication'. In many areas, phosphorus appears to be the nutrient most directly responsible for rapid increases in phytoplankton productivity, when added to lake water as a result of human activity. Even in areas remote from human habitation, airborne transport of sulphuric acid, nitrogen compounds and a variety of organochlorines leads to detectable changes in lake chemistry and fish yield. A particularly troublesome effect of airborne pollutants is the acidification of rivers to the point where migratory species of trout and salmon can no longer breed in them.

1.9.2 The depth of the water column

If a lake basin is relatively shallow, a large proportion of the water column has enough light to support photosynthesis and the volume of water below the thermocline is relatively small. When the lake stratifies, the organic matter produced in surface waters sinks below the thermocline, decomposes and produces a relatively high concentration of nutrients. When wind-driven mixing (usually in conjunction with autumn cooling) brings the nutrients into the surface waters, the next burst of phytoplankton production will be a strong one. Lakes of this type are often eutrophic.

One disadvantage of this type of lake is that the decomposing organic matter below the thermocline may cause a strong reduction in dissolved oxygen, or even create totally anaerobic conditions. In that case, the deep benthic fauna is restricted to a few species such as chironomid larvae or tubificid worms capable of surviving anaerobic conditions for short periods. Benthic animals not having this capability are restricted to the shallower zones of the lake. On the other hand, anaerobic conditions at the sediment surface stimulate the release of nutrients into the water column above.

If a lake is deep in relation to its surface area, the sinking organic matter from surface production will release nutrients that are distributed in a large volume of water. When the lake mixes, relatively low concentrations of nutrients enter surface waters and plant productivity is low. Such lakes are frequently oligotrophic. It must be remembered that this tendency is modified by the geochemistry of the surrounding drainage basin. However, mountainous terrain having deep, narrow lake basins is often also deficient in soil nutrients.

Depth of the water column affects marine productivity in an analogous manner. As Table 1.2 shows, the average rate of production per unit area of continental shelf is nearly three times as high as that of the open ocean. Especially shallow areas, such as the well-known fishing 'banks'; the Dogger Bank, the Grand Banks and George's Bank, may have productivity that is twice as high as the average for continental shelves. Estuaries are extraordinarily productive. In part this is a function of the 'estuarine trap' mechanism discussed earlier, but their shallowness is also a factor.

1.9.3 Development of fringing communities

Related to the matters dealt with in the previous section is the question of fringing communities. These are the areas shallow enough for attached plants to grow on the bottom. In general, beds of emergent or submerged aquatic plants, together with their epiphytes, are much more productive than the equivalent area of water containing only phytoplankton. Table 1.2 shows marine algal beds and reefs as being on average 20 times as produc-

Table 1.2 The magnitude of aquatic productivity in different environments. From Whittaker (1975)

Habitat	Net primary productivity (g dry wt m^{-2} year^{-1})		Secondary productivity (g dry wt m^{-2} year^{-1}) (mean)	Plant : animal productivity ratio
	Range	Mean		
Open ocean	2–400	125	8	1 : 0.06
Upwelling areas	400–1000	500	27	1 : 0.05
Continental shelf	200–600	360	16	1 : 0.04
Algal beds/reefs	500–4000	2500	60	1 : 0.02
Estuaries*	200–3500	1500	34	1 : 0.02
Lakes and rivers	100–1500	250	5	1 : 0.02
Swamps and marsh	800–3500	2000	16	1 : 0.008
Non-aqueous habitats	0–3500	760	0–20 (6)	1 : 0.003–0.02 (0.008)

* Excluding their fringing marshes.

tive per unit area as the open ocean. The corresponding communities in freshwater habitats are labelled 'swamps and marshes' in which the productivity is given as 16 times that of the open ocean. Any factor that leads to greater development of these fringing communities will lead to greater overall productivity for the water body.

For a lake, the shape of the basin is a major factor. Many glaciated lake basins have margins that drop off steeply to deep water, providing little opportunity for the development of fringing communities. Over long periods of time, wave action may erode platforms on these steeply sloping margins and allow limited development of fringing communities. On the other hand, a lake formed in a shallow depression in an area of rich soil may have very extensive fringing communities and a relatively small area of open water. The plant productivity will be very high. However, plant material from marshes and swamps has a high fibre content and is relatively indigestible to many animals. It dies, becomes detritus, and has to undergo a long process of conditioning by microorganisms before it becomes nutritious. In the process, much of the material is consumed by the microorganisms, and the efficiency of conversion to animal tissue is very low (Table 1.2, column 5). Algal beds and reefs are converted with a higher efficiency because much algal tissue has a low fibre content and can be assimilated directly by animals.

For a given type of lake, whether its shoreline is gently sloping or steep, the amount of fringing community is greater if the shoreline is convoluted than if it is straight. This principle applies equally to marine coastlines. Those with many estuaries, bays and inlets will have a better development of salt marshes, seaweed beds or mangrove forests than those that are relatively straight.

1.9.4 The gradients of rivers

It is obvious to the casual observer that there is a tremendous difference between a river that runs swiftly over steep terrain and one that meanders slowly across a nearly flat plain. The former presents difficult conditions for the growth of aquatic plants. Productivity within the river bed is usually minimal, and most of the organic matter found there consists of leaves fallen from overhanging trees or other kinds of terrestrial vegetation washed into the channel. The invertebrate fauna is usually dominated by the larvae of a few insects such as caddises and blackflies that are adapted to life in a swift current. These streams are the favoured spawning grounds for trout and salmon, but few other kinds of fish inhabit these reaches.

A slowly-moving lowland river can closely resemble a lake. It may have well-developed fringing communities in which the water movement is so slow that phytoplankton and zooplankton, similar to those found in lakes, may be abundant. While the bottom of an upland river will consist of coarse rocks and gravel, all else being washed away, the bed of a lowland river may be soft and muddy, with a rich variety of molluscs, worms, crustaceans and insects. Many more kinds of fish, such as cyprinids and members of the perch and pike family, will be found in these areas. Taking rivers as a whole, the unidirectional nature of the current exerts a limiting effect on the development of plankton in open waters, for organisms living in the headwaters today may be carried to the ocean in a matter of days. Many diatoms grow attached to the hard bottom in the upland reaches, and a proportion of these will be dislodged and carried in the open water. In lower reaches, a true 'potamoplankton' develops, with diatoms and rotifers being especially important. In the open water the crustacean plankton is relatively scarce, probably because the residence time in the water is too short for most species to complete their life histories, but in the fringing communities and in the river sediments, cladocera and copepods can often be found in great numbers.

1.10 THE BIOSPHERE AS A WHOLE

1.10.1 Aquatic algae and global warming

The fixation of carbon dioxide by aquatic algae has recently taken on a new, global significance. During the decade of the 1980s we have become much more aware of the biosphere as an entity. Contamination of the earth and the atmosphere by human activity is giving rise to problems that are truly global in scope. Holes in the ozone layer over the polar regions are thought to be attributable to photochemical reactions with chlorofluorocarbons, which are widely used in refrigeration and in the manufacture of packing and insulating

materials. Food webs, even in remote Arctic and Antarctic regions are known to be contaminated with DDT, polychlorinated biphenyls (PCBs) and other chlorinated hydrocarbons. The acid effluents of industrial plants are falling as acid rain and causing the acidification of lakes and rivers far from the point of origin. In addition to all these problems, it is now clear that industrial emissions are causing a rise in the concentration of CO_2 in the world's atmosphere and that this is likely to lead to a global rise in mean temperature. A consequence would be a rise in sea level resulting from thermal expansion of sea water and melting of the polar ice caps and also major changes in the patterns of distribution of temperature and rainfall.

Plant life takes up CO_2 in photosynthesis, but the respiration of plants and animals returns a proportion of it to the atmosphere. Clearing tropical rainforest and burning the trees is thought to be a factor contributing to the global rise in atmospheric CO_2. How important is the uptake of CO_2 by aquatic algae? Table 1.3 shows that aquatic habitats occupy about twice as much of the earth's surface as terrestrial habitats, yet present estimates are that net primary production in aquatic habitats is only about one-third of the total for the biosphere. However, this production is potentially important in the global balance of carbon, because a significant proportion of it sinks to the ocean floor or into the water column at great depth, from which it is unlikely to be returned to the atmosphere within the next century. The amount of carbon taken from the atmosphere and removed from circulation by deep lakes and the oceans is a significant item in the global budget of atmospheric CO_2.

Table 1.4 shows that of all the water on the

Table 1.3 Comparison between terrestrial and aquatic production: world totals. Data from Whittaker (1975)

	Surface area (10^6 km^2)	Approx. volume of biosphere* (10^6 km^3)	Net plant production (10^9 tonnes year^{-1})	Animal production (10^6 tonnes year^{-1})
Terrestrial	145	14.5	110.5	867
Aquatic	365	1445	59.5	3067
Ratio	1 : 2.5	1 : 99	1 : 0.54	3.54 : 1

* Based on average sea depth of 4000 m and assuming an average terrestrial inhabited zone 100 m deep.

Table 1.4 Percentage of the earth's water occurring in different aquatic systems. From Wetzel (1983)

Oceans	97.6
Ice	2.1
Exchangeable ground and soil water	0.3
Freshwater lakes	0.01
Saline lakes	0.01
Atmospheric water vapour	0.001
Rivers	0.0001

earth, well over 97% is in the ocean basins. The total aquatic production, estimated at 59.5×10^9 tonnes year^{-1}, is therefore predominantly marine. It has a carbon content of the same order of magnitude as the global emissions of CO_2 from the burning of fossil fuel. Biological oceanographers are therefore concerned to know whether this estimate of carbon fixation is accurate, what proportion of it is taken out of contact with the atmosphere each year, and how these numbers are likely to change when global warming takes place.

1.10.2 The production of food from aquatic habitats

An interesting feature of Table 1.2 is that there is an order of magnitude difference in the rate of secondary production per unit area of different habitats, and also nearly as great a difference in the efficiency with which photosynthetic production is converted to secondary production. The lowest efficiency of conversion is for swamps and marshes. As we saw earlier, most of the vascular plant production in these habitats is not readily digestible by animals and receives a lengthy period of processing by microorganisms before being consumed by invertebrate animals. We call this the detritus food web, and it is inefficient because the microorganisms convert much of the organic carbon produced by the plants to CO_2 or methane. Detritus food webs are also a prominent feature of salt marshes, seaweed beds and estuarine mudflats, and of many lakes and rivers (for review see Mann, 1988). It is therefore not surprising that they, too, have relatively low efficiencies of conversion of plant material to animal tissue.

On the continental shelves, especially in tem-

perate and arctic waters, a large proportion of a year's phytoplankton production is concentrated in the spring bloom. The zooplankton is unable to consume this material as fast as it is produced, so that a large proportion sinks to the bottom and is buried in the sediments. Upwelling areas have more continuous primary production, so there is a year-round population of consumers that uses the phytoplankton rather more efficiently. Finally, in the open ocean, about two-thirds of the total area is in tropical and subtropical regions where there is a permanent thermocline. There is a chronic shortage of upwelled nutrients and the phytoplankton depends heavily on the nutrients excreted by the zooplankton living in the same water mass. The zooplankton and phytoplankton populations remain more nearly in balance so that phytoplankton production is used relatively efficiently by the consumers.

When we examine the statistics for world fish catches, we find that by far the greatest proportion comes from continental shelves and upwelling areas. While catches from lakes and rivers are crucially important in the diet of peoples living remote from the sea, they account for only 10–20% of the world total. Why is it that, within the marine environment, the open ocean with its vast area and relatively high ratio of secondary to primary production produces such a small proportion of the world total? Under the relatively stable, stratified conditions found in tropical and subtropical regions year-round, and in temperate regions in summer, the average concentration of nutrients is low, and most of the primary production is by small cells, the nanoplankton and picoplankton, which have a favourable surface area : volume ratio. These in turn are consumed by microzooplankton, which are preyed upon by mesozooplankton, and so on. The result is that in the open ocean there are more steps between phytoplankton and fish than there are in coastal waters. Since there are losses by respiration and excretion, etc. at each stage of the process, the amount of fish production per unit area of the ocean is much lower than on the continental shelves.

Of those fish that are produced in the open ocean, the greatest biomass is in the form of small, bony fish living by day at depths of 200–1000 m. They are not particularly tasty and are too widely dispersed and too small to be economically harvested and marketed. Tuna and swordfish, on the other hand, are luxury foods that command a high price, so they are actively pursued across the world's oceans. They live in the upper 100–200 m of the water column and prey on smaller fish.

On the continental shelves, we find that in northern temperate regions the planktonic food webs lead to a relatively small number of commercially important species, such as herring and mackerel, and a relatively small number of bottom-feeding species such as cod, haddock and the flat-fish. In tropical regions, the species diversity is much greater for both pelagic and benthic species. This appears to be a reflection of the generally nutrient-low conditions and relatively greater stability of tropical food webs, as mentioned in connection with open water.

For about two decades after the Second World War the world catch of fish increased steadily as more areas were exploited and larger and more powerful vessels were used. In 1973, the total catch of fish and shellfish was nearly 70 million tonnes, but during the ensuing years that total has increased little and in several years has been less. It appears that there is not much scope for increased exploitation by traditional methods. When we compare the situation with that on land, we could say that man is still at the 'hunter-gatherer' stage in his exploitation of the sea.

In an effort to develop a parallel with agriculture, a great deal of effort is now being put into the development of aquaculture. In the tropics, for example, there is an old tradition of labour-intensive cultivation of shrimp and milkfish, and in China and Japan there is extensive cultivation of seaweed for human food. At the present time, in Europe and North America, new intensive methods are being developed for the rearing of luxury seafoods that command a good price, such as salmon, trout, oysters and mussels; moreover, the search is on for profitable ways of cultivating lobsters and halibut. In the shellfish rearing operations, the animals take natural phytoplankton as food, but salmon and trout are often held in nets

and fed with fish products derived from less valuable species. Whichever method is used, there is a prime requirement that the water be free of harmful pollutants. For this reason, careful management of coastal waters becomes a necessity. Not only must the effluents of cities be treated to minimize pollution, but care must be taken to see that intensive aquaculture operations do not themselves contaminate the waters to the detriment of neighbouring fish or shellfish farms. It is in this context that a detailed knowledge of the interactions between the physics of water movement and biological processes becomes extremely important.

FURTHER READING

Elton, C. (1927) *Animal Ecology.* Sidgwick and Jackson, London. (Reprinted in 1966 by Science Paperbacks and Methuen & Co, London.)

Fenchel, T. (1974) Intrinsic rate of natural increase: the relationship with body size. *Oecologia* **14**: 317–26.

Fenchel, T. (1987) *Ecology of Protozoa.* Science Tech Publishers, Madison, Wisconsin and Springer-Verlag, Berlin.

Likens, G.E. (1975) Primary production of inland aquatic systems. In: Leith, H. & Whittaker, R.H. (eds) *Primary Productivity of the Biosphere*, pp. 185–202. Springer-Verlag, Berlin.

Mann, K.H. (1972) Case history: The River Thames. In: Oglesby, R.T., Carlson, C.A. & McCann, J.A. (eds) *River Ecology and Man.* Academic Press, New York.

Mann, K.H. (1988) Production and use of detritus in various freshwater, estuarine and coastal marine systems. *Limnol. Oceanogr.* **33**: 910–30.

Purcell, E.M. (1977) Life at low Reynolds number. *Am. J. Phys.* **45**: 3–11.

Stockner, J.G. (1988) Phototrophic picoplankton: an overview from marine and freshwater ecosystems. *Limnol. Oceanogr.* **33**: 765–75.

Strickler, J.R. (1984) Sticky water: a selective force in copepod evolution. In: Meyers, D.G. & Strickler, J.R. (eds) *Trophic Interactions within Aquatic Ecosystems*, pp. 187–242. American Association for the Advancement of Science, Washington DC.

Taub, F.B. (ed) (1983) *Ecosystems of the World 23: Lakes and Reservoirs.* Elsevier, Amsterdam.

Wetzel, R.G. (1983) *Limnology.* 2nd edn. Saunders, Philadelphia.

Whittaker, R.H. (1975) *Communities and Ecosystems.* 2nd edn. Macmillan, New York.

PART 2
AQUATIC ECOSYSTEMS

2
Ecology of Water Columns

I. VALIELA

2.1 PRIMARY PRODUCTION

This chapter deals with how organic matter is produced in aquatic systems, and how the rates of production are controlled. We will also be concerned with the various fates of the organic matter that is produced, and will examine some of the consequences of some of the important steps in the production, consumption, loss and degradation of organic matter.

The organic matter that supports biological activity in water columns of fresh and sea water is mainly synthesized by organisms that carry out photosynthesis. Photosynthetic organisms capture light energy by means of certain pigments such as chlorophylls, and use this energy to fix carbon dioxide into organic compounds. Photosynthesis is a complex series of reactions that can be summarized as

$$CO_2 + 2H_2A \xrightarrow[\text{pigments}]{\text{light}} CH_2O + 2A + H_2O. \quad (2.1)$$

In oxygenic photosynthesis, carried out by plants and algae, the electron donor (H_2A) that is split is water. Some bacteria carry out anoxygenic photosynthesis, and make use of other electron donors such as H_2S.

Producer organisms are called 'autotrophs' because they produce their own organic matter, in contrast to 'heterotrophs' which use organic matter produced by other organisms. Organic matter does not, of course, just consist of carbon, hydrogen and oxygen. Many other elements are also involved, so that a more complete equation summarizing autotrophic activity is

1300 kcal light energy + 106 mol CO_2
+ 90 mol H_2O + 16 mol NO_3^- + 1 mol PO_4^{3-}
+ trace amounts of mineral elements
= 3.3 kg biomass + 150 mol O_2
+ 1287 kcal heat. (2.2)

There is another process by which some organisms fix CO_2, called chemosynthesis. This process is carried out by certain bacteria that can use chemical energy contained in inorganic compounds, instead of light energy. Chemosynthesis may be important in certain localized environments, such as deep sea vents where reduced compounds are released by geochemical processes, or in coastal environments with reduced layers of water or reduced sediments. In general, however, rates of chemosynthetic production are much smaller than rates of photosynthetic production.

2.2 CONTROL OF PRIMARY PRODUCTION BY BOTTOM-UP MECHANISMS

Rates of photosynthetic primary production by phytoplankton vary greatly in different waters (Table 2.1), and at different times. The large variation in photosynthetic rates suggests that there are factors that differ from place to place and time to time, which determine the evident differences in photosynthetic activity.

There are several methods by which we measure production in aquatic systems. The most frequently applied procedure measures the rates at which a radioactively-labelled inorganic carbon source is fixed by producer organisms. Table 2.1 (left column) provides a summary of net primary production rates in various aquatic environments. The

Table 2.1 Average primary production, producer biomass, turnover time and chlorophyll in some major aquatic environments. Data from Whittaker & Likens (1975)

	Net production (g dry wt m^{-2} year^{-1})	Biomass (kg dry wt m^{-2})	Annual P/B	Chlorophyll (g m^{-2})
Marine water				
Open ocean	125	0.003	42	0.03
Upwellings	500	0.02	25	0.3
Continental shelf	300	0.001	300	0.2
Near-shore reefs	2500	2	1.3	2
Estuaries	1500	1	1.5	1
Fresh water				
Wetlands	3000	15	0.2	3
Lakes and streams	20–8000	0.02	20	0.2

ranges of net production are similar in fresh water and coastal marine water. Oceanic waters support far lower net production than other environments.

The rates of primary production we now have available (summarized in Table 2.1), especially those from oceanic waters, are currently being re-evaluated. It has been argued, from considerations of the chemistry of different masses of oceanic sea water, that net primary production ought to be higher than we have measured in the open oceans.

A distinction must be made between gross and net primary production. The total amount of photosynthesis achieved by a plant or alga during a certain time period (e.g. 24 hours) is the gross production, but during that time the organism is also carrying out respiration, breaking down some of the organic compounds synthesized in photo-synthesis to release energy. The difference between gross production and respiration in that time period is net production. One of the methods used to measure primary production uses radioactive carbon; it measures something close to net production.

The extent to which biomass is produced may be limited by any of the terms on the left side of equation 2.2, or by light supply. Below we will examine the evidence for control of primary production by two of the principal control mechanisms, light and nutrients. These mechanisms affect mainly primary producers, that is, the 'bottom' of food webs, and their effects may have repercussions that move 'upward' through the various trophic steps. Such controls have been called bottom-up controls, in contrast with top-down controls that may be exerted by consumers on top of the food webs and whose consequences cascade down the trophic steps of a food web.

2.2.1 The role of light & photosynthesis

The intensity of light, that is the rate at which photons strike a given surface, varies over space and time. One of the most common demonstrations of the effect of light is to examine depth profiles of light and of photosynthesis down a water column. Light energy is absorbed by water itself, by organic substances dissolved in the water and by particles. As a result, light intensity diminishes exponentially down the water column (Fig. 2.1). Profiles depicting photosynthetic rates down the same water columns often have a characteristic shape: photosynthetic rates increase down to a certain depth, and then decrease beyond that (Fig. 2.1).

Exposure to high light intensities, as occurs near the water surface, tends to inhibit photosynthesis. This is thought to occur because of destructive photo-oxidation reactions by light that cannot be absorbed by the photosynthetic apparatus, as well as damage by the accompanying ultraviolet radiation. Deeper in the depth profile, water has absorbed more of the light energy, the inhibition is reduced and photosynthesis increases. At some point in the profile however, the diminished intensity of light at greater depths can only support lower rates of photosynthesis, and hence production rates are lowered.

The vertical scale of curves, such as the one

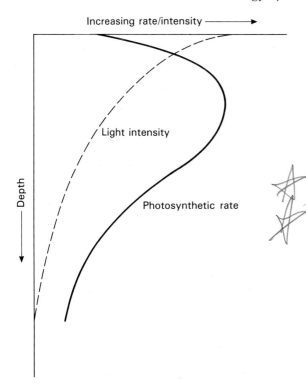

Increasing rate/intensity ──────➤

Depth

Light intensity

Photosynthetic rate

Fig. 2.1 Generalized sketch of vertical profiles of light availability and corresponding rate of photosynthesis in water columns in which phytoplankton are uniformly distributed in the vertical dimension. Non-uniform distributions may change the photosynthesis pattern.

depicted in Fig. 2.1, varies. Photosynthesis may occur to depths of over 100 m in very clear oceanic or lake water. In very nutrient-rich waters photosynthesis may take place over only the upper few centimetres, because the large mass of cells cuts down light penetration. The rate at which photosynthesis takes place may vary, even under adequate light intensities; in nutrient-rich waters, the maximum value of photosynthesis may be much higher than in nutrient-poor waters, but the depth of light penetration may be much reduced. This is only one of the many ways in which light interacts with nutrients in setting the rate of photosynthesis.

2.2.2 The role of nutrients

The fundamental importance of nutrients is that the rate at which they are supplied may determine the rate of primary production. The applied importance is that often it is man-originated enrichments

that change the nature of aqu
Knowledge of nutrient limitati
fundamental to understanding h
tems function, and also of appli
eutrophication cannot be remedi
know what element limits rates of producer growth.

The basic and applied importance of potential limitation of producer activity by nutrients has resulted in a large body of literature, in which at least three meanings of 'nutrient limitation' are used. First is the empirical notion that if after nutrients are added, net primary production increases, the productivity of the *system* is nutrient-limited, regardless of the changes in species composition that often result from the enrichment. We might refer to this as limitation of potential net production. This is the most common understanding of the concept, and the one most often used when dealing with eutrophication.

The second concept of nutrient limitation is evident in papers concerned with limitation of the growth rate of phytoplankton taxa currently present in the water column. The argument has been put forward that cells found in nutrient-poor oceanic waters contain molar ratios of C:N:P of 106:16:1 (a ratio discovered some time ago by the eminent oceanographer A.C. Redfield), and since in the lab, cells at Redfield ratios show maximal growth rates, oceanic phytoplankton must grow at maximal rates and may not be nutrient-limited at all. Of course, we might expect that species living in any habitat are, by definition, well-adapted to be successful there; if additional nutrient supplies were to become available, other species might replace the original ones, and the new species may be more productive. Therefore, potential net primary production could be nutrient-limited, even if the growth of the cells now in the water is not. This physiological–adaptive notion is most relevant for physiological understanding of the status of phytoplankton present in the water.

A third notion is limitation of net ecosystem production (net ecosystem production = gross primary production − total respiration in the ecosystem), which is an estimate of the actual rate at which biomass accumulates (or does not, as the case may be) in an ecosystem. Nutrient supply may affect gross production, but the relation of nutrient supply to respiration rates (and hence to total oduction) is probably more indirect, so the link to eutrophication (nutrient enrichment) is not clear cut. This approach is most relevant for whole-system dynamics, accumulation of materials over long time periods and geochemical studies.

We will use the first meaning of nutrient limitation in this chapter.

Nutrient limitation of primary production

Supply of nutrients determines the rate of growth of specific populations of autotrophs, as described by equation 2.2, if any of the nutrients are present in limiting amounts. Different autotrophic taxa require somewhat different combinations of nutrients (more silicon for diatoms, for example, since diatoms have silica tests surrounding the cells), so perhaps different elements may restrict growth in different places and times.

In freshwater systems, most studies show that the supply of phosphorus largely determines the rate of primary production. This evidence comes from experiments in which algal cultures or mixed assemblages from natural waters were enriched with specific nutrients, and from experimental enrichment of whole lakes. These manipulative approaches are corroborated by correlative data that show that standing crops of phytoplankton (expressed as amount of chlorophyll) increase as rates of phosphorus loading to freshwater loading increase.

The results of experimental manipulations are evidence of a cause-and-effect relationship, and therefore provide the more compelling results. Nonetheless, both kinds of data support the case for phosphate limitation of freshwater aquatic production.

In coastal marine waters nitrogen has been thought to be the primary limiting element, based on enrichment experiments done in containers, and on consideration of correlative data on the ratio of nitrogen to phosphorus (N:P) in the water, relative to the Redfield ratio. The classic enrichment experiment was one designed to study whether nitrogen or phosphorus limited phytoplankton growth in New York Harbor and neighbouring waters (Fig. 2.2). Phytoplankton from stations off New York City were placed in containers and nitrogen and phosphorus were added to separate containers. The results show that near the harbour, where nutrient concentrations were high because of contamination with waste water (Fig. 2.2a and b), there was no response by the phytoplankton to the addition of nutrients (Fig. 2.2c). Further offshore, where nutrient concentrations were lower, the phytoplankton grew in response to the addition of nitrogen but not to the addition of phosphorus (Fig. 2.2c). These results suggested that in coastal waters that were not markedly enriched, the phytoplankton were nitrogen-limited.

The results of enrichment experiments are corroborated by correlational data that show that for whole coastal systems, average chlorophyll concentrations become greater as loading of nitrogen into the systems increases (Fig. 2.3). There is considerable scatter to this relationship (note that a log scale is used). The correlational data are less compelling than the experiments, since it is difficult to attribute cause and effect in the relationship, and also because the loading rates are rough approximations.

In open ocean situations, enrichment experiments have had ambiguous results. It may be that trace elements are more important than macronutrients for oceanic phytoplankton, due to either their low supply or perhaps their toxicity.

Unlike in lakes, it has been difficult to do whole-system enrichment experiments in marine waters. This is unfortunate, since this is the approach that provides the strongest kind of experimental evidence, avoiding the artifacts possibly created by enclosing algae in small containers. There is new evidence, at least for some coastal waters, that the identity of the limiting nutrient may change seasonally, or may be greatly altered by the rate of water renewal. The data available suggest that nitrogen is the principal nutrient that limits potential net production in marine coastal waters, but perhaps its limiting role is taken over by phosphorus during certain times of year, or in situations where salinity is low.

Nutrient uptake and growth

The rate of growth of phytoplankton is related to nutrient supply in the water in an indirect way.

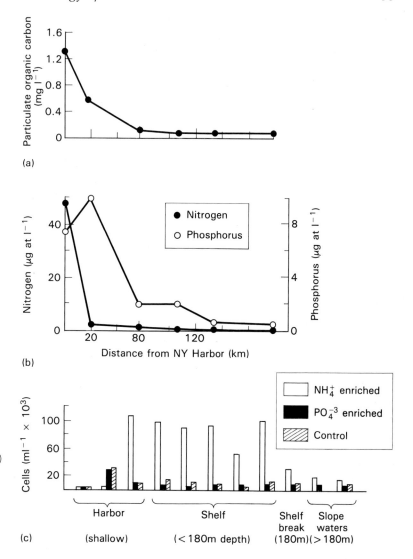

Fig. 2.2 Particulate organic carbon (a) and nutrients (b) in a transect from New York Harbor to offshore. (c) Enrichment experiments. Results of nutrient additions to samples of phytoplankton from the transect. From Ryther & Dunstan (1971).

The relationship involves mechanisms of uptake by cells, and internal pools of stored nutrients.

Nutrient uptake by cells takes place at discrete active uptake sites on cell membranes. Uptake rates increase as ambient nutrient supply increases, but there are fewer and fewer sites available as concentrations of nutrients increase. The discrete uptake sites become saturated at some higher nutrient concentrations. The relationship is therefore one of decreasing response to concentrations, tending to an asymptote (Fig. 2.4a).

The amount of the nutrient in question that is stored within the cells is in turn controlled by the rate of uptake of cells (Fig. 2.4b). Although cells may store nutrients in excess of current needs, there seems to be a feedback between cell content of that nutrient and the rate with which a given nutrient may be taken up by the cell.

Lastly, the internal pool of nutrients seems to control the rate at which a phytoplankton population may grow (Fig. 2.4c). The relation of ambient nutrient supply and growth of cells is thus at least a three-step process: ambient nutrient concentration determines uptake rates, uptake leads to a certain

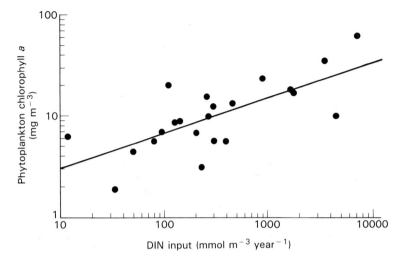

Fig. 2.3 Concentrations of phytoplankton chlorophyll *a* in the midregion of various estuaries, in relation to input of dissolved inorganic nitrogen (DIN). From Nixon & Pilson (1983).

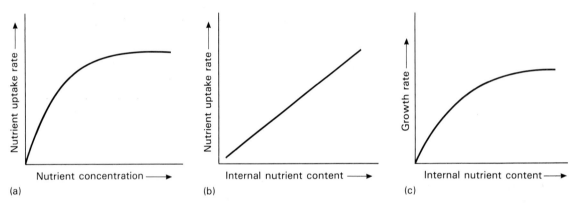

Fig. 2.4 Diagrammatic sketch of the relationships between ambient nutrient concentration, uptake rate, internal nutrient content and growth rate of phytoplankton.

cell content of the nutrient and cell content sets the rate of growth.

2.2.3 Seasonal variation in primary production

Production by phytoplankton varies over the year in many different ways, each responding to the particular combination of controlling factors present in each locality. One cycle that has been described many times is that for temperate marine waters in the northern hemisphere (Fig. 2.5), a cycle that serves as an example of how light and nutrients may combine to set rates of net primary production.

During late autumn and winter, nutrients may accumulate in the water column (Fig. 2.5b) because uptake by phytoplankton is low, and nutrients are renewed by upward movement of nutrient-rich deeper waters, by freshwater sources, or by regeneration in bottom sediments, thus leading to greater concentrations in the water. The Swedish pioneer of oceanography H.V. Sverdrup long ago suggested that the reason why uptake by phytoplankton may be low is that the phytoplankton population may be mixed vertically, below the critical depth (that depth below which respiration for the population of phytoplankton in the water column exceeds photosynthesis) (Fig. 2.5c), so that photosynthesis does not exceed respiration.

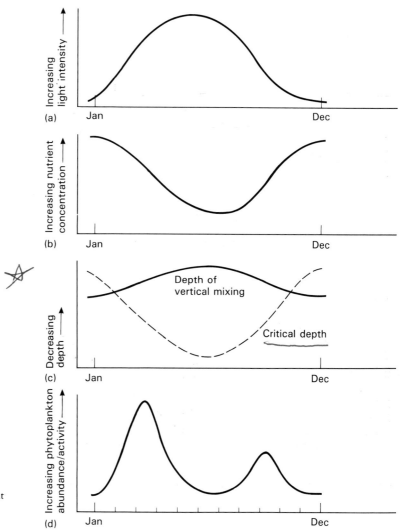

Fig. 2.5 Simplified scheme of the seasonal cycle of phytoplankton for northern hemisphere temperate waters. (a) Light intensity; (b) nutrient concentration; (c) mixing; (d) phytoplankton abundance/activity.

Since this is the case, during winter there is little phytoplankton growth. At some point during late winter or early spring, where the weather improves enough (while light intensity increases also, Fig. 2.5a), the phytoplankton populations are held in an upper layer of water that receives sufficient light. It also seems that photosynthesis is less responsive than respiration to changes in temperature. Hence, even in the cold water of late winter and spring, photosynthesis may occur at a reasonably fast rate, while respiration may be held to a low rate by the low temperatures. This differential response may actually increase the critical depth and result in a bloom.

Regardless of how the bloom is caused, phytoplankton use the plentiful nutrient supply, the spring bloom takes place (Fig. 2.5d) and nutrient concentrations in the surface waters become lower. Phytoplankton populations built up during late spring have, by summer, depleted nutrients (nitrate is most often the first nutrient to be so depleted, hence the conclusion that coastal net production is nitrogen-limited) (Fig. 2.5b and d).

During the spring bloom, phytoplankton cells

may double in number each day, an exceedingly fast growth rate. Few grazers could ever keep up with such a rate of provision of new cells, so that grazers are unlikely to be able to control the growth of phytoplankton; hence blooms take place. In summer, however, warmer temperatures create a warm, shallow upper layer that does not mix vertically, and hence nutrients can be depleted by phytoplankton within this upper layer. After nutrients are depleted, photosynthesis and growth rates of phytoplankton slow down (Fig. 2.4). It is this time of year when grazers of various kinds are most likely to have an effect on the producers, and may in fact lower phytoplankton abundance.

Recent research shows that zooplankton grazers appear to have internal clocks that trigger inactivity or resting stages. In late summer or early autumn, grazers may become inactive and/or move to greater depths, thus there is a release of grazing pressure on the surface phytoplankton. This may be the reason why at times there is a small autumnal peak in phytoplankton growth (Fig. 2.5d). Of course, it may be that autumn also brings storms that prompt vertical mixing, and hence renew nutrient supply to the upper layers of the water column.

The pattern just described should not be taken as the template for plankton seasonal cycles everywhere. There are many exceptions to the pattern. In shallow marine water, for example, there may be sufficient repeated vertical mixing for depletion of nutrients from the upper layer not to occur; storms may also temporarily disrupt the water column, even in summer in deeper waters, and bring up nutrient-rich water, creating a short-term bloom.

During the last 10 years or so, new methods and novel experiments and observations applied to, and about, aquatic primary production have built on or challenged generalizations and ideas that were thought to be reasonably well-established. This renewal of knowledge is deepening insight as to how aquatic systems operate. We are also in a time when it is harder to generalize about ecosystems, because explanations have become more and more linked to specific features of given environments. Hence, it is increasingly difficult to describe generalities of patterns — for example, a 'typical' seasonal cycle — since each local system has its own peculiarities rather than fitting into

a 'type'. Local idiosyncrasies and specific time courses in each site create a different ecological situation. The generality that remains is in the processes involved in functioning, and in whatever control mechanisms exist. These processes of production, respiration, grazing, predation, nutrient uptake and release, decomposition, hydrodynamics, etc. occur everywhere to some degree, and the combination determines how each ecosystem functions. To understand aquatic systems, therefore, it is useful to focus not on descriptive typologies but rather on processes and control mechanisms.

2.2.4 Spatial heterogeneity

We have discussed our very simplified version of the temperate seasonal cycle, principally to show that bottom-up controls such as light and nutrient supply (plus top-down mechanisms such as grazing, to be discussed below) are principal controls of net primary production. We need to emphasize that local hydrographical features are of major importance in mediating how these control mechanisms work. The effects of light and nutrients are indeed important, but what we can see of production in waters of the world differs sharply from what we might predict, based merely on the basis of variation of light and nutrients.

Examine a map of distribution of annual net primary production over the world's oceans (Fig. 2.6). Although light clearly has a role in starting the seasonal cycle, and in setting overall rates of primary production, it is evident that if light were the dominating control, we would certainly not see this pattern of primary production.

We have not emphasized temperature as a control factor, although temperature certainly affects virtually all biological transformations, and hence must be important to some extent. Nonetheless, if temperature was critical, we would see quite another geographical pattern.

What we find in Fig. 2.6 is primarily the result of nutrient provision by hydrographical mechanisms. High annual net production occurs where there is some physical process that creates enough mixing to bring nutrient-richer deeper waters to the surface, where phytoplankton can then use them. These upwellings, fronts and shear zones

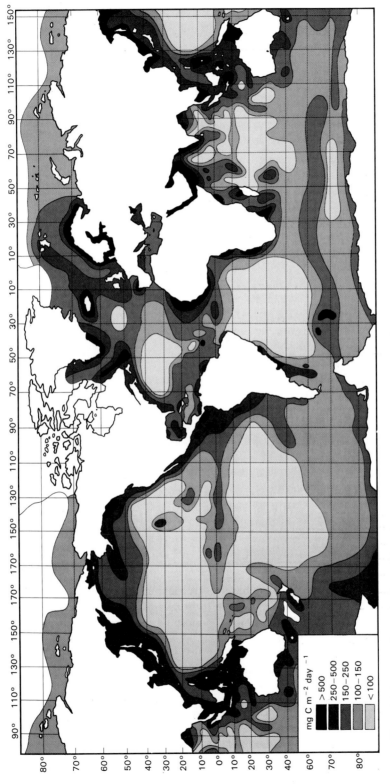

Fig. 2.6 Primary production in the world's oceans.

mg C m^{-2} day^{-1}

>500
250–500
150–250
100–150
<100

have been discussed in Chapter 1. All of these hydrographical processes act to deliver nutrients to surface waters, and hence promote higher levels of primary production.

A second major point to be noted from Fig. 2.6 is that rates of primary production are very heterogeneous over space, because of local differences in the suite of control mechanisms that affect them. We will find similar spatial heterogeneity at every spatial scale, from the many kilometre scale of the globe (Fig. 2.6) to microscales of centimetres.

The importance of heterogeneity of primary production at small spatial scales is being recognized as a major area needing research. For example, one of the major controversies today is about whether we have adequate measurements of rate of primary production in open water columns. There are a number of technological problems with current methods, and also geochemical arguments regarding the appearance of oxygen in excess of what could be produced by phytoplankton if the rates measured with present carbon isotope techniques are correct. In addition, and more relevant here, it has been pointed out that if the spatial distribution of primary production were to have very scattered, low volume but very productive 'hot spots', it would be very unlikely for us to be able to have sampled these small volume 'hot spots' using our present sampling methods; we could therefore be underestimating global net primary production because of the pattern over smaller scales.

The spatial mosaics we find in almost all environments are turning out to be the result of major processes acting on units of the environmental mosaic, each unit with perhaps slightly different histories. Thus, it is becoming clear that natural environments consist of a mosaic of patches, not a uniform 'average' assemblage of organisms. The spatial structure now recognized has altered our view of the structure of environments, as well as changing our approach to studying them.

2.3 THE FATE OF ORGANIC MATTER SYNTHESIZED BY PRODUCERS

Organic matter synthesized by aquatic primary producers is a transient substance, subject to many transformations. Herbivores eat live algal and plant biomass, detritivores and microbes consume organic matter after senescence and death of the producer and water flow and gravity transport organic matter away from the photic zone or site of production. In the following sections we examine some principal types of organism that consume organic matter, and describe some major properties of these consumers.

2.3.1 Consumption and transformations in aquatic food webs

Classical views of water column communities

For many years water column food webs were thought of as consisting of a series of well-defined trophic levels through which energy flowed. In the lowest level, phytoplankton cells (mainly diatoms and dinoflagellates in the sea; diatoms, chlorophytes and desmids in fresh water) fixed carbon, and hence were the primary producers that supported the upper trophic levels of secondary producers. The principal grazers, such as copepods and cladocerans, ate phytoplankton and were in turn consumed by larger predators, such as insects or fish. In certain lakes and oceans, larger fish or non-baleen whales, seals or man were at the very top of the food web.

It was thought that perhaps only 10% of the energy flowing into one trophic level was transferred to the next; as a result it was calculated that a maximum of five trophic steps could be sustained in open water food webs. So much energy was dissipated by respiration of organisms at each trophic level that by the fifth trophic step, virtually no energy remained to be passed on to a hypothetical sixth trophic level.

These notions of trophic levels were always recognized to be an oversimplification. For example, many taxa feed at various trophic levels, it was hard to know what to do with detritivores, energy transfer efficiencies did not always turn out to be 10% and microbes were ignored. Nonetheless, the trophic levels idea was still used as the anchor for explanations as to how aquatic communities functioned.

The classic concept of food webs said that energy flowed up the trophic levels; organic matter that was not consumed was degraded by bacteria and fungi. The bacteria and fungi were thus the principal regenerators of the nutrients that were sequestered in the organic matter synthesized by primary and secondary producers. These regenerated nutrients, plus externally supplied nutrients, kept the system going. Control of the system could occur via nutrient supply to the producers, or by action of the top predators in determining how many prey existed.

One application of the idea of trophic levels was that of bioaccumulation. Studies showed that contaminants, such as chlorinated hydrocarbons or heavy metals, accumulated in food webs. Concentrations of pesticides, for example, could be fairly low in phytoplankton, but increased in zooplankton, and were highest in piscivorous fish or in birds. This bioaccumulation came about because the pesticides contained in food ingested by consumers were not degraded in the bodies of the consumers and accumulated there. As the contaminants moved up the steps of the food web, the concentrations of toxic materials eventually reached damaging thresholds, and consumers near the top of the food web began to show toxic effects.

An example of damaging bioaccumulation is the effect of chlorinated hydrocarbon pesticides on ospreys, birds of prey that feed exclusively on fish. Osprey populations in the eastern USA were decimated during the 1960s, because chlorinated hydrocarbons used as pesticides found their way to water bodies, and moved up aquatic food webs. The chlorinated hydrocarbons accumulated in osprey tissues and interfered with calcium metabolism. As a result egg shells were so thin that few eggs managed to survive long enough to hatch.

Actually, the matter of bioaccumulation in aquatic environments is not so clear-cut. It turns out, for example, that the concentration of petroleum, chlorinated hydrocarbons and other synthetic compounds, such as polychlorinated biphenyls (PCBs) in fish is more related to fat content of the fish tissues than to trophic level of the fish. Moreover, it has been shown that concentration of contaminants such as PCBs increase during the life of a consumer such as lake trout. Bioaccumulation in aquatic food webs is therefore not only a function

of the trophic position of consumers, but also depends on other factors.

Types of organisms of the classical food web

Plankton organisms drift passively, carried along by motion of the water masses they inhabit. This is not to imply lack of motility; many of these organisms are capable of active swimming. By and large, however, their swimming speeds are small compared to the movement of water masses.

Classic marine phytoplankton (Fig. 2.7) include many species of diatoms, dinoflagellates, coccolithophorids, a few cyanobacteria and a rich assortment of other taxa. Freshwater phytoplankton contains diatoms, dinoflagellates, chlorophytes, desmids and some other taxa.

Marine zooplankton (Fig. 2.8, Table 2.2) include species of protozoans, coelenterates, ctenophores, chaetognaths, annelids, molluscs, arthropods (particularly the crustaceans) and chordates. The crustaceans were thought to be the most abundant marine group, especially the copepods. Crustaceans range widely in size, from small naupliar stages less than 1 mm long, to large euphausiids several centimetres long. Crustaceans feed as herbivores, carnivores, or parasites.

Freshwater zooplankton is dominated by crustaceans (copepods and cladocerans), rotifers and some insects (Table 2.2). While the taxonomic diversity of phytoplankton of fresh water is similar to that of marine water, the species richness of freshwater zooplankton is dwarfed by the great variety of taxa in marine zooplankton. The difference may stem from the greater area, depth, antiquity and geological continuity of oceanic planktonic environments, although no direct evidence for this claim exists. Large groups of taxa that are common in the sea are missing in fresh water: there are no freshwater foraminifera, radiolarians, acantharians, ctenophores, chaetognaths, echinoderms, appendicularians and there are many fewer types of molluscs.

Nektonic organisms are large enough, and swim fast enough, to be able to overcome water motions. This category includes a series of very diverse taxa (Fig. 2.9). In the sea there are large crustaceans such as free-swimming pelagic crabs, krill, molluscs

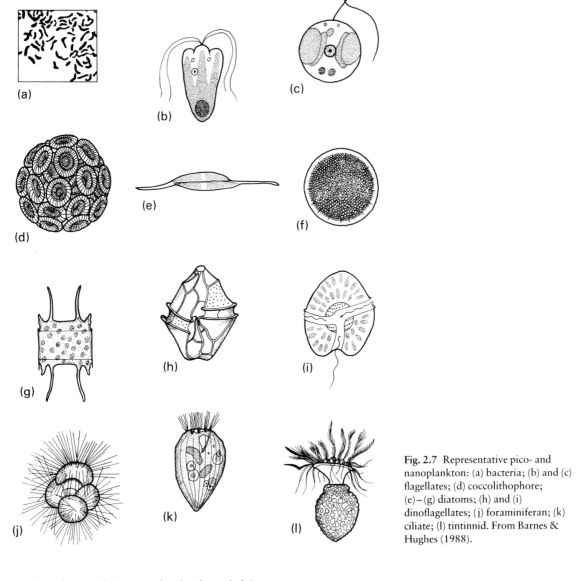

Fig. 2.7 Representative pico- and nanoplankton: (a) bacteria; (b) and (c) flagellates; (d) coccolithophore; (e)–(g) diatoms; (h) and (i) dinoflagellates; (j) foraminiferan; (k) ciliate; (l) tintinnid. From Barnes & Hughes (1988).

(such as the nautilus), many kinds of squid, fish, sea turtles, sea snakes, seals, whales and penguins. Perhaps marine birds, such as albatrosses, alcids, petrels, shearwaters, boobies, cormorants, tropic birds and frigate birds, could also be loosely called 'nekton'.

Freshwater nekton are somewhat less diverse. There are a variety of crustaceans, insects, fish, amphibians, reptiles and some freshwater mammals and birds.

New views of aquatic communities

A number of new findings and developments have very substantially changed our notion of the structure and functioning of aquatic food webs. These recent ideas include the discovery of new kinds of producers, the realization that size of organism carries with it clear constraints about ecological function, development of new methods of measurements of production and new observations

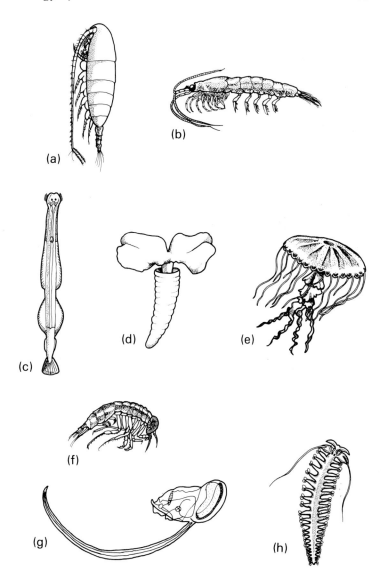

Fig. 2.8 Representative micro- and mesoplankton: (a) copepod; (b) euphausiid; (c) chaetognath; (d) pteropod; (e) jellyfish; (f) hyperiid amphipod; (g) appendicularian; (h) polychaete. From Barnes & Hughes (1988).

about the activity and abundance of microbes and metazoa of the water column.

One of the major areas of new knowledge that has changed during the last decade is knowledge about the kinds of microbes and invertebrates found in the water, especially in the sea. The new information can be conveniently summarized by referring to organisms that can be classed into various size categories.

Picoplankton (organisms measuring 0.2–2 µm) consist principally of bacteria, plus a very few eukaryotic organisms. New direct counting methods using stains and fluorescence microscopy have shown that there are from 10^5 to $>10^6$ bacterial cells ml^{-1} in sea water; these numbers turn out to be two orders of magnitude greater than earlier estimates, obtained using counts of colonies grown on agar plates. The earlier method greatly underestimated total microbial abundance because the specific media used were very selective, and so allowed a minority of the bacteria to grow and be counted.

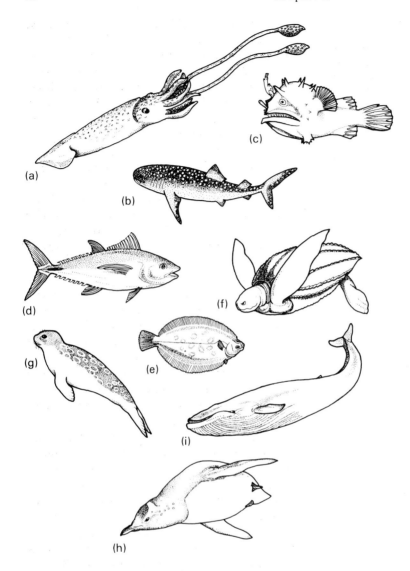

Fig. 2.9 Representative nekton: (a) squid; (b) shark; (c) deep sea fish; (d) tuna; (e) flatfish; (f) sea turtle; (g) seal; (h) penguin; (i) whale. From Barnes & Hughes (1988).

Another recent discovery is that small, picoplanktonic, photosynthetic cyanobacteria are nearly ubiquitous and may carry out a substantial part of primary production in nutrient-poor waters (Table 2.3). The contribution of picoplankton to net primary production and to biomass vary widely, but both can be large.

Another group of newly discovered picoplankton are small (0.6–0.8 μm) prochlorophytes that can be abundant (reaching up to 10^{11}–10^{15} cells ml^{-1}) in the lower illuminated layers of the open ocean. These cells are most easily detected by the (only recently available) technique of flow cytometry, rather than microscopically. These very small organisms contain chlorophyll *a*, *b* and *c*, and may be relatives of *Prochloron* (the putative organism that entered into symbiotic relationships with other organisms) and are the origin of chloroplasts of higher plants. Very little else is known about these potentially ecologically important organisms.

Nanoplankton (2–20 μm) include a variety of organisms, some of which are photosynthetic, others grazers and carnivores. They range in abundance up to 10^4 cells ml^{-1}, as determined by

Table 2.2 Representative zooplankton taxa present in either marine or fresh waters. From Lehman (1988)

	Marine water	Fresh water
Protista		
Mastigophora	*	*
Sarcodinea	*	*
Foraminifera	*	0
Radiolaria	*	0
Ciliata	*	*
Tintinnina	*	0
Metazoa		
Cnidaria		
Hydrozoa	*	+
Scyphozoa	*	0
Ctenophora	*	0
Rotatoria	+	*
Mollusca		+†
Gastropoda		
Heteropoda	*	0
Pteropoda	*	0
Annelida		
Polychaeta	*	0
Arthropoda		
Crustacea		
Branchiopoda	+	*
Cladocera	+	*
Copepoda	*	*
Calanoida	*	*
Cyclopoida	*	*
Malacostraca		
Mysidacea	*	*
Amphipoda	*	+
Euphausiacea	*	0
Decapoda	*	0
Insecta	0	*
Chaetognatha	*	0
Chordata		
Appendiculata (Larvacea)	*	0
Thaliacea		
Cyclomyaria (Doliolida)	*	0
Desmomyaria (Salpida)	*	0
Diverse larvae of benthic invertebrates, particularly in coastal regions	*	0

*, Common; +, present, but rare or represented by few species; 0, absent or extremely rare. †, In Old World lakes, larvae of *Dreissensia* (Bivalvia) can be common seasonally.

the new fluorescence methods. Many juvenile stages of zooplankton of the next size category, the microplankton, fall into the nanoplankton size class.

Microplankton (20–200 µm) include those organisms that comprised the 'classical' phytoplankton (diatoms, flagellates, etc.), plus ciliates, radiolarians and acantharians. Copepods, cladocerans and larvae of larger animals are also in this size range, which includes most of the 'classical' zooplankton. Typically, one to 10 microplankton organisms will be found in 1 ml of water.

Larger organisms (mesoplankton (>200 µm) and nekton) include a wide variety of taxa, such as some types of very large copepods, shrimp, crabs, squid, fish, whales and many others. Their abundance is usually lower than that of the smaller organisms, but it has been generally difficult to obtain good estimates of the population density of these organisms.

There is one very heterogeneous group of animals, largely mesoplankton, that were virtually ignored in the classic ideas of marine food webs. Marine waters contain many kinds of gelatinous animals of very diversified body design, including ctenophores, coelenterates and urochordates. These vary in size from a few centimetres to a few metres in length. Although gelatinous animals were known to exist, there were no methods available to assess how numerous they might be.

Just as new techniques have pointed out that microbes are more abundant in the sea than thought previously, new ways to collect samples of animals from the sea have shown that the old idea of copepods being the principal zooplankton needs some revision. Oceanographers used to largely conduct 'remote sensing' research, collecting samples using nets and sampling devices dragged by wires a long distance away. Studies using SCUBA and minisubmarines in open ocean waters have shown that there is a plethora of very delicate gelatinous organisms in the water column that were destroyed or undersampled by nets, and hence were for the most part ignored by marine ecologists.

The delicate gelatinous organisms may be as abundant as copepods, and as important trophically in the sea. The biomass of certain gelatinous free-swimming coelenterates, ctenophores and chordates may be at least as large as copepod biomass, the group classically thought to be the most prominent zooplankton taxon. There are still insufficient data with which to evaluate their role, but they

Table 2.3 Rates of primary production and biomass of picoplankton in marine and freshwater ecosystems. From Stockner (1988)

	Primary production		Biomass	
	Rate (mg C m^{-3} h^{-1})	% of total primary production	Chlorophyll concentration (mg m^{-3})	% of total chlorophyll
Marine water	1–31	1–90	0.5–1	1–90
Fresh water	1–8	16–70	0.3–1	0.2–43

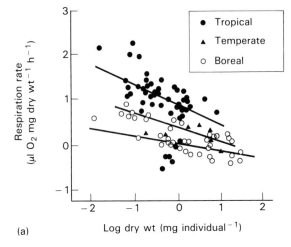

(a)

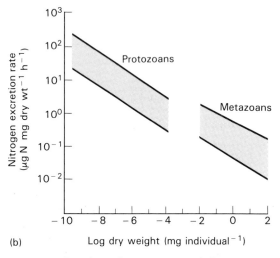

(b)

Fig. 2.10 Effect of size of organism on metabolic processes. (a) Weight-specific respiration of zooplankton. (b) Weight-specific release of nitrogen by planktonic protozoans and metazoans. Lines depict upper and lower extremes of measured values. From Ikeda (1970) and Caron (1991).

are surely quantitatively significant as consumers of surface-produced organic matter. Measurements of secondary production by these gelatinous forms are not widely available as yet.

The importance of size in aquatic food webs

In Chapter 1 and in the previous section we discussed terms such as pico-, nano-, micro- and mesoplankton, and specified size ranges of organisms in these size categories. Part of the reason such emphasis has been placed on size classes is that often different sampling and handling methods have to be applied to measure the abundance of organisms in these different classes. More fundamental, however, is that size is one of the principal criteria underlying how aquatic food webs are structured; for example, many metabolic processes run at rates in part determined by the size of the organism.

There are many important physiological and ecological correlates to size. Small organisms tend to carry out life activities at faster rates relative to larger organisms, for a variety of reasons. One is that the surface area to volume ratio of rounded objects is proportionally greater in objects of small diameter. Rates of biological processes are often controlled by mechanisms involving transport of substrates across surfaces, so that small size may favour relatively faster transport. Variables that are aggregates of several such transport processes, such as respiration rate, are very much affected by the size of the animal; the smaller the organism, the larger the weight-specific respiratory rate (Fig. 2.10a). Growth and generation times are among

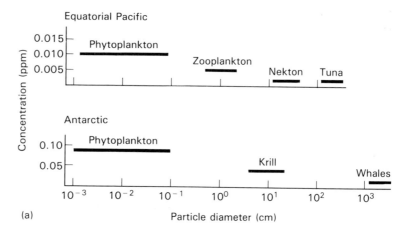

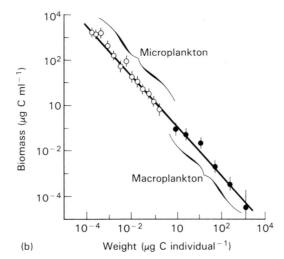

Fig. 2.11 Relationship of organism size to abundance. (a) Estimates of abundance of some major groups of different size in two oceanic environments. (b) Biomass of surface plankton in the North Pacific Gyre in relation to weight of the organisms. From Sheldon *et al.* (1972) and Rodriguez & Mullin (1986).

the many variables that are faster for smaller organisms, and so are many other processes, such as release of nutrients by consumers and nutrient uptake by producers. The weight-specific rate of release of nitrogen, for example, is higher for smaller organisms (Fig. 2.10b). Picoplankton may be responsible for 50–80% of nitrogen uptake in aquatic systems.

As another example of the importance of size of organism on ecological function, consider that if net primary production of an oceanic or lake water column is partitioned into fractions carried out by different sized producers, most of the activity is attributable to the smaller size fractions, as is most of the chlorophyll (see Table 2.3). In coastal waters,

the dominance of picoplankton is not as marked, but is still important.

There are many other size-dependent phenomena. Who eats what depends on the sizes of who and what, and since abundance is inversely proportional to size, size and abundance must concern consumers. Above we noted that organisms of increasing size are progressively more scarce. Empirical measurements and theoretical models predict such a negative relationship. Examples of the trends are given in Fig. 2.11. Over a very large spectrum of sizes of organism, estimates of abundance decrease as size increases in major water column groups in equatorial Pacific and Atlantic oceans (Fig. 2.11a). In more recently acquired

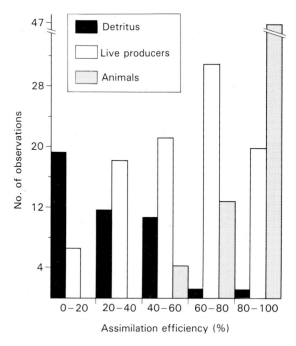

Fig. 2.12 Frequencies of assimilation efficiencies for a wide variety of animals feeding on detritus, on live producers and on animals. From Valiela (1984).

data from plankton taxa in the North Pacific Gyre, there is a very clear negative relation of reduced biomass as weight of the individual micro- and macroplankton increases (Fig. 2.11b).

There is a large body of literature that describes how predation rates are affected by the size of available prey. There are involved expressions of optimal foraging theory, in which the claim is made that predators tend to choose to feed on prey whose yield as food is high in relation to the energy or time spent by the predator in chasing, subduing and eating prey. Prey size is the principal criterion that affects both yield and expenditures.

Being a successful consumer, however, is more complicated than merely finding food of the right size. Consider detritivores and carnivores in relation to abundance and quality of their food supply. Carnivores use food items that have relatively high assimilability (Fig. 2.12) and are of high nutritive quality. Their problem is that, as discussed earlier, there is a fairly well-specified negative relationship between abundance and size of potential prey

organisms (Fig. 2.11). The larger a predator, the less numerous its suitable prey. There are often too few suitably large food items around, in contrast to the great abundance of detrital particles.

Carnivores have to ensure that the particle they seek is large enough to be a sufficient reward to more than compensate for the energy expended seeking, chasing, subduing and eating the item. Moreover, the prey has to be of a size such that capture by the feeding apparatus is mechanically feasible. Hence, food selection in such consumers is based on an assessment of size of the morsel relative to size of the consumer.

Most consumers will seek particles substantially smaller than themselves, but as large as their feeding apparatus and devices permit. There are some exceptional cases of consumers that are relatively very large compared to their food items, such as clams, appendicularians and cladocerans feeding on bacteria, or whales feeding on mesoplankton, but generally the size difference between carnivore and prey is not this large. There is, in almost all cases of predators, a clear relation of size of predator to size of prey (Fig. 2.13). The ratio of consumer to food particle may be about 10:1 or 100:1 in linear dimensions. Raptorial feeders have to chase prey, hence it seems reasonable to expect their prey to be generally relatively large. Suspension feeders may spend much time removing particles from large

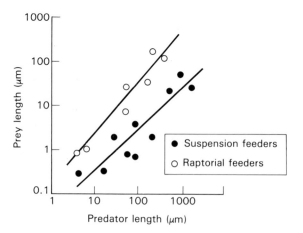

Fig. 2.13 Relation of size of raptorial- and suspension-feeding consumers and their prey. Both consumers and prey are ciliated protozoa. From Fenchel (1988).

volumes of water, but this may not be as energetically costly as hunting, so that suspension feeders may subsist on smaller prey per given size of consumer.

This is a convenient point at which to digress from the theme of size to mention that other criteria may be important in food selection. Consumers that feed on dead matter not only need to consider the size of the particle they ingest, but also its nutritive quality. Detritivores have an altogether different problem from predators. Detritivores consume particles of a very abundant material — a very large proportion of the particles in any water column, and even more so of the benthos, is detrital — yet although detritivores are very common, they have not completely overrun the waters of the earth. Their food is of relatively low quality, and their assimilation efficiency for detritus tends to be low (Fig. 2.12). Moreover, detritus still contains at least part of the chemical armoury of defensive compounds that plants and some algae use to deter herbivores. There are many compounds synthesized by consumers that reduce palatability and assimilation, and many of these compounds remain in detrital particles.

The problem that needs to be solved by detritivores (and by herbivores to some extent) is one of finding food items of sufficiently high palatability and nutritive quality. Hence we find that, in contrast to the prey size selection that is the trademark of predators, detritivores (and herbivores) show very well-developed selectivity as to which food they feed on, and use chemical cues detected by complex chemosensory organs to make decisions about food items.

An interesting side issue in the matter of chemical defences is that many sessile organisms, such as sponges, sea slugs, gorgonians and tunicates, also develop defensive chemical armouries, much like macroalgae and plants. Chemical defences are associated with a sessile or slow-moving life form, rather than with specific taxonomic groups.

The topics discussed in this section touch on a few of the ways in which size of organism is involved in the evolution and ecology of aquatic food webs. They do show that size of organism is a key criterion in determining how populations fit into aquatic food webs: smaller organisms generally have higher activity rates. We also know that although size of particle is important, it is not the sole criterion used by consumers to choose food.

2.3.2 Release, use and aggregation of dissolved organic matter

Phytoplankton, macroalgae and aquatic plants release substantial amounts of dissolved organic matter (DOM) to the surrounding water during the course of their life activities. 'Dissolved' here is an empirical category: it merely refers to organic compounds that pass through a filter of mesh size 0.45 nm. The important recent finding is that anywhere from 10 to 50% of net primary production may be released by algae and plants via exudation from live producers.

In addition, perhaps 15% of net primary production of phytoplankton may be released to the water during the course of what has been called 'messy feeding'. Grazers may break or crush algal cells or tissues, releasing fluids into the water. Further DOM may be released from dead matter during the process of decay, by the leaching of soluble compounds out of detritus.

The DOM released can be used by heterotrophs, including bacteria, fungi and many other small organisms. The more exudation of organic matter by algae, the more bacteria can grow. This may be why there is a clear relationship between bacterial production and primary production for both lakes and coastal waters (Fig. 2.14a).

Another fate of DOM released by organisms is aggregation into amorphous particles. Such particles have been termed 'marine snow', although not all marine snow particles may have such an origin. The aggregates may be nutritionally valuable for consumers, such as invertebrates and fish. Fragments of dead producers, which we can refer to as 'morphous detritus', are well-known in having poor nutritional quality. In fact, this is a principal reason why detritivores show a spectrum of rather lower assimilation efficiencies than carnivores (Fig. 2.12) and often grow poorly on diets of particulate morphous detritus. In contrast, the shapeless (amorphous) aggregates of DOM released from producers are better food, because: (i) they are made up of aggregated smaller molecular weight

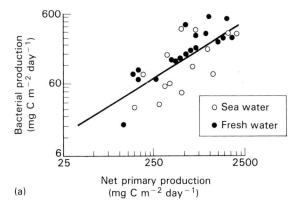

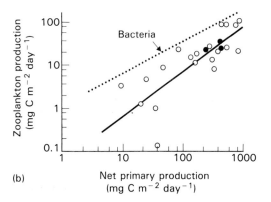

Fig. 2.14 (a) Relation of bacterial and primary production in fresh and sea water. (b) Relation of zooplankton and primary production in lakes. For comparison, the dotted line is the best fit line from the bacterial production/primary production relationship. From Cole *et al.* (1988).

compounds, which are generally more assimilable than the larger polymers that remain in the cell and tissues; and (ii) have lower content of detrimental or inhibitory secondary compounds than morphous detrital particles. Work is still needed on this pathway of organic matter, but it may be of quantitative significance for aquatic food webs.

The microbial loop

New information about the abundance of microbes that are not eaten by microplankton, about release of DOM by producers and about the abundance of non-living organic particles, has suggested the existence of a very active 'microbial loop', a substantial part of aquatic food webs not foreseen in the classic view of these webs.

The high microbial abundance revealed by the new methods of counting bacteria raised the question of how active these microbes may be, and resulted in development of methods to measure bacterial production. Although there is still some controversy about these methods, two — one based on the frequency of dividing cells and a second based on uptake of thymidine, a compound only bacteria use — have been used in many waters. These methods suggest that in the field, bacteria may have generation times of a few hours to a few days, and may have production rates equivalent to perhaps 20% of total primary production on a unit volume basis (Fig. 2.14a). Bacteria in the water column, however, are not perfect transformers; they respire, and hence their gross production is about twice that of net production, because bacterial assimilation efficiency — (organic matter assimilated/organic matter decomposed) × 100 — is about 50%. Thus if we combine assimilation plus respiration terms, about 40% of net primary production is likely to flux through the bacteria.

All the above calculations are on a per unit volume basis. The importance of bacterial production should be greater if we were to consider a per square metre basis. The deeper the water column, the greater the importance of bacteria, since phytoplankton production would, of course, only occur near the surface.

There is some information, however, that suggests that small heterotrophs may release DOM. For example, bacteria release large quantities of mucopolysaccharides, and the external 'digestion' strategy of microbes demands the release of enzymes to the water. Of course, if release of DOM by small protozoa turns out to be large, the role of bacteria may be less important than has been suggested.

The number of bacteria per unit volume of water typically lies within the range 10^5 to somewhat over 10^6 cells ml^{-1}, a remarkably narrow range. These limits occur in spite of the potential for very active growth suggested by the short generation times. Since microzooplankton and larger animals (except

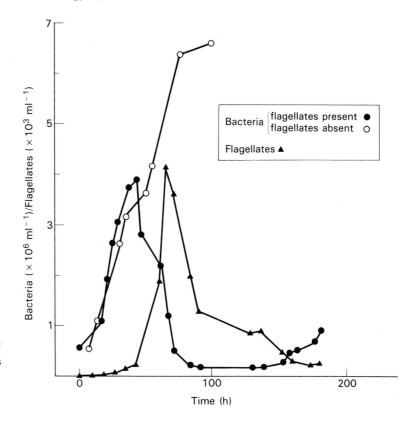

Fig. 2.15 Time course of numbers of bacteria and bacterivorous flagellates in sea water and numbers of bacteria in sea water when flagellates were removed. From Fenchel (1987).

for certain freshwater cladocerans) are unable to feed on particles as small as bacteria, there must be some other sink for bacterial production. That sink is most likely to be grazing by protozoans.

Heterotrophic nanoplankton consume bacteria (and possibly prochlorophytes) in water. Nano-flagellates may clear the water column of bacteria up to twice per day. The ability of flagellates to control the abundance of bacteria can be shown by experiments in which the abundance of bacteria and flagellates was measured in circumstances where both were present and where flagellates were removed (Fig. 2.15). As bacteria grew over the course of hours, their abundance increased, but flagellate abundance responded to the increase in food and within some tens of hours the flagellates reduce bacterial abundance. Where there were no flagellates, bacteria continued to grow and reached densities much greater than in the other treatment.

The nanoplankton, represented by the flagellates in the case we have been examining (Fig. 2.15),

are fed upon by larger ciliates and dinoflagellates. These larger groups, together with the larger phytoplankton, are the producer trophic level of the classical metazoan aquatic food web.

Food chains consisting of bacteria, small ciliates, large ciliates or small microzooplankton, transfer very little energy as carbon to larger metazoans. The data available on the transfer energy and growth of microbes are still fragmentary, but some constraints can be calculated. Even if we allow 50% transfer of energy from one trophic link to another, which is likely to be too high, only 12% of the energy passes through 3 trophic steps; only 6% through four steps. Thus, the more feeding links, the less the likelihood of transfer of matter beyond the microbial loop to the classical plankton. In nature, it is usual to find three to five steps in most microbial loop food webs.

Knowledge of what controls the structure and function of microbial food webs, and of how they may be coupled to the classical food web, is still rudimentary, but it is likely to be a function of complex relationships. For instance, feeding by fish or large zooplankton may lower the consumption of protozoans by large zooplankton sufficiently to affect the fate of bacterial production. Experiments where fish were present or absent showed that where fish were absent, 50% of bacterial production was ingested by large zooplankton. Where fish were present, only 4% of bacterial production was used by large zooplankton. In this latter situation, protozoans consumed the bacteria, so that much less carbon left the microbial loop.

The question has been put this way: is the microbial loop a link or a sink? With so many steps dissipating energy, the answer appears to be that to the extent that the microbial loop exists, it is a sink for energy and organic matter furnished by primary producers. This degradation of organic matter means, however, that the microbial loop has a critical function in that it releases (mineralizes or regenerates, in the jargon of terrestrial and aquatic scientists, respectively) the mineral elements in the organic matter. The nutrients released by the microbial loop organisms may be a critical pathway for the support of continued activity for the whole plankton. In the classic view of plankton dynamics, it was the microzooplankton that were thought to be mainly responsible for regeneration of nutrients.

The role of larger zooplankton

Zooplankton in the water column consume particles of appropriate size and nutritional quality. The proportion of primary production that is available to the classical microzooplankton is smaller than that consumed by bacteria. In the freshwater lakes that yielded data for Fig. 2.14b, zooplankton production was correlated with net primary production. Of course we cannot tell if both vary in response to a third factor, or if there is a cause-and-effect relationship, but zooplankton production was about half as large as bacterial production (Fig. 2.14b). The sum of production by herbivorous and carnivorous zooplankton averaged 12% of primary production. Since zooplankton have growth efficiencies (growth/consumption) of roughly 25%, zooplankton respiration would be about 36% of net primary production, while

bacterial respiration would be about 29%. If indeed respiration is related to rates of mineralization, these rough calculations suggest that perhaps 65% of primary production is consumed in water columns. The data also suggest that both bacteria and larger zooplankton have important roles in regeneration of nutrients from primary organic matter synthesized by producers in water columns. This is clearly different from the ideas associated with the classical food web, mentioned earlier.

A considerable portion of food ingested by zooplankton is not assimilated and becomes packaged into faecal pellets. The study of production and sinking of faecal pellets (plus cast moult skins, mucus released by gelatinous animals and other organic materials that make up the remaining portion of marine snow and detritus) have played an important part in forging the new views about the functioning of the pelagic community.

In the early 1960s, oceanographers found that concentrations of organic particles were rather constant below the illuminated top layer of the sea. These vertically uniform concentrations presented an illogical paradox. We knew that there were many zooplankton consuming particles up and down the oceanic water column, and we knew that the sea floor supported many benthic organisms. If these animals depended on producers near the sea's surface, it was reasonable to expect that there would be a decreasing gradient of particulate organic matter down the depth profile instead of a homogeneous vertical profile concentration. The steepness of the gradient would reflect the relative abundance of consumers in each water column.

The answer to the paradox came on re-examination of the sampling techniques and spatial scale of measurements. The early samples of water were taken with collecting bottles of a few litres of volume, and the sample entered the bottle through a narrow opening. There was little chance that such a device would adequately sample the rain of faecal pellets and other large particles from above, the unexpected source of organic matter that supported the communities of consumers below the upper layers of the sea.

It was only when sediment traps with a much larger sampling area were deployed at different depths over long periods that a better assessment of

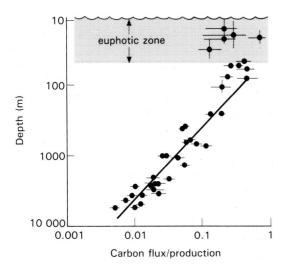

Fig. 2.16 Carbon flux standardized to production in the photic zone in relation to depth, for a series of observations on a wide variety of areas. The euphotic zone is the upper illuminated layer. From Suess (1980).

the actual flux of organic matter was obtained. The frequency of such particles in any one litre of sea water is very low; but accumulated over suitably large intervals of space and time, this turns out to be the major pathway of surface-produced organic matter to the depths of the ocean and to the sea floor.

Data obtained from sediment traps (Fig. 2.16) show the gradients we expect for a resource pro-

duced near the surface and consumed by organisms as the particles fell downward toward the sea floor. The descent of particles is reasonably fast and seems to be proportional to particle size (Fig. 2.17).

Since pellet size is a function of animal size, the delivery of particles downward is in part calibrated by the size distribution of zooplankton near the surface. If the size of the zooplankton near the surface were small enough, it may be that no organic matter reached the sea floor, and no benthic community could be supported. In fact, in the Middle Atlantic shelf off southeastern USA, the zooplankton near the surface layers, although abundant, are quite small. Correspondingly, the benthic fauna below are scarce. The link between these two observations may be that the faecal pellets produced in this area are small, sink slowly and are largely consumed before they can reach the sea floor.

The transport of organic matter vertically downward by fast-sinking organic particles may be the explanation for the surprizingly widespread distribution of man-made polychlorinated biphenyls, industrial chemicals first manufactured in the mid 1940s. Such compounds can be incorporated into particles such as faecal pellets, and they can in this way be transported downward. Until the fast sedimentation of faecal pellets was identified, there was no ready explanation of how these compounds found their way to great depths, where waters had last been near the surface thousands of years ago.

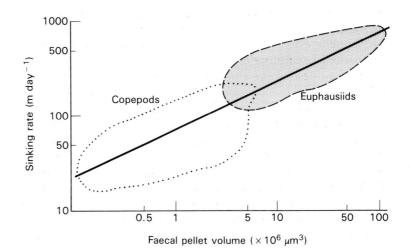

Fig. 2.17 Relation of sinking rate to volume for faecal pellets of copepods and euphausiids in the Ligurian Sea. From Small *et al.* (1979).

There is, therefore, a complex intertwining of the relationship between size distribution and structure of a community at the surface, and the whole ecosystem dynamics below, as well as in the transport of biogeochemically important compounds. We have indeed come a long way from the classic phytoplankton–copepod–fish trophic web.

The new view of the plankton community can be summarized in a schematic representation that includes both size and trophic transfer (see Fig. 1.3). A considerable portion of the production by autotrophs (organisms shown on right-hand column of Fig. 1.3) is released as DOM, which may be consumed by bacteria, which in turn are consumed by ciliates and other small fauna. These also release DOM, which is available to the bacteria. The larger zooplankton (the classic zooplankton trophic level) eat both producers and consumers, and in turn are eaten by larger organisms.

The scheme of Fig. 1.3 does not include aggregation of DOM, and consumption of such aggregates by consumers of all sizes; it also does not indicate that most consumers produce particles. Missing is an indication that some of the larger particles, such as faecal pellets, moults, etc., may sink down the water column or may be consumed by other organisms. These are additional components of planktonic food webs that will require attention.

2.4 TOP-DOWN CONTROLS OF WATER COLUMN COMMUNITIES

Earlier sections emphasized that bottom-up controls, primarily nutrients and light, whose effects are mediated by hydrography, powerfully influence net primary production in the water column, and hence on the rest of food web. We have also seen that grazers and predators can effectively consume their selected food items. Is there evidence that such consumption is sufficient to constitute a top-down control on aquatic communities?

2.4.1 Top-down control in freshwater water columns

In fresh water there is strong evidence that top predators can completely restructure the food web of the ecosystem in which they live. The purposeful or accidental introduction of a new predator into a freshwater environment often results in wholesale changes in the structure of the community: this has happened in many instances. Among the best known examples are the introduction of alewives into a lake in Conneticut in the USA, invasion of the Great Lakes by lampreys and the entrance of Nile perch into Lake Victoria in East Africa. We will examine one other example, that of the introduction of a piscivorous cichlid fish, native to the Amazon River, into Gatun Lake in Panama.

Cichla ocellaris was introduced to Gatun Lake when rains caused an overflow from an artificial impoundment where the fish were being kept. As *Cichla* spread over the lake during succeeding years, its voracious feeding devastated populations of native fish. Before the introduction of *Cichla*, or in areas where there are no *Cichla*, the lake's food web is complex, and 14 species of fish occur, 12 of which are abundant (Fig. 2.18a).

There are direct first-order effects of the introduction of *Cichla*. Predation by *Cichla* decimated populations of several fish species, including *Melaniris*, a key species that served as food for three other species of predators (Fig. 2.18a). Where *Cichla* is present, the food web is therefore much simpler. Seven fish species remain, but only *Cichla* and *Cichlasoma* are abundant (Fig. 2.18b).

There are indirect, second-order effects. *Cichlasoma* became more abundant because predation by *Cichla* apparently removed a number of species that previously fed on the fry of *Cichlasoma*. Another second-order effect involves the young *Melaniris* that fed on microzooplankton, especially on the cladoceran *Ceriodaphnia cornuta*. Predation by *Melaniris* was sufficiently increased to favour the occurrence of a horned form of *C. cornuta*; pointed extensions on the body of the horned form of the cladoceran apparently made them less susceptible to being eaten by *Melaniris*. After *Cichla* eliminated *Melaniris*, the proportion of horned *C. cornuta* diminished. Apparently the normal form of the cladoceran has a reproductive advantage in the absence of predation pressure.

A third second-order effect involves tarpon and black terns (Fig. 2.18a). Tarpon feed on *Melaniris*. The terns follow the tarpon, and feed on the *Melaniris* brought to the water surface by attacks

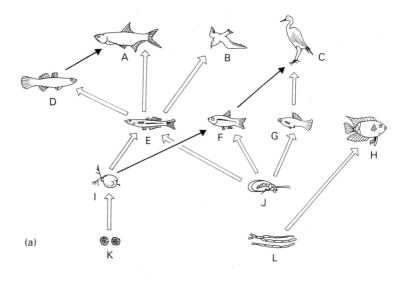

(a)

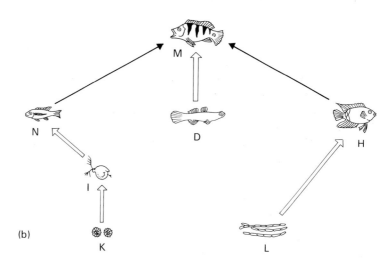

Fig. 2.18 Food webs of Gatun Lake
(a) in the absence of *Cichla* and (b)
with *Cichla* present. Thickness of lines
indicates relative importance of food
item. A, *Tarpon atlanticus*; B, the tern
Chlidonias niger; C, herons and
kingfishers; D, *Gobiomorus dormitor*;
E, *Melaniris chagresi*; F, several
characinid fish species; G, several
poeciliid fish species; H, *Cichlasoma
maculicauda*; I, zooplankton; J,
terrestrial insects; K,
nanophytoplankton; L, filamentous
green algae; M; adult *Cichla ocellaris*;
N, young *Cichla*. From Zaret & Paine
(1973).

(b)

of the tarpon. Both tarpon and terns (and other predators such as kingfishers and herons) disappeared from areas with *Cichla*.

There are presumed third- or fourth-order effects of the introduction of the new predator. Many of the fish species eaten by *Cichla* in Gatun Lake feed on insects (Fig. 2.18a). Data on insect abundance in the lake are lacking, but mosquitoes, which have aquatic larvae, have increased in areas with *Cichla*. Before the introduction of *Cichla*, 95% of malaria victims in the Canal Zone of Panama carried *Plasmodium vivax* as the disease organism; *P.*

falciparum was the agent for the remaining 5% of the cases. Since *Cichla* entered the area, more than 60% of the malaria cases carried *P. falciparum*, and two serious outbreaks of malaria have ocurred near *Cichla* areas. These correlations by no means establish cause and effects, but are suggestive of how pervasive the effects of predators may be in fresh water.

The importance of top-down controls in freshwater food webs is therefore well-established. In fact, these principles have been put to applied use. There are cases where introduction of a fish that fed on a zooplanktivorous fish, successfully released microzooplankton from predator control. The microzooplankton then proceeded to consume phytoplankton sufficiently to greatly improve water transparency and quality.

2.4.2 Top-down controls in marine water columns

The importance of top-down controls is, in contrast to the situation in fresh water, not well-established for marine water columns. In the sea we lack the opportunity to carry out whole ecosystem, manipulative experiments that provide the strongest test of possible controls. The best evidence we can muster, to very roughly assess the role of top-down controls in marine systems, comes indirectly, from study of the effects of intensive fishing and from consideration of reproductive strategies of marine top predators.

First, the evidence from fishing fleets. Commercial fishing can remove a considerable proportion of the stock of a particular fish species, and there are many examples of such depletions. If these predators were truly controlling marine water column food webs, we would expect to see some consequences of the depletion caused by fishermen. Unfortunately, there is little evidence of such a first-order effect. For example, the halting of intense fishing during the Second World War in the North Sea did not result in a great increase of fish abundance, nor in stunting of the fish population present at the conclusion of the war. It may be that we simply have not critically set out to measure the effects of disappearance of fish, but it appears that by and large, marine water column food webs are mostly affected by bottom-up effects. Perhaps top-down controls are of little consequence, or their effects are overwhelmed by the vagaries of chance events.

Second, the evidence from reproductive strategies. Many pelagic fish exhibit what is referred to as a strong cohort recruitment pattern. Recruitment is far from uniform from year to year. Instead, every so often — apparently at random — the hazards that plague pelagic larvae and juveniles of fish are fortuitously relaxed. Mortality is somehow reduced and a strong cohort for that year appears. The individuals in these strong cohorts are far more numerous than those for cohorts of other years. For example, the 1962 and 1967 cohorts of haddock in the North Sea were 20 times as numerous as those of other years. The 1958 autumn age class of herring in the Gulf of Saint Lawrence was 44 times, and the spring 1959 class was 60 times, as large as the smallest age class recorded in 1969. We are therefore dealing with a widespread recruitment pattern that results in widely varying entry of individuals into the ecosystem from one reproductive season to the next.

If the existing population of fish were controlling their food supply, by definition they would have reduced the abundance of the food species. Entry of a very large number of new individuals into the area, as per the strong cohort pattern, could be expected to result in competition for food, and this would result in increased mortality and reduced growth. Instead, strong cohorts survive and grow better than average cohorts. In fact, entire fisheries often depend on harvests of the strong cohorts over several years. These cohorts make up a major part of marine fish populations. The explanation may be that the abundance of marine fish stocks is poised at levels far below the potential which can be supported by food supply. Perhaps the enormous, chance-driven mortality of larvae is a bottleneck that prevents a closer coupling of recruitment to available food supply.

Marine fish produce very large numbers of very small eggs per female. Perhaps only a few out of every million eggs survive to become an adult: the larvae suffer tremendously high mortality rates. In part, the mortality is the result of their small size. Small larvae have to grow their way up the food web, and are exposed to a gauntlet of different size predators as they grow. This is a particularly serious problem for marine fish, because there are more of these size-related feeding steps in the sea than in fresh water.

In addition, the habitats of marine fish are not bounded by land. Larvae, being planktonic, can, by chance events, drift into areas inappropriate for their survival. This is a common occurrence, and because of the many chance factors that affect the outcome, it may be that the number of fish that survive and enter the adult population one year are not numerically related to the number of fish in the parent generation. Thus the effect of food supply on the adults is not transmitted to the next generation. If this lack of coupling of abundance to food is the case, it does make sense that the strong cohort pattern can occur.

There are some marine top predators that, in contrast to marine fish, produce relatively few and large young, and reduce their mortality by parental care. These are the marine mammals and birds. There are no direct data on whether these taxa can control marine food webs. We have, instead, only some glimpses of circumstantial evidence.

Industrialized whaling in the Southern Ocean reduced stocks of whales by an order of magnitude by the mid 1960s. If whales controlled the Antarctic food web, we would expect that the whaling would have led to a very significant increase in abundance of the organisms — krill, squid and fish — that whales eat.

Unfortunately there are no census data with which to assess the abundance of animals in the southern ocean. We do have, however, some ancillary evidence that suggests that increased availability of krill, squid and fish may have occurred. First, the populations of other top predators that feed on the same food items as whales appear to have increased. Seals and penguins seem to be expanding their range and abundance in the southern ocean. Second, growth and reproductive success of those individual whales that have survived have improved. These improvements suggest a better 'standard of living', or more available food, now that whale populations are lower.

These observations suggest, in a general and circumstantial way, that abundance of marine mammals and birds may be closely coupled to the abundance of their food resources. They could therefore control prey populations, and hence the marine food web. It is difficult to demonstrate how or why fish are unable to match mammals and birds. Perhaps it is a consequence of the evolutionary decision to employ the 'few large' versus the 'many small' reproductive strategy, with the resulting differences in exposure to mortality and chance events.

FURTHER READING

Atlas of the Living Resources of the Seas. (1972) FAO, Rome.

Barnes, R.S.K. & Hughes, R.N. (1988) *An Introduction to Marine Ecology.* 2nd edn. Blackwell Scientific Publications, Oxford.

Caron, D.A. (1991) *The Evolving Role of Protozoa in Aquatic Nutrient Cycles.* NATO ASI Conference in Marine Science. Springer-Verlag, Berlin.

Chisholm, S.W., Olson, R.J., Zettler, E.R., Goericke, R., Waterbury, J.B. & Welschmeyer N.A. (1988) A novel free-living prochlorophyte abundant in the euphotic zone. *Nature* 334: 340–43.

Cole, J.J., Findlay, S. & Pace, M.L. (1988) Bacterial production in fresh and salt water ecosystems: a cross-system overview. *Mar. Ecol. Progr. Ser.* 4: 1–10.

Davis, C.S. (1987) Components of the zooplankton production cycle in the temperate ocean. *J. Mar. Res.* 45: 947–83.

Fenchel, T. (1987) *Ecology — Potential and Limitations.* Ecology Institute, Oldendorf/Luhe, Germany.

Fenchel, T. (1988) Marine plankton food chains. *Ann. Rev. Ecol. Syst.* 19: 19–38.

Hecky, R.E. & Kilham, P. (1988) Nutrient limitation in freshwater and marine environments: a review of recent evidence on the effects of enrichment. *Limnol. Oceanogr.* 33: 796–822.

Howarth, R.W. (1988) Nutrient limitation of net primary production in marine ecosystems. *Ann. Rev. Ecol. Syst.* 19: 89–110.

Ikeda, T. (1970) Relationship between respiration rate and body size in marine plankton animals as a function of temperature and habitat. *Bull. Fac. Fish.* (Hokkaido Univ.) 21: 91–112.

Lehman, J.T. (1988) Ecological principles affecting community structure and secondary production by zooplankton in marine and freshwater environments. *Limnol. Oceanogr.* 33: 931–45.

Nixon, S.W. & Pilson M.E.Q. (1983) Nitrogen in estuarine and coastal marine ecosystems. In: Carpenter, E.J. & Capone, D.G. (eds) *Nitrogen in the Marine Environment,* pp. 565–648. Academic Press, New York.

Pomeroy, L.R. & Wiebe W.J. (1988) Energetics of microbial food webs. *Hydrobiologia* 159: 7–18.

Rodriguez, J. & Mullin M.M. (1986) Relation between biomass and body weight in a steady state oceanic ecosystem. *Limnol. Oceanogr.* 31: 361–70.

Ryther, J.H. & Dunstan, W.M. (1971) Nitrogen, phosphorus and eutrophication in the coastal, marine environment. *Science* 17: 1008–13.

Sheldon, R.W., Prakash, A. & Sutcliffe W.H. (1972) The size distribution of particles in the ocean. *Limnol. Oceanogr.* 17: 327–40.

Small, L.F., Fowles, S.W. & Uulu, M.Y. (1979) Sinking rates of natural copepod fecal pellets. *Mar. Biol.* 51: 233–41.

Stockner, J. G. (1988) Phototrophic picoplankton: an overview from marine and freshwater ecosystems. *Limnol. Oceanogr.* **33**: 765–75.

Suess, E. (1980) Particulate carbon flux in the ocean-surface productivity and oxygen utilization. *Nature* **288**:260–3.

Valiela, I. (1984) *Marine Ecological Processes.* Springer-Verlag, New York.

Whittaker, R.H. & Likens, G.E. (1975) The biosphere and man. In: Leith, H. & Whittaker, R.H. (eds) *Primary Productivity of the Biosphere*, pp. 305–28. Springer-Verlag, New York.

Zaret, T.M. & Paine R.T. (1973) Species introduction in a tropical lake. *Science* **182**: 449–55.

3
Ecology of Coastal Ecosystems

I. VALIELA

3.1 INTRODUCTION

In this chapter we first briefly review the types of organisms that occur in the coastal zone and mention some distinctive geological and hydrographical features; we then devote our attention to comparisons that highlight how coastal ecosystems differ from deeper aquatic systems. After these considerations, we will examine some general mechanisms that control the structure and function of coastal communities.

3.2 PROPERTIES OF COASTAL ENVIRONMENTS

The material in this chapter deals primarily with marine coasts, but we can briefly point out some contrasts with lacustrine shores. The coasts of freshwater lakes differ from those of marine coasts in that variation in water level in the sea shore is determined primarily by tides. While in some lakes seiches may impose a regular rhythm to changes in water level, the pattern of high and low water creates an intertidal habitat in marine coasts that is lacking in lakes. The water transport involved in tidal exchanges in marine coastal waters provides a different set of conditions for life, and enhances transport of ecologically important materials in and out of coastal environments.

Another important difference between lakes and the marine coast, of course, is salinity of the water. Unlike lakes, estuaries and coastal waters often contain various mixtures of fresh and salty water that create physiologically and ecologically challenging circumstances for organisms. The often-changing mixtures of fresh and sea water create difficult osmotic gradients that greatly affect coastal organisms. Different taxa solve the osmotic problem in different ways. Estuarine vascular plants, for instance, have evolved salt glands that exclude or secrete salts. Since exclusion of salts cannot be perfect, many marine producers have also developed the ability to synthesize organic compounds, such as betaines, prolines and dimethylsulphoniopropionate (DMSP), that serve to maintain osmotic balance in their fluids.

We should note parenthetically that the discovery of DMSP in relatively high concentrations in estuarine vascular plants, in macroalgae such as *Ulva* and in phytoplankton, has attracted interest recently because DMSP degrades to dimethyl sulphide (DMS). Dimethyl sulphide from marine sources may be a major mechanism providing sulphur to the atmosphere and may be involved in the acidification of precipitation. Of more importance still, it may influence cloud cover on a global scale by providing 'cloud condensation nuclei'.

Estuarine animals also need to solve the osmotic problem posed by the changing salinity of estuaries. Some species exude salts from their skin. Many other estuarine species are covered by an outer layer of mucus, which may prevent salt transport into the animal. Fish and other animals also respond to osmotic differences by physiological changes involving drinking, transport of ions across gill surfaces and control of the amount of water passed via urine.

Differences in salinity among water masses have very significant ecological consequences because of differences in density of water that result from differences in salinity. Differences in salinity emphasize or change the stratification of estuarine and coastal water columns compared to those of lakes. The results are important for organisms in water masses with differing salinity, because their access to resources — light, nutrients, oxygen, food organisms — may be curtailed by the resulting stratification.

Some of the chemical differences between lakes and coastal marine waters have non-obvious second- and third-order consequences. For instance, fresh water has much lower concentrations of sulphate than sea water. The rate of sulphate reduction in freshwater sediments is therefore much lower than that in coastal sediments, and the importance of sulphate reduction as a path for the microbial degradation of organic matter is much less. As a result, freshwater sediments release much less sulphide gas than coastal sediments. This is the reason why coastal salt marshes smell of rotten eggs, while freshwater wetlands release odourless methane. Freshwater wetlands, incidentally, are the major source of atmospheric methane, one of the important atmospheric greenhouse gases. The release of sulphides from sulphate reduction — plus the release of DMSP from marine producers already mentioned — mean that marine waters are more significant as a source of sulphur input to the atmosphere than fresh waters.

We should not overemphasize the differences between lakes and coastal waters, however, since there are many similarities. Most of the important ecological processes discussed in this book have roles in both freshwater and marine settings.

Although there may be differences in details, most ecological processes that function in coastal ecosystems (shelves, bays, estuaries, lagoons, fjords, reefs, salt marshes, mangrove swamps, etc.) are the same as those operating in other environments. In each of these kinds of environments, however, there is a somewhat different hierarchy among the suite of processes that are most important in determining the function of each ecosystem.

3.3 BIOLOGICAL COMPONENTS OF COASTAL ENVIRONMENTS

Phytoplankton are generally the major producers in the shallow waters of coastal ecosystems. The taxa involved are similar to those found in offshore waters, and the suite of mechanisms that can set the rate of primary production are the same. The exact identity of the specific factor responsible for actually limiting the rate of net primary production may differ in coastal waters compared to open oceanic waters, as already discussed in Chapter 2.

The major taxa of zooplankton and nekton of coastal areas are not, in general, different from those of offshore waters although, of course, the species differ.

In coastal environments in which the water column is such that the photic zone reaches the bottom, macrophytes (vascular plants and macroalgae) may become the dominant producers. Macroalgae (in the kingdom Protista) and higher plants (kingdom Plantae) constitute the major distinguishable types of vegetation of seagrass meadows, salt marshes, mangrove swamps, kelp forests and rockweed beds.

Soft sediments in estuaries, or along shores protected by island bars or barrier beaches, support seagrass meadows. In temperate latitudes, the seagrasses are mainly in the genera *Zostera*, *Posidonia*, *Cymodocea* or *Halodule* (Fig. 3.1) and *Ruppia* in brackish waters. *Thalassia* is the most common tropical seagrass. Seagrasses tend to be restricted to a narrow band of subtidal depths, because they tend to be, as mentioned earlier, light-limited, and because they do not grow well intertidally. Seagrasses occur primarily in protected estuarine areas and even more commonly in lagoonal or bay situations, widely distributed over the world's shores (Fig. 3.1). Seagrass beds are important nurseries and sources of food and cover for a great variety of marine species. Seagrasses also tend to stabilize unconsolidated coastal sediments, and so have several important functions in the shallow subtidal zones of lagoons, bays and estuaries.

The intertidal range of temperate protected shores supports salt marsh vegetation (Fig. 3.2), such as *Spartina*, *Juncus*, *Carex*, *Salicornia*, *Distichlis*, *Halimione*, *Suaeda* and many other genera. In latitudes where there is no danger of frost, the same ecological habitat is used by mangroves. Mangrove genera include *Rhizophora*, *Avicennia*, *Conocarpus* and others.

Whether the intertidal shore is fringed by salt marsh or by mangrove vegetation depends on latitude (Fig. 3.2). It is not clear why mangroves are

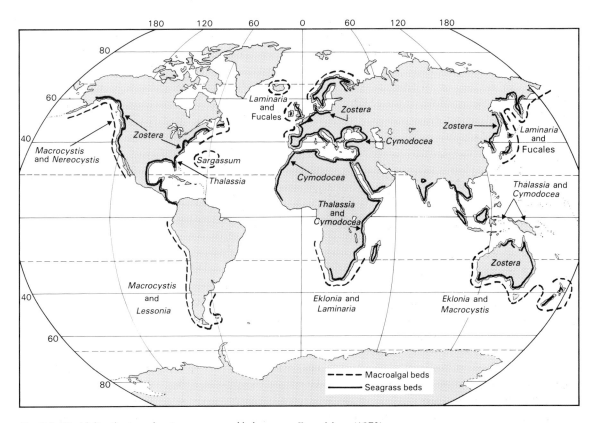

Fig. 3.1 World distribution of major seagrass and kelp genera. From Mann (1972).

restricted to latitudes roughly north and south of 35°. It is suspected that mangrove seedlings do not survive frost. Salt marsh plants do not grow well under the canopy of mangrove trees, since they cannot tolerate shading. These restrictions may explain the narrowness of the zone of overlap between salt marsh and mangrove vegetation.

Seagrass meadows and coastal wetlands, such as salt marshes and mangrove swamps, contain vegetation whose role in the ecology of the coastal zone should not be underemphasized. These vascular plants provide food and shelter for a wide variety of coastal organisms. The juveniles of many commercially important crustacean and fish species make special use of vegetated coastal environments. In fact, the harvest of commercial coastal fisheries in the southeast of the USA is directly proportional to the acreage of salt marsh habitat on the shore. Many of the species taken by coastal fisheries have

some stage in their life histories that feeds or finds protection in coastal wetlands. In addition, these wetlands are critical stepping stones in migratory routes for waterfowl and serve as nesting and overwintering areas for many migratory birds.

Coastal wetlands are characterized by high rates of biological activity. Rates of primary production in wetlands are relatively high (see Table 2.1), and there are correspondingly high rates of degradation of organic matter. The high rates of production and decay and the rapid exchange of water created by tides, create the potential for active transport of organic matter and nutrients in and out of coastal wetlands.

While the extent and direction of transport varies from one estuary to another, the trend is for wetlands to export materials to deeper adjoining waters. These exports can be thought of as subsidies that one unit of the mosaic of communities in

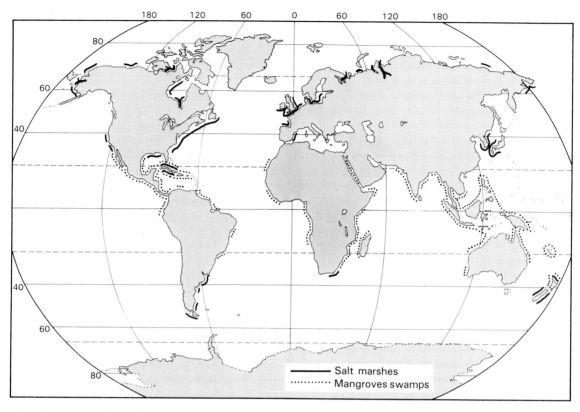

Fig. 3.2 World distribution of salt marshes and mangrove swamps.

the coastal zone provides to another adjoining unit. For example, organic matter is borne by brackish water, leaving salt marsh estuaries on the coast of Georgia, USA, as it moves offshore. Where the estuarine water mixes with the coastal water, there is a depletion of oxygen in the receiving water mass. This depletion is the result of enhanced respiration of organisms in offshore water, prompted by the organic matter provided by salt marsh estuaries. Similarly, the export of organic particles provides food for coastal animals, and the export of reduced nitrogen compounds, such as ammonium and organic nitrogen, convey energy-rich nitrogen compounds to nitrogen-limited coastal phytoplankton and to energy-limited bacteria.

Vascular plants dominate seagrass meadows, salt marshes and mangrove swamps. These producers contain lignocellulose complexes that make much of the tissues not readily available to consumers,

as already noted in Chapter 2. Vascular plants also contain substances that reduce palatability to consumers. Grazing by herbivores on live plants, therefore, results in consumption of perhaps less than 10% of annual net production of coastal vegetation. The remainder is either degraded *in situ* by microbes, or gets transported by tidal waters. The water of salt marshes and mangrove swamps typically contains many detrital particles.

During decay of detrital particles, some dissolved organic matter is lost from the detritus. Once the DOM is in the water, bacteria, in some as yet unexplained fashion, cause aggregation of the DOM into amorphous particles. These amorphous particles are numerous in the soup that tides flush in and out of salt marshes and mangrove swamps. The amorphous particles, more like flakes than particles are then further colonized by other microorganisms, and are eaten by detritivorous consumers. We discussed the dilemma of food quality for detritivores in Chapter 2. In coastal waters, detritivores are especially abundant, and so is detritus. We would predict that if amorphous particles were nutritionally better than morphous particles, that detritivores would actively select amorphous particles for feeding: this remains to be tested.

Hard substrates, in shores exposed to waves, typically support growth of macroalgae. Subtidal kelp beds or forests are common in colder waters (Fig. 3.1). Kelp are giant brown algae, most frequently *Laminaria*, *Eklonia* or *Macrocystis*. Intertidal macroalgae, often called rockweeds, cover rocky shores the world over: *Fucus*, *Ascophyllum*, *Postelsia* and *Ulva* are among the myriad of genera.

Coastal environments dominated by brown algae — kelp and rockweeds — show many of the same properties as coastal wetlands. Although in some circumstances grazing by sea urchins can deplete vegetation of kelp forests, it is more common for most of the producer biomass to go uneaten by herbivores. Some macroalgae contain refractory chemicals, resembling lignins, and compounds that inhibit grazing and microbial attack. In most cases, most of the annual production is, as in the case of coastal wetlands, degraded by decomposers, or exported out of the system by water flow.

Some species of green and red algae contain organic compounds that inhibit feeding by grazers and invasion by fungi. These inhibitors are often small molecular weight halogenated hydrocarbons. Most green and many red algae, however, are more susceptible to grazers than either brown algae or vascular plants. Green and red algae generally make more assimilable food for consumers while they are alive, and the smaller amount of detritus that is produced also decays much faster than detrital matter from vascular plants.

3.4 GEOMORPHIC FEATURES OF COASTAL ENVIRONMENTS

There are shorelines where the bottom is a gentle slope, and water reaches up to a straight or concave beach. Bottom sediments may be unconsolidated sands, or hard rock. These simple environments are merely shallow, landward extensions of the continental shelf. There may be straight, exposed shores, or bays less exposed to ocean waves. Although these simple environments do occur, most coastal settings have suffered much alteration by geological processes and are much more complicated.

Over most of the world's shorelines, local geology and water flow combine to produce a variety of coastal environments. For instance, in the temperate or cold zones of the world, glaciation is a primary source of sediments and landforms that create today's coastal environments. The original landforms left by glacial action are subsequently and continually modified by water flow. In one common sequence in glaciated terrain, long-shore currents move sand across indentations of the coastline and create bars. Lagoons form behind such bars, and are a common place where seagrasses, salt marsh, or mangrove vegetation become established.

Coastal environments usually have somewhat lower salinities than open ocean water. Fresh water may be borne into lagoons and bays by small streams or by flow of groundwater. In many of these common and shallow environments, the water column may be mixed vertically by wind, which overcomes stratification.

Estuaries are environments in which large rivers deliver fresh water, and in which mixing of river water with sea water is the dominant physical process. The physics of estuarine mixing is complex and is affected by geological setting, topography, flow of fresh water and several other features. River water carries dissolved salts and particles, and this loading, plus the physical mixing, leads to very active biological transformations.

The processes involved in the mixing of fresh and sea water give rise to various phenomena of ecological importance. There are often strong gradients of chemical and physical properties, such as salts and temperature, that affect organisms. There are also chemical consequences of the mixing of fresh and salty water. Salty water desorbs ions from particles, so that concentrations of certain dissolved ions increase. Passage of dissolved ions from fresh to brackish water is also often accompanied by flocculation and precipitation of certain other salts, especially those involving metals. These chemical consequences of mixing alter the delivery of essential elements to estuarine sediments, and affect the activity of sediment microflora.

Deltas occur in estuaries where the sediment load carried by the river is so large that accumulations of sediment near the river mouth grow towards the sea. Deltaic sediments tend to include more small-sized particles, such as clays and silts, than sands. The shores of the complex channels that form in deltas provide extensive protected shallow habitats that invariably support dense vegetation of vascular plants.

Fjords and firths are coastal environments in which glaciers scoured river valleys, but left a shallow sill at the mouth where the tongue of ice rose as it made contact with the denser sea water. Firths are shallower than fjords and may lack their U-shaped cross-section. Due to their glacial origins, fjords and firths are restricted to latitudes above 45° or so. The deeper fjords may have anoxic waters below the depth of the sill. These deeper waters have primarily anaerobic microorganisms as the major biological elements. The walls of fjords and firths are so steep that macroalgae and vascular plants are not important: primary production in these coastal systems is primarily carried out by phytoplankton.

Where the bottom is hard rock, but is shallow enough to allow light to support algal growth, there may be extensive kelp forests subtidally, and beds of rockweeds intertidally. These brown algae tend to be associated with cold waters, either because of latitude or because colder water upwells nearby (Fig. 3.1).

3.5 HYDROGRAPHIC FEATURES

Circulation of water in the nearshore is complex, and results from a combination of factors. There are forces produced by differences in density of fresh and sea water, and by differences in temperature of waters. The effects of these are added to, or modified by, tidal forces, wind-driven circulation and the effects of the earth's gravitational field.

Less dense or warmer water tends to remain in the surface layers of a water column; denser, colder water remains in a layer below, and the water column is therefore stratified. Winds and tides, however, may mix the water column, causing a vertical exchange between surface and deeper layers. Chapter 1 offers more discussion of estuarine circulation: here it is sufficient to say that the balance between stratification and mixing vary over time and space, and that the ecological consequences are significant. For example, upper layers of the water column that do not receive renewed supplies of nutrients from the richer layers below may become depleted of nutrients.

In estuaries, gravitational and tidal forces may be the principal hydrographical features. In other coastal environments, wind-driven circulation may be as important. In areas where winds blow in certain directions and at certain speeds along western shores of continents, wind-driven up-wellings of nutrient-rich deeper water may occur (see Chapter 1). These upwelled nutrients support some of the most productive blooms of phytoplankton found anywhere in the sea. The effects of this high production on the rest of the coastal food web is enormous, as can be seen from the fact that although upwellings constitute only 0.1% of the world's surface, they have the potential to produce up to 50% of the world's fish harvest.

The importance of hydrography is also evident in areas where water masses of different properties meet in a sharp boundary. The narrow zone where they meet, called a front, is often a site of enhanced biological productivity. Fronts may have a permanent, very large, near-global scale, such as those associated with the Antarctic Divergence, or have an ephemeral, small spatial scale of just metres, as may happen in small, shallow estuaries during heavy rains. Coastal fronts may be associated with estuarine circulation, upwellings, breaks in slope of continental shelves and with river plumes.

3.6 CONTRASTS OF COASTAL AND OPEN WATER COLUMN ECOSYSTEMS

In this section we compare some features of coastal ecosystems with those of deeper water columns. The contrasts selected provide a way to emphasize function of ecosystems and bring up aspects of system-level elemental cycles that were underemphasized in Chapter 2. We will focus on three major features of coastal ecosystems: (i) well-developed couplings to land and to the sea floor; (ii) a greater concentration of life-supporting substances and generally higher rates of ecological activities than in the open-sea water column; and (iii) the more prominent role of macroproducers.

3.6.1 Couplings of coastal environments to land and to the sea floor

During the early 1960s, the Egyptian government built the Aswan High Dam on the Nile River. Its purpose was to control floods and provide a reliable supply of fresh water to the region. In fact, in a few short years a number of unforeseen effects of the dam became evident. Sediment was trapped behind the dam: these sediments historically replenished nutrients in the coastal plain of the Nile, and supported a productive agriculture for millenia; these sediments also replenished the sediments that were eroded from the Nile Delta by the Mediterranean. Sediment particles and dissolved nutrients that were not left on the floodplain or captured in the delta of the Nile, were transported into the very nutrient-poor eastern Mediterranean, where they supported phytoplankton and a rich fishery. The consequences of the dam have included hastened erosion and retreat of the Nile Delta, marked shifts in navigation channels, lower agricultural production and, because the nearshore water was much less fertile, the collapse of the sardine fishery off the delta, presumably because of the scarcity of phytoplankton.

In Ecuador, entrepreneurs devised an intermediate technology mariculture of shrimp, which demanded low energy and little capital investment, but yielded a crop with high value in foreign markets. This involved conversion of mangrove areas into dyked shrimp lagoons, whose nutrient-rich water was pumped from the mangrove estuary. Since juvenile shrimps cannot yet be grown under culture, the practice depended on harvesting sufficient wild juveniles to grow into adults in the dyked lagoons. Soon, however, the stock of young shrimp were depleted by the collections and the culture collapsed. No one has evaluated the consequences of the substantial removal of mangrove

habitat that has resulted from this practice, but there are sure to be effects on adjoining waters, since it is known that the yield of coastal fisheries is related to the acreage of wetlands along the coast.

The rivers that drain industrialized nothern Europe are among the most nutrient-laden waters anywhere. The nutrients transported by the Rhine, Elbe and other rivers have been enough to significantly increase concentrations of nitrates and other nutrients over recent decades, as measured at a station in the German Bight of the North Sea. The nutrients provided by rivers have been sufficient to show increased concentrations, in spite of the very large volume of coastal sea water into which they discharge. In fact, the nutrients are of sufficiently high concentrations that phytoplankton are no longer nutrient-limited. This must be leading to major changes in the marine food web, and may be related to problems with toxic blooms of phytoplankton, which kill fish and shellfish.

Urbanization along the shores of New England is proceeding rapidly. Most dwellings in the New England coastal zone (except in cities) dispose of sewage via septic tanks that discharge nitrogen to the groundwater. The flow of groundwater and spring-fed streams transports the nutrients downslope, and eventually the nutrient load, particularly nitrates, reaches coastal waters. Once in the marine system, the nutrients stimulate growth of coastal producers and favour the abundance of certain opportunistic macroalgae. Large areas of shallow coastal water become covered by thick layers of benthic algae at the expense of other producers such as eelgrass. Unfortunately, an eelgrass bed is a much better habitat and nursery for fish, bivalves and other animals than a bottom covered by thick, loose, unstable layers of algae. High production of organic matter by macroalgae due to nutrient enrichment leads to more frequent anoxic events which kill animals such as shellfish, shrimp and flatfish.

These are but four sketches that vividly demonstrate the close couplings of watersheds and adjoining coastal environments. The examples also illustrate that today, human activities dominate the fate of the coastal zone.

The effects of anthropogenic activity are varied, as is evident in the four examples briefly described above, and their importance is created by the way coastal systems are coupled to watersheds. Due to the central role of nutrients in many of the couplings, in this section we choose to focus primarily on the ways in which nutrient exchanges and transformations couple coastal environments to adjoining land and to the sediment below.

Nutrient dynamics in coastal ecosystems

In discussions of the aquatic water column in Chapter 2, we mentioned how patterns and rate of primary production are largely determined by nutrient supplied to the surface layers of the ocean by hydrographical mechanisms (upward vertical advection and turbulent mixing). In much of the area of large lakes, and most of the area of the oceans (88% of the sea is deeper than 1000 m), water columns are deep enough to extend below the illuminated zone, and we can indeed conceive of the productive layers as receiving nutrients from a 'black box' of deeper water below. Estimates of such deliveries of nutrients to specific areas are rough, but it is apparent that, for example, at least for the North Pacific Gyre and the Sargasso Sea — both representative of deep oceanic areas — the major nitrogen source is transported from below (Fig. 3.3). Other gains, such as from precipitation and by nitrogen fixation, both to be discussed below, are much smaller than hydrographic deliveries. Losses of nitrogen via denitrification (also discussed below) are large only in the North Pacific Gyre, where nearby hydrographical features lead to the presence of water with the reducing conditions that favour denitrification.

Nearer shore, new nutrients are supplied to marine ecosystems principally from nearby land. Erosion and decay on land provide elements to fresh water, which in turn transports the elements to the sea. In recent decades, human population growth and development of coastal watersheds have increased the release of materials into fresh water and into the atmosphere, and have resulted in marked elemental enrichment of nearshore waters.

Anthropogenic eutrophication — the enrichment of receiving waters that results from human activities — is one of the most pervasive changes altering coastal environments all over the earth.

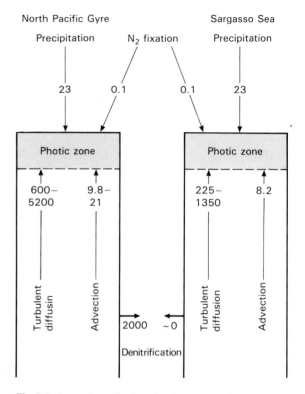

Fig. 3.3 Approximate budgets for nitrogen supply in two oceanic systems. Units are μg at $N\,m^{-2}\,day^{-1}$. From McCarthy & Carpenter (1984).

Sea-level rise, pollution with hydrocarbons, atmospheric warming and ozone depletions all have serious potential consequences for coastal environments. Eutrophication, however, is at least as pervasive as any of these threats, but is actually — not just potentially — affecting coastal ecosystems. Moreover, since eutrophication changes nutrient supply, it fundamentally alters the entire make-up of coastal food webs. Furthermore, coastal eutrophication is on the rise: by the year 2000, over half the people of the USA, for example, will live within 100 miles of the coast. This can only exacerbate the problem of eutrophication. Thus, for basic and applied reasons, we will need to know as much as possible about nutrient supply and dynamics in the nearshore.

Sources of nutrients supplying coastal marine ecosystems are varied, and transport and transformations of nutrients in these environments are very active. The supply of, say, nitrate molecules to coastal marine producers may be newly imported into the ecosystem by precipitation, fluvial transport, or human waste discharge. In addition to this 'new', externally supplied nitrogen, there is an additional supply of 'old' nitrogen, that is made available (regenerated) from decay of organic matter present in the ecosystem. Regenerated nutrients are released during decay in the same proportions as are present in organic matter:

$$106 \, CH_2O + 16 \, NH_3 + H_3PO_4 + 106 \, O_2 \rightarrow$$
$$106 \, CO_2 + 16 \, NH_3 + H_3PO_4 + 106 \, H_2O.$$
(3.1)

The regeneration of nitrogen yields ammonia (NH_3) which in water is usually converted to ammonium (NH_4^+).

The ratio of elements has already been discussed as the Redfield ratio of 106 C : 16 N : 1P. Although it is not always easy to distinguish 'new', externally-derived nutrients from 'old', internally-regenerated nutrients, it is worthwhile to consider these two sources separately.

Sources of nutrients in coastal ecosystems

'New' nutrients can be transported into coastal ecosystems via precipitation, freshwater transport and in the case of nitrogen, by nitrogen fixation carried out by microbes.

Precipitation — rain, snow and dry deposition — can provide nutrients (NH_4^+, NO_3^-, PO_4^{3-}, etc.) to waters. Land and man are still the sources of these nutrients, however, because atmospheric nitrate comes primarily from the combustion of fossil fuels, which release nitrous oxide, which in turn is converted to nitrates in the atmosphere. Atmospheric ammonium arises mainly from animal wastes from agriculture. N_2O emissions have increased continuously since 1900, and are projected to increase another 40–60% over the next 40 years; loading of precipitation-borne nitrates, therefore, can be expected to increase substantially.

Estimates of loading of nitrogen to Chesapeake Bay demonstrate the importance of precipitation-borne nitrogen (Table 3.1): 39% of the inputs of nitrogen to the bay are now being provided

Table 3.1 Estimated sources of nitrogen to Chesapeake Bay. From Fisher *et al.* (1988)

Sources	% of total nitrogen loading
Waste water	23
Atmosphere	
Ammonium	14
Nitrate	25
Fertilizer	34
Animal waste	4

Table 3.2 Budget of inputs and regeneration of nitrogen and phosphorus for Narragansett Bay, Rhode Island. From Nixon (1981)

	Annual input (10^6 g at year^{-1})	
	Nitrogen	Phosphorus
Inputs		
Fixation	20.2	—
Precipitation	2.8	0.19
Surface runoff	16.2	0.8
Rivers	235	17.3
Sewage	278	39.9
Total	552	58
Regeneration		
Water column		
Menhaden	0.8	0.1
Ctenophores	8.1	0.8
Net zooplankton	98.5	—
Benthos	264	41.1
Total	371	42

by precipitation. Of course, to a great extent the contribution by precipitation as opposed to that of other terrestrial sources, depends on the area of the water body. Precipitation, for example, provides a relatively small proportion of nitrogen to smaller systems such as Narragansett Bay (Table 3.2), or Great Sippewissett Marsh (Table 3.3).

Freshwater transport

Anthropogenic activities on watersheds generally increase nutrient content of rivers and groundwater. Downslope transport of this enriched fresh water results in loading of coastal ecosystems. Fertilizers applied to crops may run off, or leach

Table 3.3 Nitrogen budget for Great Sippewissett Marsh including major pathways of gains and losses. From Valiela & Teal (1979)

	Nitrogen (kg year^{-1})
Gains	
Groundwater flow	6120
Precipitation	380
N$_2$ fixation	3280
Total	9780
Losses	
Tidal export	5350
Sedimentation	1295
Denitrification*	4120
Total	10 765

* Modified from Valiela & Teal (1979), by not adding concurrent N$_2$ fixation to the measured rate of denitrification.

through soil and subsoil and enter fresh water; sewage is very frequently released into streams or groundwater. Both these sources are quantitatively significant in many coastal areas. Felling of forests reduces interception of precipitation-borne nutrients on land. All these mechanisms increase nutrient transport to receiving waters worldwide.

As one case history, consider the already-mentioned case of nitrate content of the waters of the German Bight in the North Sea. Between 1964 and 1984, concentrations of nitrates in winter increased from 5.6 to 16.5 µg at l^{-1}. This rather large annual increase of 0.5 µg at l^{-1} is the result of delivery of nitrates (and other nutrients) by the Elbe to the North Sea; the river's nutrients are primarily provided by sewage waters.

In shallow coastal waters, marshes and lagoons with no significant streams, groundwater may be a large contributor to the delivery of nutrients (Table 3.3). Nutrients in groundwater are also increased by disposal of waste water into the subsoil of watersheds.

Nitrogen fixation

Bacteria (including cyanobacteria) are the only organisms that can convert gaseous N$_2$, dissolved in water, into NH$_3$ and hence into organic nitrogen. This reaction (N$_2$ + 3H$_2$ → 2NH$_3$ → → organic N) is energy demanding, and is called nitrogen fixation; the gaseous nitrogen is 'fixed' into an organic form. The rate of nitrogen fixation in water varies greatly (see values in parentheses, Table 3.4). There is a tendency for richer water to support higher rates of nitrogen fixation (compare average rate for oceans to estuaries; oligo- and mesotrophic to eutrophic lakes). This effect of eutrophication may take place because waste waters also increase supplies of phosphorus, at least in fresh water, which may increase N$_2$ fixation rates. In most water columns, N$_2$ fixation furnishes an amount of nitrogen that is only a modest fraction of the quantity required to support primary production (Table 3.4, right column). In addition, the amount of nitrogen provided by fixation tends to be small relative to

Table 3.4 Rates (average (range)) and other features of nitrogen fixation in water columns. Oligo- and mesotrophic lakes receive low and intermediate nutrient loadings, while eutrophic lakes are exposed to high nutrient loadings. From Howarth *et al.* (1988)

	Nitrogen fixation rate (g m^{-2} year^{-1})	Nitrogen fixation as % of total N loading	Nitrogen fixation as % of N needed by producers
Marine systems			
Oceans	0.04 (0.02–0.09)	<0.04	0.4 (0.07–0.5)
Estuaries	0.7 (0.01–1.8)	12 (4.7–17)	2.2 (1.3–3)
Freshwater lakes			
Oligo- and mesotrophic	0.3 (0–2)	16 (0.02–81)	3 (0.03–8.9)
Eutrophic	1.7 (0.2–9.2)	28 (5.5–82)	3 (0.5–7)

other sources of nitrogen (Table 3.4, centre column) for most ecosystems. Some exceptional oligotrophic lakes do receive much of their nitrogen via fixation (Table 3.4). Fixation is also quantitatively important in salt marshes (Table 3.3).

It has been claimed that in many freshwater ecosystems, water column fixation provides 30–80% of total nitrogen which enters annually. This is considerably higher than for marine systems, where, except for salt marshes and seagrass beds, in which fixation is clearly important (Table 3.3), fixation may provide less than 5% of the nitrogen needed by producers (Table 3.4, right column). There are also many observations of abundant nitrogen-fixing cyanobacteria in freshwater systems, while nitrogen-fixing blue-green algae are not as abundant in sea waters.

Although there is large variation in fixation rate, the mean rate for fresh waters does tend to be larger than that for marine waters (Table 3.4, left column). Moreover, the relative size of inputs via fixation tend to be larger in fresh waters than in marine waters (Table 3.4, centre column).

The reasons for the claimed difference in abundance and role of nitrogen fixers in fresh and marine waters are not well-understood. At least two hypotheses have been proposed: one is that the relatively high concentration of sulphate in sea water compared to fresh water may inhibit nitrogen fixation. The sulphate ion is sterically very similar to molybdate, which is a key part of the enzyme system involved in nitrogen fixation. The cell uptake mechanisms cannot easily distinguish between the two ions, so that if sulphate is present in high concentrations, bacteria could take up sulphate rather than molybdate, and the nitrogen fixation rate will be reduced. Another hypothesis suggests that turbulence destroys the aggregates of organic matter or cells. These aggregates form and create small anaerobic zones. Since fixation of nitrogen requires such anaerobic conditions, the more turbulence, the less fixation. This hypothesis supposes that, on average, there is more turbulence in marine than fresh waters, hence the differences in fixation rates. Both hypotheses, and the underlying premise of differences in fresh and coastal marine systems, need further study.

The nitrogen fixation rate in sediments (Table

Table 3.5 Ranges of nitrogen fixation rate in sediments of different aquatic habitats. From Howarth *et al.* (1988)

	Nitrogen fixation rate ($g\ m^{-2}\ year^{-1}$)
Marine systems	
Oceans	0–0.0004
Estuaries and coastal waters	0.01–1.56
Seagrass beds	0.84–50
Salt marshes	0.24–51
Cyanobacterial mats	1.32–76
Mangrove swamps	0.03–2.6
Freshwater systems	
Lakes	0.001–0.28
Freshwater marshes	0.01–6
Peat bogs	0.05–2.1
Cypress swamps	0.39–2.8

3.5) tends to be higher than in water (Table 3.4). Sediments of the most nutrient-poor environments (oceans, lakes; Table 3.5) tend to have a lower rate than richer environments. There seem to be no major differences between the rate of nitrogen fixation in fresh and marine sediments. Sediments of freshwater and coastal environments are usually anaerobic. In sediments, potential sulphate inhibition is low, since sulphate reducers convert sulphate to sulphide, and also create other geochemical changes that favour N_2 fixation; also, sediment pore water is obviously not subject to turbulence.

The nitrogen fixation rate in sediments increases to some extent in nutrient-rich conditions (Table 3.5), but the rate is inhibited at relatively high (20–100 µM) concentrations. Where NH_4^+ is so readily available, bacteria take up NH_4^+ directly to use in protein synthesis, rather than use the energy-demanding fixation pathway to obtain the needed reduced inorganic nitrogen. Such a relatively high concentration of NH_4^+ is common in sediments, less so in water columns; nitrogen fixation may, therefore, be low in sediments and, perhaps, in water columns of eutrophic environments.

Regeneration of 'old' nutrients and losses: the couplings to sediments

In coastal ecosystems there is a very active use, release and reuse of nutrients already in the environment. Internal sources of these 'old' nutrients are quantitatively significant (see Table 3.2, for example). Much of the regeneration of organically-bound nutrients to mineral nutrients takes place in sediments. Since coastal sediments often are reducing, anaerobic microbial processes such as denitrification are important.

Nutrients contained in organic matter are released by decay and by animal excretion. We have already discussed these processes in water columns, in dealing with the microbial loop. Regeneration can occur in the water column or in sediments, and the relative importance of the release of 'old' nutrients in these two parts of aquatic systems varies. In the more shallow water columns of coastal environments, regeneration from sediments assumes a preponderant role. For example, in Narragansett Bay, over 71% of the regenerated nitrogen and 98% of the regenerated phosphorus are released from bottom sediments (Table 3.2). The shallow depth of coastal environments means that falling particles do not remain exposed to microbial and animal consumers for long in the water; most organic particles fall to the bottom and are mineralized there.

We need to emphasize that nutrients provided by regeneration cannot be thought of as an additional source of nutrients. Regenerated nutrients are already in the ecosystem, and are being recycled. This is why this source of nutrients is commonly referred to as 'old' nutrients.

Denitrification

The very active regeneration of nutrients that occurs in sediments of coastal systems has important consequences, since the anoxic sediments characteristic of coastal ecosystems allow denitrification to take place. The net result may be to lower availability of inorganic nitrogen in coastal waters.

The ammonia released by decay of organic matter, or by excretion from animals, is nitrified under aerobic conditions:

$$NH_3 + \tfrac{3}{2}O_2 \rightarrow HNO_2 + H_2O$$
$$HNO_2 + \tfrac{1}{2}O_2 \rightarrow NO_3^- + H^+. \qquad (3.2)$$

The microorganisms that carry out these reactions do so to obtain energy stored in the reduced nitro-gen atoms. The nitrate produced by these reactions can then be used as an electron acceptor by organisms that oxidize organic matter under anaerobic conditions:

$$C_6H_{12}O_6 \text{ (organic matter)} + 24\ HNO_3 \rightarrow$$
$$6\ CO_2 + 9\ H_2O + 3\ N_2O \rightarrow$$
$$30\ CO_2 + 42\ H_2O + 12\ N_2.$$
$$\qquad (3.3)$$

This series of reactions releases nitrous oxide (N_2O, one of the key atmospheric gases, which reacts with ozone) or N_2. Both these gases are unavailable to producers, so that the process converts nitrate, a form of nitrogen available to producers, to N_2, which is unavailable to them.

The loss of nitrogen by denitrification from coastal waters is proportional to the nitrogen inputs to the system (Fig. 3.4). The greater the nitrogen loading to coastal waters, the more nitrate that is made available to denitrifiers and the greater the rate of denitrification.

Denitrification may remove up to 50% of nitrogen that enters aquatic ecosystems (Fig. 3.5). Estuarine systems seem to best support denitrification; there 40–55% of inputs are lost as N_2, regardless of the nitrogen-loading dose. In fresh water (rivers or lakes), losses via denitrification are smaller, 10–35%, and are lower at the highest

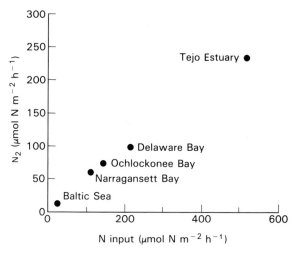

Fig. 3.4 Plot of denitrification rates against external nitrogen input rates in estuaries. From Seitzinger (1988).

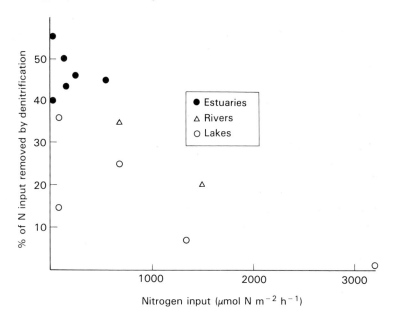

Fig. 3.5 Per cent losses by denitrification in relation to external nitrogen loading to various aquatic environments. From Seitzinger (1988).

nitrogen loading rates (Fig. 3.5). These observations agree with our discussion of nitrogen limitation of producers in coastal waters, in Chapter 2: nitrogen is likely to be in short supply in coastal waters, less so in fresh waters. If denitrification is such an active process in estuaries, why do we have a eutrophication problem? Note that in Fig. 3.5, the proportion of loss by denitrification becomes lower as nitrogen loading becomes higher. Denitrification cannot be thought to be the sole solution or control for eutrophication, for losses by denitrification are unable to keep up with increasing loading.

In addition to direct loss of nitrogen, the domination of coastal nutrient budgets by regeneration in sediments may lead to a relative loss. Degradation of organic matter in sediments — or anywhere else — releases nutrients in Redfield ratios, but the subsequent release of nutrients from sediments to the water column may not. The ratio of $N:P$ released from sediments of 12 different estuaries, for example, averaged about 8.4 (half the $16:1$ Redfield ratio expected). Although the explanation is not certain, it may be that some of the NH_4^+ regenerated from organic matter is converted to NO_3^-, which is then denitrified to N_2. If, indeed, benthic regeneration is the dominant source of nitrogen in estuaries, the lower $N:P$ released from sediments may help explain why typically there are low $N:Ps$ in water during summer, when denitrifiers are active, and why growth of producers is nitrogen-limited in coastal waters.

3.6.2 Fertility of coastal environments

As coastal ecosystems are so exposed to nearby sources of nutrients, it is not surprising that concentrations of almost any substance are greater than in deeper-water environments. In fact, there are pronounced chemical concentration gradients away from coasts and sediments. For instance, the concentration of organic particles (including phytoplankton cells), and of nitrogen and phosphorus decreases with increasing distance from New York Harbor (Fig. 2.3).

The differences in chemistry between coastal and offshore waters further translate into parallel differences in rates of biological activity. Almost all ecological processes take place at a faster rate on a per square or cubic metre basis in coastal waters than in oceanic waters.

Comparisons of net phytoplankton production between coastal and open ocean waters are difficult because of the disputed nature of the methods used to measure primary production (see Chapter 2),

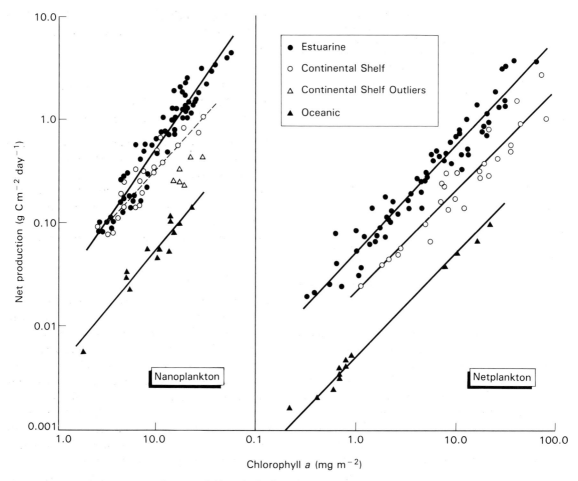

Fig. 3.6 Net production rate as a function of chlorophyll *a* for nanoplankton and netplankton size fractions. From Malone (1980).

and because there is a wide range of standing crops in both kinds of ecosystem (see *x* axis, Fig. 3.6). To allow for differences in standing crops, we can plot net production against chlorophyll, as in Fig. 3.6. For any one concentration of chlorophyll, it is apparent that there is greater productivity in estuaries than in shelf waters, and that both these environments support about one order of magnitude greater phytoplankton production than oceanic waters.

The gradients in net production are similar for nano- and netplankton. The means and ranges of net production and chlorophyll for both of these size classes of phytoplankton are also rather similar

(Fig. 3.6). The only notable exception is that in oceanic waters, nanoplankton are more important than netplankton, since both average production and average chlorophyll of netplankton are one order of magnitude lower than those of nanoplankton.

3.6.3 Prominence of macroproducers

Another major difference between open ocean and coastal systems is the relative importance of macrophytes. While floating macroalgae such as sargasso weed, a brown alga, can be locally abundant in certain nutrient-poor, warm oceanic

Table 3.6 Ranges of net primary production by phytoplankton and macrophytes in various coastal environments. From Malone (1980) and Mann *et al.* (1980)

	Net primary production ($g \, C \, m^{-2} \, day^{-1}$)
Phytoplankton	
Estuaries	0.2–1.5
Bays	0.1–1.0
Lagoons	0.2–0.7
Fjords	0.1–1.2
Shelf	0.4–0.9
Upwelling	0.3–1.0
Macrophytes	
Seagrasses	0.2–18.5
Mangroves	0.0–5.7
Salt marshes	0.2–36.5
Kelp forests	0.4–9
Rockweeds	<12

waters, it is phytoplankton that are the dominant producers, in general, in open waters. Nearer the coast, in shallower water, macroalgae and vascular plants are more prominent and may be the key species of producers. For example, in Nova Scotia Bay, kelp produced, on average, $603 \, g \, C \, m^{-2} \, year^{-1}$, while phytoplankton produced about $200 \, g \, C \, m^{-2} \, year^{-1}$. Algal macrophytes thus contributed 75% of the total production of the bay.

Net primary production by producers in different coastal environments is very variable (Table 3.6). Net production by phytoplankton is much less than $10 \, g \, C \, m^{-2} \, day^{-1}$ (Table 3.6, Fig. 3.6). Rates of production achieved by macrophytes are frequently more than $10 \, g \, C \, m^{-2} \, day^{-1}$.

It is difficult, however, to meaningfully compare absolute ranges of net primary production among different aquatic environments, because the biomass produced may have quite different turnover times (Table 2.1). In phytoplankton, the biomass present may turn over daily; in mangrove trees, the annual net production is but a small fraction of the biomass present in leaves, roots, trunks, etc. Even within kelp, there may be wide differences in the ratio of net production (*P*) to average biomass (*B*). In *Macrocystis*, *P/B* is about 1, on an annual basis. In *Laminaria*, *P/B* ranges from 2 to 7; a stand of *Macrocystis* therefore turns

over slowly, much like a forest, while stands of *Laminaria* grow and turn over quickly, more like a weed patch.

3.7 CONTROLS OF PRODUCTION AND ABUNDANCE IN COASTAL ENVIRONMENTS

The controls of local abundance and distribution of coastal communities dominated by vascular plants and macroalgae are very complex and often species- and site-specific. There are, of course, certain physiological limits for certain species, but there are also overall bottom-up and top-down controls of the components of coastal systems.

3.7.1 Top-down controls

Top-down controls have been shown to be important in a series of experiments carried out in hard-bottom coastal systems. If predators, such as sea otters, or disease are absent, the abundance of sea urchins in kelp beds increases (Table 3.7). High abundance of urchins in turn coincides with low abundance of the kelp on which urchins feed.

Manipulative experiments on the rocky intertidal zone have further shown that seastars and predatory snails control abundance of invertebrates. Moreover, when predation or herbivore pressure is intense, only a few species of food survive; when predation pressure is lower, more species of prey are present. In the rocky shore, consumers (i.e. top-down controls) not only seem to determine abundance of prey and of macroalgae, but also the species composition of the counts. For example, when few herbivorous snails are present in wave-protected tidal pools of the rocky New England shore (Fig. 3.7a), algal food species preferred by snails proliferate. There is a maximum number of species of algae at intermediate grazer densities. At high grazer densities, only the non-palatable Irish moss survives; at very low grazer densities, the competitively dominant species such as *Ulva* outgrow other species and hence only a few algal species exist. At intermediate grazer densities, the competitively dominant species (which happen to be, in this case, preferred food for the grazers also) are prevented from overgrowing the less preferred

Table 3.7 Frequency of key species of urchin and kelp in 1-m² quadrats, from three sites in Alaskan kelp forest, with no otters and with otters present for shorter and longer time intervals. Data from Duggins (1980)

	Frequency of species (mean no. m⁻² ± standard deviation)		
	No otters	Otters present <2 years	Otters present >10 years
Urchin species			
Strongylocentrotus franciscanus	6 ± 7	0.03 ± 0.03	0
Strongylocentrotus purpuratus	4 ± 13	0.08 ± 0.06	0
Kelp species			
Annual species*	3 ± 7	10 ± 5	2 ± 5
Laminaria groenlandica	0.8 ± 5	0.3 ± 0.6	46 ± 26

* Annual kelp include *Nereocystis leutkeana*, *Alaria fistulosa*, *Cymathere triplicata* and *Costaria costata*. These algae are opportunistic species that temporarily use space free from urchin grazing.

algae. In exposed shores (Fig. 3.7b) Irish moss cannot grow, and species richness is simply inversely proportional to the abundance of grazers. Local conditions thus mediate the outcome of ecological processes.

In the sea otter–urchin–kelp example, ephemeral kelp species are dominant for a period, and are eventually replaced by competitively dominant kelp species, as long as the top predators are present for sufficiently long periods (Table 3.7, compare centre and right columns). Thus, there are important second-order interactions between carnivores, herbivores and algal food species.

3.7.2 Bottom-up controls

Resources (bottom-up controls) have been shown to control production by assemblages of coastal phytoplankton. Coastal phytoplankton growth, as we saw earlier, increases as nutrient supply increases. Nitrogen is the principal nutrient limiting growth of coastal phytoplankton, but there is increasing evidence that phosphorus may also be limiting at some times of the year. Phosphate may limit phytoplankton during summer, when N : P in coastal waters tends to be lowered.

Nutrients also limit macrophyte production. There is ample experimental, and other, evidence that nitrogen supply, mediated by the oxidation state of the sediments, is responsible for control of net production in salt marsh grasses. Seagrasses seem to be mainly limited by light, not by nutrients; they thrive in sediments that provide sufficient nutrients, but the amount of light reaching seagrass beds is critical in determining both net production and depth distribution of seagrasses.

Nutrient supply limits growth of many macroalgae. For instance, growth of green algae such as *Ulva* or *Cladophora* depends on supply of nitrates or ammonium. Where urbanization of coastal watersheds provides added nitrates from waste water, such as in Venice Lagoon, or in many estuaries and bays all over the world, a thick layer of algae develops on the bay floor. This occurs as long as the bottom is shallow; in deeper waters the bottom is below the illuminated zone and macroalgae cannot grow; there phytoplankton dominate. Once again, there is the pattern of local contingencies playing a role, modifying the effects of general ecological processes.

3.7.3 A case history: controls in a salt marsh ecosystem

It is evident, then, that nutrients, light, grazers and predators can play key roles in determining how much production takes place and how abundant producers are. It is also clear that other specific

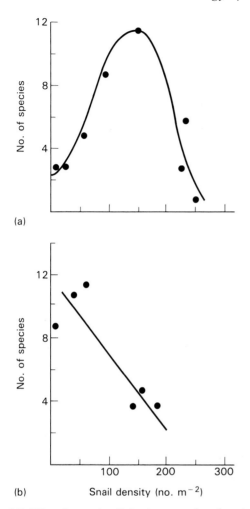

Fig. 3.7 Effect of grazer (snail) density on number of species of rockweeds of two different habitats, (a) tide pools and (b) rock surfaces, of the rocky shore of New England. From Lubchenco (1978).

benthic microalgae on the bottom of tidal creeks in salt marshes, are controlled by resources or consumers. We have already mentioned the bottom-up control of production by salt marsh grasses; there is little evidence of top-down control of salt marsh vegetation. Here, we will therefore focus on the unicellular algal populations that grow on the sediment surface.

The abundance of benthic microflora in salt marsh creek bottoms peaks in late winter and spring (Fig. 3.8a). The increase is probably triggered by the increasing supply of light. This seasonal bloom is similar to the spring bloom of phytoplankton in water columns. The actual amount of producer biomass is determined by availability of nutrients, since experimental addition of nitrogen and phosphorus increase biomass (see the difference in chlorophyll between fertilized and control plots, Fig. 3.8).

But what role do top-down controls play? The major predators are killifish and crabs. The abundance of these groups has a strong seasonal pattern (Fig. 3.9): they are most abundant during the summer, and migrate to deeper waters during cold months. Their effect, therefore, should be primarily seen during the summer. In fact, abundance of macroinvertebrates (>0.5 mm in length) decreases when fish abundance increases. As the weather warms, there is indeed a sharp reduction of producer biomass; by September, chlorophyll in the benthos is low (Fig. 3.8). This effect is clearly related to the activity of consumers, since chlorophyll in sediments protected by cages that prevent entrance of fish and crabs continues to be high through the summer (Fig. 3.8a, interrupted lines).

The smaller animals (meiofauna, organisms <0.5 mm in length) in creek bottoms also showed seasonal patterns (Fig. 3.8b). As in the case of microalgal chlorophyll, the meiofauna increased markedly inside cages, evidence of the role of the larger predators in controlling their abundance, also. Many of these small species eat microalgae, so that in the absence of meiofauna, we may expect to see an increase in abundance of the producers, as in the sea otter–urchin–kelp system. It turns out, however, that in this case the killifish are not exclusively carnivorous and can eat microalgae, so that reduction of the abundance of microfaunal 'grazers'

local contingencies are important. We should not assume, because of the lack of evidence, that top-down controls are only important in hard substrates, and bottom-up controls only in soft sediments. There are too few studies in which we can compare the relative roles of resource and consumption as control processes, and evaluate the local conditions as well. One such study has been carried out in a salt marsh ecosystem, which we can use as a case history.

Experimental manipulations were carried out to examine whether plants on the marsh surface, and

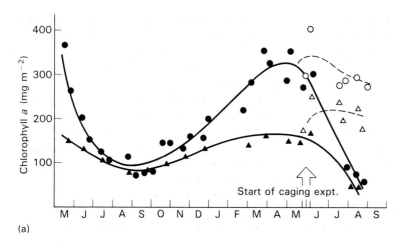

(a)

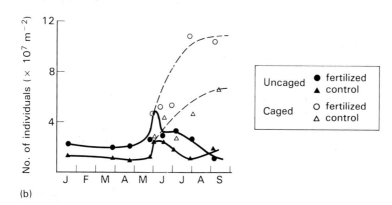

(b)

Fig. 3.8 Seasonal abundance and results of exclosure experiments from (a) benthic microalgae and (b) benthic meiofauna in Great Sippewissett Marsh. From Foreman (1989).

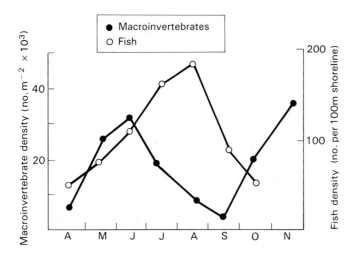

Fig. 3.9 Seasonal pattern of abundance of benthic macroinvertebrates and fish in salt marsh tidal creeks. From Wiltse *et al.* (1984).

does not result in an increase of producers. Once again, details (in this case, the food habits of a key species) of the local scene modify the general effects of consumers.

In this shallow benthic system, there is a clear seasonal shift from resource (light and nutrient) control of producers during cold weather to predator control during warm months. Both bottom-up and top-down controls are active, and are therefore important in determining what the community looks like.

Predators that exert control on abundance of their prey organisms, as the marsh killifish do in summer, also may be self-controlling, since they are ensuring themselves of a scarce food supply. For example, for any size of fish, growth rates of

killifish during August–September are lower than during June–July. This is probably due to the much lower abundance of macro- and meiofaunal benthic invertebrates during late summer. By consuming prey, the fish limit their own food supply, and the result is a reduction in their growth rate. Although it is unwise to generalize from this one case history, it does show how clearly coupled components of coastal ecosystems are; how generalized control processes operate; and how the local conditions alter consequences. Specifically, it also suggests some general relationships, such as the more primary production, the greater the growth of fish.

Due to the variety of possible controls on various features of coastal ecosystems, it has been difficult to find general predictive relationships. For

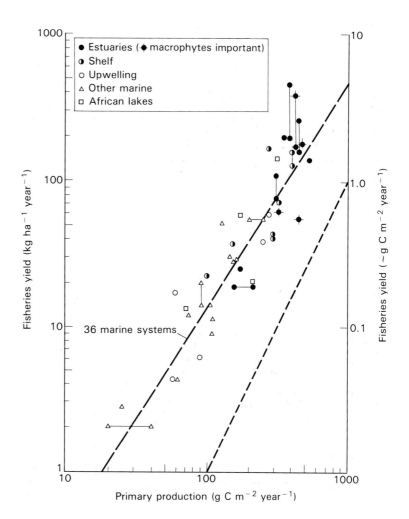

Fig. 3.10 Fisheries yield per unit area as a function of primary production per unit area in estuarine and other coastal marine systems and large lakes. From Nixon (1988).

instance, it is not at all clear how the yield of the
very important coastal fisheries is related to the
major ecological processes discussed. The ultimate
dependence of fish growth on the activity of producers in coastal systems is undeniable, but the
translation of primary production to fish yield is
necessarily only approximate, because it is affected
by at least one additional trophic step, by local
features of concentrations and site and by inade-
quacies in the quality of the data. Nonetheless,
there is some relation between fish abundance and
overall primary production in coastal and freshwater ecosystems (Fig. 3.10). Note that the graph is
on a log–log basis, which improves the look of the
data. In spite of all this, coastal systems in which
annual primary production is greater do appear to
support a larger fishery yield. Incidentally, coastal
marine systems seem, for some unknown reason, to
support a much greater yield of fish at any level
of primary production than temperate freshwater
lakes (shown only by a best-fit line in Fig. 3.10).

Coastal systems where macrophytes are important cluster on the top right of Fig. 3.10: both yield
and production are higher in those circumstances.
One reason for this relationship is that macrophyte-
dominated coastal waters are often richer in resources, and provide better and more protected
habitats for juvenile fish. This relationship is
one major reason which justifies conservation of
macrophyte-dominated coastal environments.

FURTHER READING

Duggins, D.O. (1980) Kelp beds and sea otters: an experimental approach. *Ecology* **61**: 447–53.
Fisher, D., Ceraso, J., Mathew, T. & Oppenheimer, M.
(1988) *Polluted Coastal Waters: the Role of Acid Rain.*
Environmental Defense Fund, New York.

Foreman, K.F. (1989) *Regulation of benthic microalgae and
meiofaunal productivity and standing stock in a salt marsh
ecosystem: the relative importance of resources and predation.* PhD Thesis, Boston University.
Howarth, R.W., Marino, R. & Lane, J. (1988) Nitrogen
fixation in freshwater, estuarine, and marine ecosystems: 1.
rates and importance. *Limnol. Oceanogr.* **33** (part 2):
669–87.
Lubchenco, J. (1978) Plant species diversity in a marine
intertidal community: importance of herbivore food preference and algal competitive ability. *Am. Nat.* **112**:
23–39.
Malone, T.C. (1980) Size fractionated primary productivity
of marine phytoplankton. In: Falkowski, P.G. (ed) *Primary
Production in the Sea*, pp. 301–19. Plenum Press,
New York.
Mann, K.H. (1972) *Ecology of Coastal Waters.* Blackwell
Scientific Publications, Oxford.
Mann, K.H., Chapman, A.R.O. & Gagné, J.A. (1980)
Productivity of seaweeds: the potential and the reality.
In: Falkowski, P.G. (ed) *Primary Production in the Sea*,
pp. 363–80. Plenum Press, New York.
McCarthy, J.J. & Carpenter, E.J. (1984) Nitrogen cycling in
near-surface waters of the open ocean. In: Carpenter, E.J.
& Capone, D.G. (eds) *Nitrogen in the Marine Environ-
ment*, pp. 487–512. Academic Press, London.
Nixon, S.W. (1981) Remineralization and nutrient cycling
in coastal marine ecosystems. In: Nielson, B.J. & Cronin,
L.E. (eds) *Estuaries and Nutrients*, pp. 111–38. Humana
Press, New Jersey.
Nixon, S.W. (1988) Physical energy inputs and the compara-
tive ecology of lake and marine ecosystems. *Limnol.
Oceanogr.* **33** (part 2): 1005–25.
Radach, G., Berg, J. & Hagmeier, E. (1980) Long-term
changes of meteorological, hydrographic, nutrient and
phytoplankton time series at Helgoland and at LV-ELBE1
in the German Bight. *Cont. Shelf Res.* **10**: 305–28.
Seitzinger, S. (1988) Denitrification in freshwater and coastal
marine ecosystems: ecological and geochemical significance. *Limnol. Oceanogr.* **33**: 702–24.
Valiela, I. & Teal, J.M. (1979) Nitrogen budget of a salt
marsh ecosystem. *Nature* **280**: 652–56.
Wiltse, W.I., Foreman K.F., Teal, J.M. & Valiela, I. (1984)
Effects of predators and food resources on the macrobenthos of salt marsh creeks. *J. Mar. Res.* **42**:
923–42.

4

Ecology of Deep-water Zones

B.T. HARGRAVE

4.1 INTRODUCTION

The terms profundal (Latin *profundus* — deep) and abyssal (Greek *abyssos* — bottomless) refer to the deepest parts of lakes and oceans, respectively. The profundal zone in a lake extends from below the thermocline to the deepest depths of the basin (Fig. 4.1). Water contained within this zone is called the hypolimnion (Greek *hypo* — under, *limn* — lake)*. The abyssal environment of the ocean, which occupies areas deeper than 2000 m, is the single largest habitat type on earth and it represents 84% of the world ocean area. The deepest regions of the ocean (hadal zone), from 6000 to 10 000 m, account for about 1% of the area. Since sea water covers 70% of the earth's surface, abyssal zones occupy over 50% of the world.

4.2 STRUCTURE OF DEEP-WATER COMMUNITIES

4.2.1 Types of organism

Profundal and abyssal populations are composed of a broad range of sizes of organism. Bacteria and fungi (<10 μm) and microfauna (10–50 μm), such as colourless zooflagellates, ciliated protozoa and amoebae, form the basis of food webs in deep water. Fungi predominate where leaf litter or macrophyte debris are important sources of organic matter. In general, however, bacteria colonize detritus and sediment particles. Bacterial cells produce extracellular enzymes to facilitate uptake of dissolved organic matter and to solubilize

particulate organic substrates. The carbohydrate-rich exudates are in the form of polymers that cement sediment particles together.

Meiofauna (animals 50–1000 μm in length) are the smallest metazoan fauna in the sediments and consist of groups such as nematodes and gastrotrichs. Meiofauna in marine sediments include diverse taxa, including harpacticoid copepod crustaceans, and some families (Turbellaria and Gnathostomulida) that are also able to withstand long-term exposure to H_2S. Meiofauna usually comprise a large fraction of benthic biomass in oligotrophic and deep-sea sediments, where the biomass of macrofauna may be reduced because of low rates of food supply.

Macrofauna (animals >1000 μm in length) in lake profundal and ocean abyssal sediments, are similar at the level of phyla and include annelids and molluscs. They may occur within the sediment (infauna), at or on the sediment surface (epifauna), or move by swimming above the bottom (hyperbenthic). Many macrofauna species, particularly tube-dwellers, which must circulate oxygenated water for respiration within their tubes, cannot withstand prolonged periods of oxygen deprivation. Burrowing species may be less adversely affected by brief exposure to anoxia. Annelids (tubificid and oligochaete worms) are numerous in organically-rich lake sediments, since their resistance to anoxia allows colonization of reduced deposits.

Different classes of macrofauna are present in freshwater and marine deep-water sediments. Tubificid and naiadad oligochaete annelids, and larvae of dipteran insects (chironomids), are the most common classes in profundal lake sediments — numbers may reach hundreds or thousands

* These terms are further explained in Chapter 5.

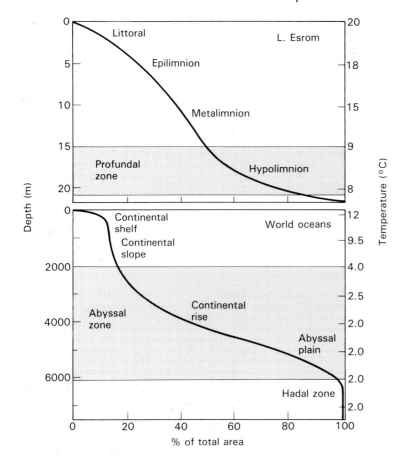

Fig. 4.1 Hypsographic curves displaying the cumulative percentage of total area and average temperatures as a function of water depth for Lake Esrom and for the world oceans. Terms on curves are used to describe depth zonation in lake and ocean basins. Data from Jonasson (1984) and Heirtzler (1982).

of individuals per square metre. Crustaceans (amphipods) and molluscs (unioid bivalves) are also numerous.

On the other hand, polychaetes predominate in muddy, abyssal ocean deposits, with a high diversity and broad size range of species. Smaller numbers of crustacean and molluscan species occur with some groups such as protobranch bivalves that are exclusively marine. Echinoderms — such as crinoids (sea lilies), brittlestars and holothurians (sea cucumbers) — brachiopods (lamp shells) and glass sponges are also restricted to marine waters where they are often the most visible fauna on the surface of abyssal sediments.

Crustaceans (mysids, amphipods, isopods) and predatory fish, such as eels, are the most common mobile fauna swimming near the bottom in deep lakes. In abyssal waters, large gammarid amphipods of the family Lysianassidae are ubiqui-

tous at depths greater than 1500 m, where they are often scavengers on carcasses of other animals. Cephalopod molluscs (squid and octopus) and predatory fish such as the rat-tail (*Coryphanoides*) are the top predators in abyssal food webs and are the largest predators in abyssal environments.

4.2.2 Physiological and behavioural adaptations

Low rates of metabolism, growth and reproduction are characteristic of most deep-water species — all are adaptations to their environment of increased hydrostatic pressure, decreased food supply and low temperature, relative to shallow water. Marine barophilic bacteria have been shown to have optimum growth when cultured at *in situ* pressure and temperature (Yayanos *et al.*, 1979). While most deep-water organisms can tolerate changes in hydrostatic pressure, they are often very sensitive to changes in temperature. For example, pressure effects on the respiration of deep-water crustaceans are more pronounced with temperatures above those usually found in deep-water environments. Some species, such as the deep-sea amphipod *Eurythenes gryllus*, have constant rates of respiration over a wide range of pressure (0.1–32.8 MPa), but pleopod activity and oxygen uptake decrease dramatically above 3°C and death occurs above 6°C (George, 1979).

Many mesopelagic fish species have swim bladders, which are used to maintain buoyancy. Internal gas pressure is maintained at depth through gas diffusion or secretion, and the size of the bladder may be used to sense depth. Buoyancy is also achieved in vertebrate and invertebrate species by lipid storage and by increasing tissue water content above levels typical of more shallow-water species. These adaptations reduce energy expenditure in a food-limited environment.

Adaptations to reduced or episodic food supply in deep water are evident in the number of fish species with extensible jaws that permit swallowing of large prey. Crustaceans and molluscs often have elastic or elongated digestive tracts, compared to related species in shallow water, that permit food storage and prolonged digestion time. The large number of carnivores or scavengers, and relative scarcity of filter-feeding species in deep-water environments is generally thought to be due to the lack of small particles that would normally be consumed by filter feeders.

The lack of penetration of sunlight to deep-water environments does not mean that there is always a complete absence of light. At ocean depths greater than 500 m, many more organisms, from bacteria to fish, produce light (bioluminescence) by chemical reactions (chemiluminescence) that convert chemical energy to radiant energy. Reactions are enzyme based, and caused by the oxidation of the protein-like substance luciferin in organs under nervous control. Luminous symbiotic bacteria may also exist within light-emitting organs.

4.2.3 Feeding mechanisms

Particle feeders in deep water can be classified on the basis of where they obtain their food. Sus-

pension feeders, such as chironomid larvae, bivalve molluscs and sponges use mucous nets, gills with ciliate surfaces, or other filtration processes to concentrate finely-suspended particulate matter. Some profundal species, such as the small clam *Pisidium* that lives in soft mud in lakes, filter suspended bacteria from the sediment (interstitial) pore water.

Meiofauna have evolved a variety of feeding methods by developing specialized mouth parts to browse surfaces, shred larger pieces of organic debris, or prey on other fauna with piercing and cutting mandibles. Benthic infauna are usually deposit feeders or carnivores, and many species such as benthic molluscs, polychaetes and echinoderms feed by ingesting particles from the sediment surface.

Tentaculate deposit feeders are common in marine muds but rare in freshwater sediments. Some tentaculate polychaetes have invaded lakes, but no freshwater groups have evolved this feeding mechanism, which allows selective feeding on particles chosen because of their size or organic content. The absence, in lakes, of tidal motion that transports fresh particulate matter across the sediment surface could explain the absence of this feeding mode in lake benthos.

Infaunal species that do not feed within the sediment have a body structure to filter particles from suspension (e.g. sponges) or may extend appendages or build tubes above the interface to serve the same purpose. Larger epifaunal species, such as crustaceans, are mobile and able to move freely on or over the bottom. These form a major part of the diet of many benthic (demersal) feeding fish. The largest individuals, carnivorous squid and fish, represent the top trophic level in abyssal zones. Eels are equivalent predators in profundal environments.

4.2.4 Biological zonation

Early studies of benthic communities used species descriptions, or number and weight of organisms per unit area, as a basis for comparing different aquatic environments. Faunal associations, or assemblages of species described as 'level bottom communities' in lake and marine sediments were named after the dominant macrobenthic species present in a particular area. This descriptive approach was extended by calculations of indices of species diversity, where the total number of species in a sample — or the distribution of numbers of individuals among species — was compared in different areas.

Diversity in marine macrobenthic fauna has been graphically illustrated by plotting numbers of individuals against numbers of species within a faunal group. Sanders (1968) used the rarefaction method to show that the number of species increases rapidly with sample size for groups of fauna such as polychaetes and molluscs, in continental slope and abyssal plain ocean transects, and in tropical shallow waters. Much smaller increases occur in samples from more shallow marine areas, or estuaries where diversity is lower.

High total numbers of species in deep lakes such as Lake Baikal, and in marine abyssal sediments, have been thought to be due to evolutionary adaptation of species to the physically stable environment ('stability–time hypothesis'). Biological processes may structure the environment and allow species to evolve into separate niches. Recent studies have emphasized that biological factors, on a variety of scales, create disturbances and form microhabitats for different deep-sea benthic species (Jumars, 1989).

4.2.5 Vertical distribution of fauna

We can recognize patterns of zonation of species distributions over depth in lake and ocean basins. In shallow-water littoral or upper continental shelf areas, suspension feeders (which filter organically-rich particles from suspension) predominate. At greater depths, high suspended-matter concentrations may be maintained in some areas, where nepheloid layers (near-bottom water enriched with suspended particles) are present but the organic content is low. In general, however, suspended particle concentrations are very low, and deposit feeders are more abundant in the soft sediments that typically occur in deep water.

Factors which determine food supply to the benthos often also affect the type of bottom substrate. Areas of high current velocity usually have

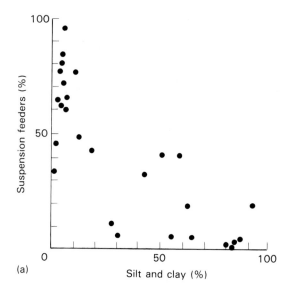

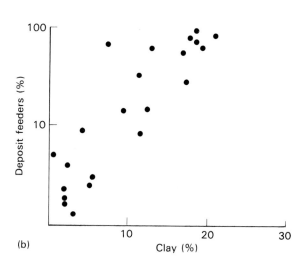

Fig. 4.2 Proportions of (a) suspension- and (b) deposit-feeding macrofauna in benthic populations from marine sediments with varying amounts of fine particles. From Menzies *et al.* (1973).

coarse sediments, such as gravel or sand, and suspension-feeding fauna are abundant (Fig. 4.2). Low current velocity, on the other hand, allows the accumulation of sediments, such as clays and organic particles with a small size, that favour colonization by burrowing infauna and deposit feeders.

A sharp decrease in benthic biomass occurs below the thermocline in many lakes, but changes

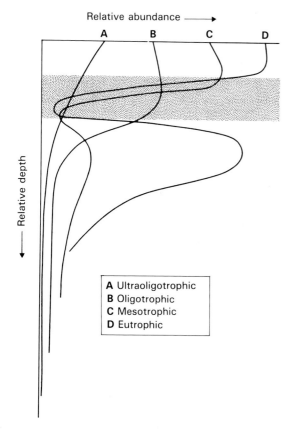

Relative abundance ⟶

A   B   C   D

Relative depth

A Ultraoligotrophic
B Oligotrophic
C Mesotrophic
D Eutrophic

Fig. 4.3 Diagrammatic representation of idealized profiles of abundance of benthic macrofauna in lakes of different trophic status. Dissolved oxygen is usually present in deep water of all lake types except in eutrophic lakes that undergo periods of low dissolved oxygen during summer stratification. Shaded area represents thermocline depth zone. From Brinkhurst (1974).

are dependent on trophic status (Fig. 4.3). For example, oligotrophic lakes have a low biomass of macrofauna and values (of a few g m^{-2}) decline more or less continuously with increasing depth. Large changes occur with depth across the thermocline in productive lakes that are not observed in oligotrophic basins. The biomass of profundal benthos in eutrophic lakes may increase just below the thermocline, before declining to low values in the deep hypolimnion. A consistent pattern of decreasing biomass with depth occurs in the ocean with the steepest gradients under upwelling areas of high productivity.

These complex patterns are due primarily to

interactions between food and dissolved oxygen supply, and in some cases sediment-type distributions. The seasonal reduction of dissolved oxygen in deep water, below the thermocline in eutrophic lakes, is probably the most important factor in limiting biomass of benthic fauna. In oligotrophic and mesotrophic lakes, where dissolved oxygen in the hypolimnion is not totally consumed, but overall rates of organic matter input to the benthos are rather low, levels of benthic biomass in the upper profundal zone below the thermocline may be as high as in littoral regions.

More eutrophic conditions increase the food supply to fauna at all depths, but respiration by decomposers in deep water and sediment lowers the dissolved oxygen available to benthic fauna. Consequently, only species tolerant of low oxygen concentrations can survive. In lakes that do not become anaerobic, biomass in the profundal zone below the thermocline can reach high levels, due to high rates of organic matter production and sedimentation. Low biomass within the depth zone of thermocline formation usually coincides with the transition from communities of suspension-feeding benthos above the thermocline, to profundal fauna dominated by deposit feeders.

The permanent absence of dissolved oxygen prevents growth of macrofauna, and only anaerobic bacteria, protozoa and some species of meiofauna can survive. This could explain the sharp decrease in biomass of macrobenthos observed below the thermocline in many lakes, and in marine sediment under upwelling areas. Bacterial cell numbers in profundal sediments of eutrophic lakes are many times greater than in oligotrophic lakes, reflecting the greater availability of sedimented organic matter for bacterial oxidation (Wetzel, 1983).

We can also observe the combined influence of dissolved oxygen availability and food supply in the high numbers of bacteria and fauna that occur close to the sediment surface (Fig. 4.4). The sediment–water interface is the zone of concentration of freshly-settled particulate matter, thus it is not surprising that most infauna species are in greatest abundance near the sediment surface. Deep burrowing species are adapted to utilize lower concentrations of organic matter and

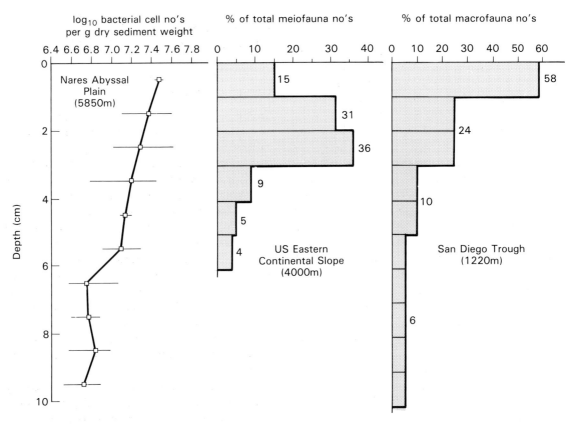

Fig. 4.4 Vertical distribution of numbers of bacteria (by direct cell counts after staining), meiofauna and macrofauna (separated from sediment by sieving) in three deep-ocean locations. From Jumars (1989).

associated microbial populations that occur in deeper sediment layers.

4.2.6 Benthic food webs

Combining species into feeding groups ('guilds') is useful for identifying factors that may structure food webs in deep water. We recognize that the relative abundance of infauna (mostly deposit feeders) and epifauna (suspension feeders and scavengers) could determine energy transfer from the benthos. Polychaetes and bivalve molluscs are relatively digestible prey for benthic predators, such as demersal fish, while suspension-feeding epifauna such as sponges and tunicates contain little energy. Energy flow through these contrasting types of communities may be fundamentally

different, based on the potential for consumption by predators.

Benthic predators may be of any size, feeding on prey of equal or smaller size. In marine deep water, echinoderms, such as brittle stars, and many crustaceans, such as lysianassid amphipods and decapod crabs, are scavengers. They are readily attracted to bait placed on the bottom. In lakes, species such as dipteran larvae, predatory molluscs, benthic and demersal fish are carnivorous and attack living prey. Some species, such as sea cucumbers (Echinoderms) indiscriminantly ingest surface sediment. Many freshwater amphipods also feed non-selectively on surface sediment and organic debris. Small nematodes, harpacticoid copepods and polychaetes are ingested, along with sediment and detritus particles. These small animals may be the main source of nutrition for larger macrofauna.

Meiofauna may serve as integral links in benthic food webs between microorganisms and macrofauna. In some cases, they may be a sink for energy, with a major ecological role in mineralizing organic matter. Groups of meiofauna have evolved specialized mouth parts for piercing and sucking, shredding and rasping, to selectively exploit different fractions of living and non-living sediment organic matter. Some of the biomass which they produce may be consumed by macrobenthos that unselectively ingest sediment, but no macrobenthic species appears to feed selectively on this size class of fauna. The biomass of meiobenthos is often higher in subsurface layers than at the sediment surface (Fig. 4.4). This may be an adaptation to avoid predation by macrofauna in surface layers. It would also explain why the distribution of several groups of meiofauna is restricted to subsurface, anoxic layers avoided by macrofauna (Section 4.2.5).

Two divergent evolutionary trends in body size have occurred in deep-water environments. The evolution of larger body size (gigantism) with increasing depth is most clearly seen in groups such as the crustaceans, where the largest amphipods and isopods in the world are found in the profundal zone of large, deep lakes and in abyssal regions of the ocean. For example, evolution has occurred in Lake Baikal where isolation over a long time period has allowed unique species to form. These organisms show the same physiological adaptations to depth as occur in the ocean, but the species are restricted to this one lake.

The opposite evolutionary trend, a decrease in body size over depth, occurs in other taxa in the ocean. Small body size, which reduces energy requirements for biomass maintenance, is an adaptation to limited food supplies in the deep sea. In some species, males are smaller than females since they do not have to store energy to produce eggs. Competition for limited food resources and predation are two important factors controlling community composition. Small size may also allow individuals to escape detection by large predators.

4.3 FOOD SUPPLIES TO DEEP-WATER POPULATIONS

4.3.1 Sedimentation of products from primary production

Stratification in all water bodies, whether seasonally induced or permanently maintained by salinity differences between water mass layers, affects the rate and pathways of transfer of particulate organic matter to deep water. When a thermocline is present, the upper, mixed layer is separated from deep water by gradients in temperature and salinity. Particulate matter is consumed and decomposed in warmer water above the thermocline, which serves as a barrier to downward transport by gravitational settling. Material that does sink has been processed by grazers and microbial decomposers, so that the readily available fractions of organic matter have been consumed. Organisms in and on the bottom, the benthos, subsist on this material, and that accumulated within surface sediment layers.

When stratification breaks down, caused by seasonal cooling at the surface, mixing of the water column replenishes the bottom layers with oxygen. Dissolved nutrients, released from sediments and accumulated in deep water during stratification, are mixed upwards and, if this enriches the photic zone, phytoplankton blooms occur. Living algal cells, which settled to the bottom during spring and summer, may also be swept into surface waters to cause autumn phytoplankton blooms on lake overturn. The input of this freshly-produced organic matter can stimulate the seasonal growth of the benthos. For example, the combined availability of oxygen and input of fresh organic matter, arising from the breakdown in stratification in Lake Esrom, resulted in about 60% of the annual growth of the dipteran midge larvae occurring during 1 month in autumn (Jonasson, 1984).

Since depth and primary production are key factors controlling the rate of food supply to deep-water communities, attempts have been made to relate these variables to particulate organic matter sedimentation. Traps have been suspended at various depths to collect faecal pellets (material egested

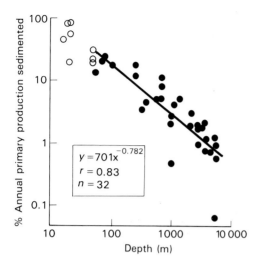

$$y = 701x^{-0.782}$$
$$r = 0.83$$
$$n = 32$$

Fig. 4.5 Relationships between annual phytoplankton production, total water column depth and sedimentation of organic carbon in various offshore ocean areas. In the regression equation, x = depth, y = % annual primary production sedimented to the bottom, r = correlation coefficient and n = number of observations. From Hargrave (1984).

by zooplankton) and aggregates of detrital debris from other planktonic populations. In some cases, after phytoplankton blooms, algal cells may settle directly to deep water, without being consumed by planktonic grazers. These rapidly sinking particles are thought to be the primary energy supply to the benthos.

Seasonal changes in organic matter sedimentation have been measured over depth in lakes and in many ocean areas, from continental shelves to the deep sea. Rates of deposition are positively correlated with phytoplankton production but negatively related to water depth. Sedimentation of organic carbon has been calculated as the percentage of primary production at various depths (Fig. 4.5). Although, on an annual basis, only a small proportion of organic matter produced in the photic zone is sedimented, particulate matter containing freshly-produced organic matter from surface plankton blooms is transferred from the upper layer to deep water in periods of a few days to weeks.

4.3.2 Horizontal transport

The presence of nepheloid layers in many lake basins, abyssal plains and deep ocean trenches shows that factors in addition to vertical settling can be involved in the supply and removal of particulate matter at the sediment–water boundary (Fig. 4.6). There is speculation that the enrichment is caused by turbulence near the sea bed, which keeps particles in a disaggregated state. Sinking rates are reduced because of small particle size, maintained by high levels of turbulence. Because of these processes, particulate matter sedimentation measured in the littoral zone of lakes, or on a continental shelf margin, is usually higher than would be predicted from data obtained in the central part of a lake or ocean basin.

Resuspension, caused by near-bottom water motion, is generated where internal waves impinge on the shelf slope. In large lakes, long-standing waves and seiches are created when wind stress is reduced and the whole water mass rocks back and forth as the equilibrium level is regained. Downslope transport, through sediment slumping in areas of steep bottom gradients, is another mechanism that moves deposited material into deeper water. The process is called sediment 'focusing' in lakes. Periodic high currents at depths greater than 4500 m ('benthic storms'), at the base of oceanic continental slopes, also bring particulate material into suspension and make it available for horizontal transport.

All of these processes add to and remove from bottom sediment, at a particular depth, and it is the balance of transport that ultimately controls the organic matter available to organisms. Filter-feeding benthic macrofauna are able to remove suspended particles from water near the sea bed, if current velocities are not too great. Deposit-feeding benthos, however, may be deprived of food if organically-rich fine particles do not settle and accumulate.

4.3.3 Chemoautotrophy and chemolithotrophy

In basins where water exchange is restricted and oxygen is depleted, as in the hypolimnion of a stratified lake, anaerobic metabolism predomi-

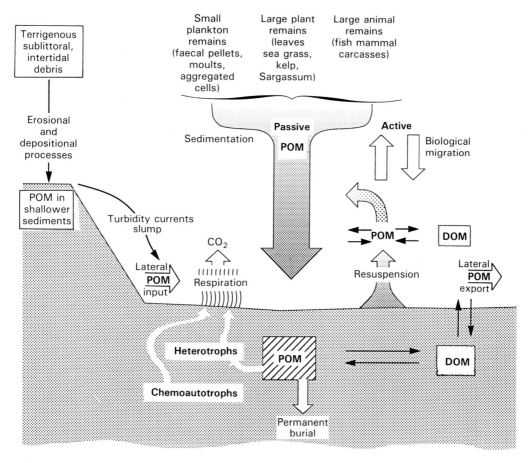

Fig. 4.6 Pathways and processes that transport particulate organic matter and material between and within the water column and deep sediments in lake and ocean basins. DOM, dissolved organic matter; POM, particulate organic matter. From Rowe & Staresinic (1979).

nates over aerobic respiration. In dark environments, chemolithotrophy (the use of inorganic and organic compounds as sources of energy) is a unique property of communities of anaerobic bacteria. Groups of microorganisms that oxidize hydrogen, nitrogen, sulphur and iron, flourish where concentrations of these reduced substrates are sufficient to support their metabolism. Sulphur-oxidizing bacteria utilize abundant hydrogen sulphide or thiosulphate in hot, mineral-rich water flowing from hydrothermal vents, and they are abundant in the deep, anoxic waters of the Black Sea. These microorganisms generate substances that can be used by other bacteria that undergo anaerobic (as opposed to aerobic) respiration. They use sulphate, nitrate, oxidized metals (such as manganese and iron) or carbon dioxide as electron acceptors, thus permitting respiration in the absence of dissolved oxygen (Table 4.1).

Heterotrophic fixation of CO_2 in the dark by chemoautotrophic bacteria can also produce new particulate organic matter in eutrophic waters and sediments, where organic substrates are plentiful. Mid-depth supply of particulate matter by this form of microbial production occurs in permanently stratified lakes and enclosed seas, such as the Black Sea, at the interface between anoxic deep water and surface layers.

A dramatic example of the use of dissolved, reduced inorganic compounds for energy supply in

Table 4.1 Bacterial metabolic reactions that oxidize organic carbon under different oxidation–reduction conditions. From Nealson (1982)

Process	Electron acceptor	Reaction products	Eh range (mV)
Aerobic respiration	O_2	CO_2, H_2O, biomass	+500 to +800
Nitrate reduction	NO_3^-	NO_2	+300 to +500
Metal reduction	MnO_2	Mn^{2+}	+100 to +400
	Fe_2O_3	Fe^{2+}	
Sulphate reduction	SO_4^{2-}	SH-, acetate	−100 to −400
Methanogenesis	CO_2	CH_4	−800
Fermentation	Organic carbon acids	H_2, CO_2, organic	Various

deep-water communities, comes from recent discoveries of hydrothermal-vent communities on the Galapagos Rift, 1700 m off South America (Grassle, 1986). Reduced metal and sulphur compounds are oxidized by chemolithotrophic bacteria in the heated vent water. The same bacteria also grow as symbionts in the tissues of filter-feeding clams, mussels and tube-worms (*Vestimentifera*). The communities have a high biomass, relative to the surrounding oligotrophic benthic communities.

4.3.4 Large particle sinking

In addition to fine-particle passive sedimentation, larger particles (plant remains, leaves, seaweed fragments and animal carcasses) also supply organic matter (Fig. 4.6). Since input from these sources is episodic and not spatially uniform, we have no general methods to quantify rates of supply. However, baited time-lapse cameras have been used to show that large numbers of scavenging invertebrates and fish are attracted to bait moored on the bottom. Carcasses must reach the deep sea, since abyssal populations are adapted to utilize this type of food. Similarly, many studies have shown how, despite low temperature, high pressure and low nutrients, barophilic (grow only under pressure) and psychrophilic (requiring low temperature) bacteria exist in low biomass in the deep sea and are able to rapidly utilize organic substrates, when they are available.

4.3.5 Vertical ladders of migration

Additional sources of organic matter to deep-water communities arise from two separate biological processes. Firstly, zooplankton and nekton migrate to the surface to feed at night. Vertical movements of organisms in lakes may traverse the entire water column, if stratification is not well-established. Daily migration of marine plankton has only been documented over the upper 500 m. Animals that feed near the surface will egest material at depth and increase the downward rate of transport of particles. Secondly, predators, swimming up from even greater depths, can consume individuals that migrate downwards. Trophic interactions linking predators and prey over depth have been described as a 'ladder of vertical migration'.

Vertical movements of plankton on a daily basis, in response to day–night cycles, occur throughout the water column in shallow lakes and in the upper ocean. However, large actively-swimming nekton normally thought to live near the sediment have been trapped hundreds and thousands of metres above the bottom of the ocean. High rates of faecal pellet accumulation in sediment traps placed at midwater depths have been measured in lake and ocean basins. Resuspension or lateral transport of sediment from the basin margins is often the cause, but re-ingestion of sinking material by grazers at mid-depths could also produce high faecal pellet numbers at intermediate depths.

4.3.6 Dissolved organic matter as a food source

A final source of organic input to deep-water zones may occur through uptake of compounds from the large pool of dissolved organic matter (DOM). Concentrations exceed those of particulate matter by about 10 times, but little is known about the origin or dynamics of these substances. The filter-passing material (less than 1 μm in size) has a range of molecular weights. The fact that concentrations are higher in deep water than at the surface, and even higher within the sediment than in the free water, is thought to indicate that compounds are resistant to bacterial decomposition.

Radioactive tracers have been used to follow uptake of labelled amino acids by a variety of soft-bodied marine invertebrates (Wright & Stephens, 1982). Several studies have shown that dissolved organic compounds may be a significant nutritional supplement for some marine organisms. Fresh-water animals, however, because of the requirement for osmoregulation to retain body salts, have tended to develop impermeable outer membranes, and no uptake of DOM has been measured. The existence of this additional source of food for abyssal fauna is important since organically-rich particulate food is not abundant in deep water. Some abyssal invertebrates, such as pogonophores, some bivalves and tubificid oligochaetes, do not have mouths or digestive tracts, and the use of DOM provides their only source of nutrition.

An alternative source of nutrition — by uptake of dissolved compounds from internal, rather than external, sources — is found in fauna that host endosymbiotic sulphur-oxidizing bacteria. In deep-water environments that have high sulphate concentrations, such as near hydrothermal vents, sulphate reduction by symbiotic bacteria in tissues of bivalve molluscs is thought to provide an energy supply to host animals (Grassle, 1986). In lakes, where methane accumulates, it is possible that methane-oxidizing bacteria could be symbiotic, but this has not been confirmed.

4.4 BIOLOGICAL PRODUCTION IN DEEP WATER

4.4.1 Biomass distribution

Depth and the amount of particulate organic matter sedimentation are two critical factors controlling benthic biomass in all aquatic ecosystems. Macrofauna biomass in shallow, profundal sediments of eutrophic lakes, and on marine continental shelves below the base of the seasonal thermocline, varies between 10 and 100 g fresh wt m^{-2}. If primary productivity is enhanced, through eutrophication or upwelling, and there is reduced availability of dissolved oxygen (Section 4.2.5), macrofauna biomass may be up to 10 times lower.

Benthic biomass in deep oligotrophic lakes is 10–100 times lower than these values, due to reduced organic matter sedimentation. Similarly, biomass declines logarithmically with depth in the deep sea. Values of 0.1 g m^{-2} or less have been reported at depths greater than 4000 m. The decrease in numbers of macrofauna is proportional to the productivity in overlying water, but there is no clear trend for the smaller organisms, the meiofauna. The lowest biomass of macrofauna (approximately 0.05 g m^{-2}) occurs under central gyre regions of the mid-ocean, away from continental margins.

The decrease in macrofauna biomass does not extend to the deepest depths of the ocean, and samples from 4 to 11 km depth do not follow the relationship derived for shallower samples. Deep-ocean trenches often have a higher biomass than adjacent abyssal regions. The pattern probably reflects the input of additional food for the benthos in these areas, by transport of particulate matter from continental margins.

4.4.2 Population production

Although standing crop does not give us information about growth rate, reduced benthic biomass in deep water probably indicates lower rates of productivity. Direct observations of growth in non-microbial populations require frequent sampling in one area, to follow increases in numbers or biomass over time. Few such studies have been carried out at abyssal depths in the ocean, but data has been obtained from some lakes sampled over an annual period.

Annual production (P) for different macrofauna species in shallow water has been calculated. When this is divided by the annual average biomass (B), the specific production rate (P/B) or 'turnover ratio' has been found to decrease with increasing bodyweight of individuals and lifespan (Banse & Mosher, 1980). Since maximum length of life is

Table 4.2 Comparison of estimates of productivity from biomass and *P/B* ratios calculated for various size groups of benthic organisms, in a hypothetical community, at abyssal depths greater than 4000 m. From Rowe *et al.* (1986); Schwinghamer *et al.* (1986)

Group (size)	Biomass (g m^{-2})	Organic carbon concentration (mg m^{-2})	*P/B*	*P/B**	Annual production (mg C m^{-2})
Bacteria (1–3 μm)	0.2–2	20–210	100–250	10–25	200–5250
Meiofauna (>40 μm)	0.05–0.5	2–20	2–15	1–7	2–40
Macrofauna (>250 μm)	0.01–10	4–400	0.8–5	0.4–3	2–1000
Megafauna (>1 cm)	0.02–1	0.8–40	0.4–3	0.2–1	2–40
Fish (>10 cm)	0.02–1(?)	1–40	0.2–1.3	0.1–0.7	0.1–28
Abyssopelagic plankton (up to 50 m above bottom)	0.01–0.1	0.4–0.5	0.3–2	0.2–1	0.8–5
Total	0.31–14.6	28–695			207–6463

* Indicates ratios corrected to 2°C from a relationship between turnover ratios and temperature for freshwater macrobenthic species (Johnson & Brinkhurst, 1971) and reduction of the calculated value for bacteria by 10 to reflect assumed substrate limitation.

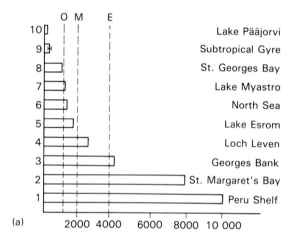

(a)

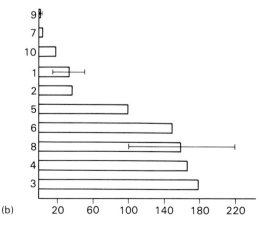

(b)

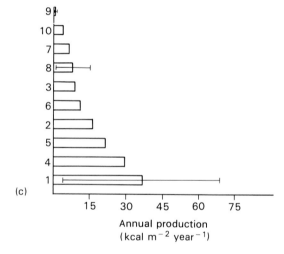

(c)

Annual production
(kcal m^{-2} year^{-1})

Fig. 4.7 Comparison of annual production by (a) primary producers (phytoplankton, attached macrophytes), (b) benthic macrofauna and (c) benthic or demersal fish. Numbers on the vertical axes refer to the ranking of locations by their relative rates of primary productivity. Horizontal bars indicate the range of values calculated from measurements at more than one site in a study area, or arising from assumptions in calculations. Dashed lines show levels of primary production equivalent to oxygen loss measured as mg O$_2$ cm^{-2} day^{-1}, considered by Hutchinson (1957) to separate oligotrophic (O), mesotrophic (M) and eutrophic (E) water bodies. Data from Hargrave & Phillips (1986) and Jonasson (1984).

difficult to determine for most species, individual bodyweight is usually used to calculate *P/B*. Various species of benthic macroinvertebrates have ratios of approximately 5 when expressed per generation. Thus, *P/B* values divided by 5 give the turnover time as number of generations per year. Meiofauna, because of their smaller body size and more rapid generation times, have *P/B* ratios on an annual basis, which are about 10 times larger than values typically calculated for macrofauna.

As would be expected from differences in levels of primary productivity, turnover ratios for a given size of benthic fauna in eutrophic waters are higher than in oligotrophic or deep-water areas. Values also decrease with decreasing temperature. Thus, characteristic low temperatures in deep-water environments (Fig. 4.1) would cause a marked reduction in specific production of benthic fauna.

Productivity based on biomass in various organism size groups, can be calculated from these relationships (Table 4.2). Ratios calculated from empirical regressions for shallow-water benthic organisms usually growing at temperatures above 10°C, will be too high for oligotrophic lakes and mid-ocean abyssal areas: low temperatures and reduced food supplies would decrease predicted turnover ratios. Therefore turnover ratios were reduced by 10 for bacteria and 2 for various sizes of fauna. These corrections yield a range of values for annual benthic community production (0.2–6 g C m^{-2}) (Table 4.2) that is reasonable for water depths where sedimentation may amount to a few per cent of annual phytoplankton production (Fig. 4.5).

Many studies of production by benthic macrofauna in lake and marine sediments have been carried out, but sampling difficulties have limited the number of observations in deep water. No studies have measured production by all size classes of organism in any lake profundal or ocean abyssal community. A summary of results from various lake and marine locations is presented in Fig. 4.7.

Ranking of sites by levels of primary production shows the general proportionality between energy fixation by primary producers and that transferred to macrobenthos and benthic fish, but the relationship is not linear. Factors such as basin size, bathy-metry, overall nutrient supply and the availability of oxygen determine the absolute proportion of energy transferred from primary producers to benthic macrofauna and fish populations.

FURTHER READING

Banse, K. & Mosher, S. (1980) Adult body mass and annual production/biomass relationships of field populations. *Ecol. Monogr.* 50: 355–79.

Brinkhurst, R.O. (1974) *The Benthos of Lakes.* Macmillan, London.

George, R.Y. (1979) What adaptive strategies promote immigration and speciation in deep sea environments. *Sarsia* 64: 61–5.

Grassle, J.F. (1986) The ecology of deep-sea hydrothermal vent communities. In: Blaxter, J.H.S. & South, A.J. (eds) *Advances in Marine Biology 23*, pp. 302–62. Academic Press, London.

Hargrave, B.T. (1984) Sinking of particulate matter from the surface water of the ocean. In: Hobbie, J.E. & Williams, P.J. leB. (eds) *Heterotrophic Activity in the Sea*, pp. 155–78. Plenum Press, New York.

Hargrave, B.T. & Phillips, G.A. (1986) Dynamics of the benthic food web in St. Georges Bay, southern Gulf of St. Lawrence. *Mar. Ecol. Progr. Ser.* 31: 277–94.

Heirtzler, J.R. (1982) Physical and chemical environments of the deep sea. In: Ernst, W.G. & Morin, J.G. (eds) *The Environment of the Deep Sea*, pp. 4–54. Prentice-Hall, New Jersey.

Hutchinson, E.G. (1957) *A Treatise on Limnology. Vol 1. Geography, Physics and Chemistry.* John Wiley & Sons, New York.

Johnson, M.G. & Brinkhurst, R.O. (1971) Production of macroinvertebrates of Bay of Quinte and Lake Ontario. *J. Fish. Res. Bd. Can.* 28: 1699–714.

Jonasson, P.M. (1984) Lake Esrom. In: Taub, F.B. (ed) *Lakes and Reservoirs. Ecosystems of the World 23*, pp. 185–204. Elsevier, Amsterdam.

Jumars, P.A. (1989) Transformations of seafloor-arriving fluxes into the sedimentary record. In: Berger, W.H., Smetacek, V.S. & Wefer, G. (eds) *Productivity of the Ocean: Present and Past*, pp. 291–311. John Wiley & Sons, New York.

Menzies, R.J., George, R.Y. & Rowe, G.T. (1973) *Abyssal Environment and Ecology of the World Oceans.* John Wiley & Sons, New York.

Nealson, K.H. (1982) Bacterial ecology of the deep sea. In: Ernst, W.G. & Morin, J.G. (eds) *The Environment of the Deep Sea*, pp. 179–99. Prentice-Hall, New Jersey.

Rowe, G.T. & Staresinic, N. (1979) Sources of organic matter to the deep-sea benthos. *Ambio* 6: 19–23.

Rowe, G.T., Merrett, N., Shepherd, J., Needler, G., Hargrave B. & Merietta, M. (1986) Estimates of direct biological transport of radioactive waste in the deep sea with special reference to organic carbon budgets. *Oceanol. Acta* 9: 199–208.

Sanders, H.L. (1968) Marine benthic diversity: a comparative study. *Am. Nat.* **102**: 243–82.

Schwinghamer, P., Hargrave, B. & Hawkins, C.M. (1986) Partitioning of production and respiration among size groups of organisms in an intertidal benthic community. *Mar. Ecol. Progr. Ser.* **31**: 131–42.

Wetzel, R.G. (1983) *Limnology.* 2nd edn. Saunders, Philadelphia.

Wright, S.H. & Stephens, G.C. (1982) Transepidermal transport of amino acids in the nutrition of marine invertebrates. In: Ernst, W.G. & Morin, J.G. (eds) *The Environment of the Deep Sea*, pp. 302–23. Prentice-Hall, New Jersey.

Yayanos, A.A., Dietz, A.S. & Boxtel, R.V. (1979) Isolation of a deep sea barophilic bacterium and some of its growth characteristics. *Science* **205**: 808–10.

5

Lakes and Oceans as Functional Wholes

D.W. SCHINDLER

5.1 INTRODUCTION

In many of their basic functions, lakes and oceans operate in a very similar way. Lakes vary by orders of magnitude in size, depth, age and salinity. They number in the millions, occurring at almost all latitudes and altitudes, and in a variety of geological and hydrological settings. As a result, we may learn much about the factors controlling aquatic communities and biogeochemical cycles by studying and comparing the different properties of lakes along gradients in size, chemical composition and other features. In making such comparisons, it is in many cases convenient to treat oceans simply as end members; that is as our largest, deepest and oldest lakes. It is therefore not surprising that many aquatic scientists have made important contributions to our understanding of both freshwater and marine ecosystems.

In all sizes and types of water bodies, the functioning of biogeochemical cycles and biotic communities involves complex interactions between physical, chemical and biological processes. These processes vary according to a water body's particular setting, as well as 'feedbacks' among the processes within the systems. The processes are similar across the spectrum of aquatic ecosystems, but differ in relative importance.

5.2 PHYSICAL PROPERTIES OF LAKES AND OCEANS

5.2.1 Thermal stratification

Most lakes and oceans have three distinct vertical layers. The upper, well-lit, well-mixed and rela-

tively warm region is known as the epilimnion to limnologists, but simply as the mixed layer to oceanographers. An intermediate layer, which shows rapid temperature decline with depth, is known as the thermocline (to oceanographers) or metalimnion (to limnologists, who consider the thermocline to be the plane of maximum rate of change of temperature with depth).

In oceans, the thermocline is permanent in most latitudes. It generally occurs between 100 and 1000 m in tropical and temperate regions. (In arctic regions, surface waters are cold and thermal stratification is absent, or even reversed — i.e. when waters are ice-covered.) In lakes, thermal stratification varies greatly. Where a thermocline is present, it ranges from a few metres in depth in the smallest lakes, to 20–30 m in the largest ones. It can be non-existent in shallow, temperate and tropical lakes, or those in oceanic climate regimes. In polar or high alpine regions, lakes can be permanently ice-covered, or ice-free for only a few weeks. Some lakes are permanently stratified, or meromictic, due to a combination of thermal and chemical features. In temperate regions, most larger lakes stratify seasonally, with mixing occurring in spring and autumn, ice-cover in winter and a thermocline for several months in summer. Shallower lakes may stratify and then destratify several times in a summer, as the result of windy periods. Such polymictic lakes are also common in the tropics, where, in some cases, even diurnal temperature changes are enough to cause mixing. Detailed descriptions of the variation in mixing patterns are discussed by Hutchinson (1957).

Finally, the hypolimnion, which is usually too cold and dark to support much photosynthetic production (except in small, clear lakes), is a term used by limnologists to refer to all depths below the thermocline. Oceanographers usually refer to this region simply as the deep ocean. In either case, deep-water temperatures decline only very slowly with depth. In lakes, typical bottom temperatures are close to 4°C — the temperature of maximum density for fresh water. In oceans, deep-water temperatures can be much lower, often approaching 1°C. In polar regions, water temperatures can actually reach −1.5 to −1.7°C without freezing, due to the depression of the freezing point by salinity. Such temperatures require special adaptations in marine organisms, but reduce metabolic requirements to miniscule values.

All lakes have epilimnions, while to have a metalimnion and a hypolimnion, a lake must have sufficient depth and sufficient seasonal weather change. These characteristics are most prominently displayed in temperate latitudes, where there is a well-defined relationship between lake size and thermocline depth. Maritime, tropical, alpine and arctic regions tend to have lakes that are more weakly stratified, and with deeper thermoclines.

The most feeble of mixing processes is molecular diffusion, the dispersal of molecules simply from collisions with one another. Of course, molecular diffusion affects the distribution of heat and chemical solutes in all layers of all lakes. In the metalimnions and hypolimnions of small lakes, where other mixing processes are very weak, the vertical gradients of temperature and various chemicals can usually be explained almost entirely by diffusive processes. An exception is in very clear lakes, where light penetration is sufficient to warm deeper layers by radiant energy, and to affect the distributions of biologically-reactive chemicals by supporting net photosynthesis. However, in the epilimnions of all water bodies, and in deeper layers of large lakes and oceans, other mixing processes are far more important than molecular diffusion: these processes are generally referred to as advective. They result from turbulent energy imparted by wind or pressure changes to the surface of lakes. Most of such energy is dissipated in the epilimnion, causing the steep temperature gradients of the thermocline.

In most lakes, advective processes exert mixing energy many orders of magnitude greater than diffusion, allowing dispersal of heat and chemical

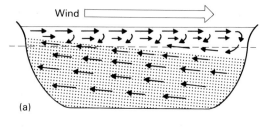

(a)

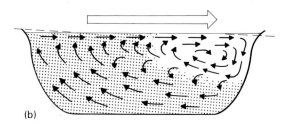

(b)

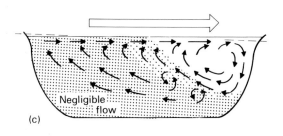

Negligible flow

(c)

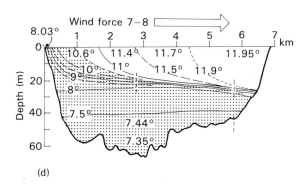

(d)

Fig. 5.1 Diagram to show the effect of strong, steady wind in a lake basin. (a)–(c) represent stages in the development of a tilted thermocline. Arrows show velocity and direction of flow; the original hypolimnion is shaded. (d) shows the temperature distribution (°C) resulting from a wind force of 7–8 (Beaufort Scale) along the longitudinal axis of Lake Windemere (northern basin), for about 12 hours. The ratio of length to depth is exaggerated 50 times. From Mortimer (1954).

solutes rapidly enough to have major effects on the interchange of substances between layers.

One phenomenon that drives advective mixing between layers is seiche activity. Seiches are set up by the effect of wind on the water surface. For example, a wind blowing across a water body from east to west will tend to 'pile' water at the western (or downwind) end of the lake. This extra weight on the surface will, in turn, tend to force deeper water to move in the opposite direction, to compensate for the weight difference. In extreme cases, the 'piling' of surface waters can be enough to tilt the thermocline by metres in small lakes, or even tens of metres in large ones (Fig. 5.1). In some cases, this actually allows cold, possibly nutrient-rich, deep water to be entrained into the surface water of a lake or ocean. When the wind stops, these layers rush back in the direction from which they began, starting an oscillation known as an internal seiche, that continues for days. As a result, there is strong horizontal motion, and considerable vertical mixing due to turbulence in the shear zones between layers. In general, horizontal movement in lakes is much more rapid than vertical mixing. As a result, although horizontal dimensions of most lakes are much greater than vertical distances, the chemical characteristics of smaller lakes are much more a result of exchanges with sediments occurring at the same depth than they are of exchanges with deeper sediments, i.e. most chemical exchanges between water and sediments in small- to moderate-size lakes can be visualized as occurring horizontally, not vertically; as shown by the fact that chemical concentrations are generally uniform horizontally, but highly stratified in a vertical direction. The operation of mixing processes may, for practical purposes, be visualized as a series of rapidly-moving horizontal conveyor belts, transporting substances between the water column and the sediment surface at the same depth in the lake. The relative vertical movement is small, perhaps no larger in relative magnitude than the vibrating of a conveyor belt.

5.2.2 Effects of mixing and thermal stratification on chemical and biological properties

The epilimnion and hypolimnion (or deep ocean) differ greatly in their chemical and biological, as well as in physical, properties. In general, epilimnions are: (i) where nutrients are fixed photosynthetically; and (ii) where most of the organic material for supporting biological life in all layers is produced. Deeper layers are zones that accumulate the products of decomposition, including the nutrients necessary to sustain productivity in surface waters. The processes that transport nutrients among layers are, therefore, of considerable interest. The aforementioned three regions of lakes and oceans differ vastly in their mixing characteristics, chemistry and biological communities. In very deep lakes, there may be zones within the hypolimnion: shallower depths in the hypolimnion are typically referred to as the profundal, while great depths (which may extend to over 1000 m in deep lakes) are referred to as abyssal. These are the same two terms used for the shallower and deeper parts of the deep ocean.

The effects of horizontal water movement also play an important role in governing the chemistry and productivity of water bodies. In large lakes and oceans, where the horizontal distance from a given point in the water column to sediments can be extremely great, horizontal movement explains the major features of chemical stratification. Much of the vertical mixing (upwelling) in large water bodies is due to the horizontal movement of surface water, which is blown offshore, allowing nutrient-rich deep waters to rise into the well-lit surface regions. The coasts of Peru and northern Chile, and the southwest coast of Africa, are examples of such regions, which are usually the most productive parts of oceans. We will discuss these processes in more detail later, when describing El Niño events.

The most important mechanism moving biologically-reactive chemicals from surface waters to the hypolimnion is particle settling. For a steady-state ocean, the downward flux of chemicals with sedimenting particles, plus a relatively small movement of chemicals due to downwelling and other mixing events, must equal the sum of upwelling and other processes moving chemicals upwards, plus the input to the oceans from other sources, such as rivers and the atmosphere. This particle settling, and the relative horizontal movements of surface and deep waters, are important in determining the nutrient chemistry of various parts of the ocean.

The surface ocean can be viewed as a thin, warm

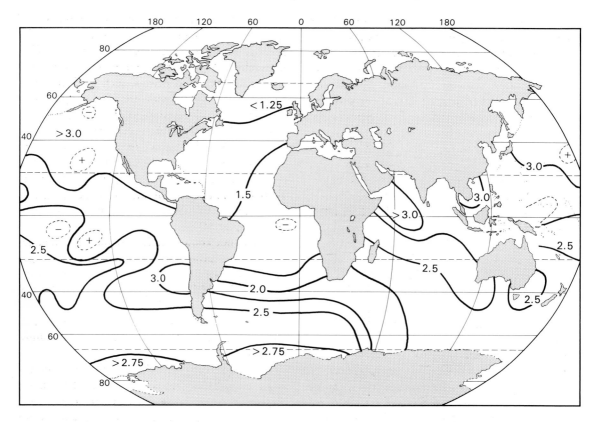

Fig. 5.2 Distribution of phosphates at a depth of 2000 m in the oceans of the world. Contour intervals 0.25×10^{-6} mol l^{-1}. From Broecker (1974).

epilimnion

skin over a vast body of dark, cold deep water. The skin is generally thickest in the tropics, and thins toward the poles. It disappears entirely in the Arctic and Antarctic, leaving cold deep water in contact with the atmosphere.

More than half of the replenishment of ocean deep water occurs between Greenland and the British Isles. It slowly flows down the Atlantic to the south, around Africa, through the Indian Ocean, and north through the Pacific, reaching the Arctic once again the Bering Strait. En route, it is replenished by some flow from cold waters of the Antarctic. The flow of water into the deep ocean is balanced by upwelling at several locations around the globe. The 'youngest' deep water in the oceans thus occurs in the North Atlantic, and the oldest deep water in the North Pacific.

As deep water is slowly moved horizontally through the oceans, it receives a continuous rain of nutrients, moved from surface water by biological uptake and particle sedimentation. As a result, as deep water moves southward through the Atlantic and then northward through the Pacific, it becomes progressively more enriched in nutrients and depleted in oxygen, which is consumed in the decomposition of organic matter. Consequently, there is considerable horizontal variation in the nutrient concentration at a given depth in the deep ocean; vertical nutrient and oxygen profiles for the Pacific are much more dramatic than those for the Atlantic (Fig. 5.2). Such horizontal circulations occur to a lesser degree in great lakes.

Differences in chemical composition analogous to those that occur spatially in the oceans, are those that occur seasonally in lakes. At spring overturn, lake water is relatively homogenous chemically and thermally, much as it is in polar seas. Over the course of the summer, there is a gradual transfer of nutrients from surface water to below the thermocline. While surface water may receive additional inputs from precipitation, inflows and pollution sources, in general there is a progressive increase in concentration of nutrients in the hypolimnion, due to particle sedimentation and decomposition. In autumn, when lakes are again thermally homogenous, the cycle begins once more. Of course, the time scales are very different — months in lakes, millenia in the oceans.

Tides, Coriolis force imparted by the earth's rotation and changes in atmospheric pressure are other major sources of energy which promote physical mixing in oceans and large lakes. Their effects are immeasurably small in small lakes.

To coastal biotic communities (including humans living near the sea shore), tides are particularly important; the regular diurnal, semi-diurnal and seasonal patterns of the tides can cause water levels to change from a few tenths to several tens of metres at regular intervals. Tides are produced by the gravitational pull of the sun and moon on the earth. The regular ebb and flood of the tides also produces currents in areas near the continental margins. The regular submersion and exposure of intertidal zones has resulted in some interesting and bizarre evolutionary adaptations in the organisms which occupy them. In areas near river mouths, periodic fluctuations in salinity make the environment even less stable, and require even greater adaptation of organisms. The combination of freshwater runoff and tidal exchange help to flush pollutants from coastal areas at a rapid rate. Even so, the eutrophication, and pollution with toxic materials, of coastal waters is assuming high importance, due to the concentration of human population and industry in coastal areas.

5.3 GAS EXCHANGE AT THE AIR–WATER INTERFACE

The wind's effect on water surfaces is probably the most conspicuous form of mixing energy to most of us, for it determines the size of surface waves. Maximum surface waves can range in height from a few millimetres on small ponds, to tens of metres in more windswept regions of the oceans. In addition to imparting energy for mixing, as discussed above, surface waves affect a number of properties important to the functioning of lakes and oceans, such as their ability to exchange gases with the atmosphere. Gas exchange can be important to the growth of photosynthetic algae in ecosystems where carbon dioxide and nitrogen are in short supply, as we will discuss later. It also determines many of the features that are important to global chemical cycles, such as the exchanges of greenhouse gases like CO_2, methane and nitrous oxides (see Chapter 6).

We do not know exactly how the wind modifies the surface properties of waters to affect gas exchange. Several types of model have been developed, which allow gas exchange to be predicted from measurable properties. Some of the models are quite complex; however these seem to perform no better on average than simple models.

The simplest gas exchange model in common use is the so-called stagnant film model. This model depicts two well-mixed reservoirs, the atmosphere and the epilimnion, which exchange gases with each other through a thin 'skin' called the stagnant boundary layer (z), through which gases pass only by molecular diffusion (Fig. 5.3). As wind velocity increases, the thickness of the boundary layer is visualized as decreasing, decreasing the distance through which gases must move to pass from one reservoir to the next. The thickness of the boundary layer, and the gradient in concentration between the two reservoirs, determine what the rate of exchange of a given gas will be.

Under constant conditions, the rates of exchange of different gases are assumed to be in proportion to their coefficients of molecular diffusion, which are determined in the laboratory and are assumed to vary only with temperature. The diffusion coefficients (D) are known for most of the common gases (Table 5.1). The model for net flux of gas (by weight) between the surface water and the atmosphere may be written mathematically as

$$F = D\{(gas)top - (gas)bottom\}/z$$

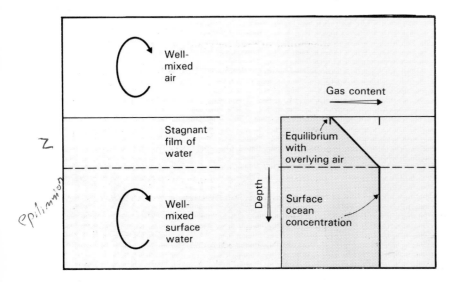

Z

epilimnion

Fig. 5.3 Gas exchange model. A thin film of 'stagnant' water separates the well-mixed overlying air from the well-mixed underlying surface water. Gases are transferred between air and water only by molecular diffusion through this film. The concentration of gas within the film, grades from that corresponding to equilibrium with the air at the top, to that found in the surface ocean at the base. The film thickness varies inversely with the degree of agitation of the air–water interface. In the ocean this film averages tens of micrometres in thickness. From Broecker (1974).

where F is the flux of gas per unit area and z is the depth in metres. The flux is usually expressed in $mol\,m^{-2}\,year^{-1}$.

In practice, what is measured by scientists is the mass transfer of gases, D/z. This is usually done by measuring the change over time in the concentration of a gas in both water and air. If the theory is correct, measurement of any gas allows us to calculate fluxes of other gases for which diffusion coefficients are known. It is therefore customary to select gases for study which are easy to measure. For example, the radioactive gas radon-222 is continuously produced in water by the radiodecay

of radium-226, weathered from soils and bedrock. The concentration of radon in the atmosphere is negligible; it decays with a half-life of only 3.5 days. As a result, in order to calculate gas exchange, we only need to measure the change of radon in surface water with time. This can be done with great precision, as radon can be stripped from large volumes of water and trapped by freezing or adsorption of the gas in a small chamber that can be radio-assayed. Alternatively, the rate of gas transfer can be measured using a gas that is present in the atmosphere in much larger amounts than in water, as carbon-14 was after the atmospheric testing of atomic devices in the 1950s and 1960s. Obviously, if D/z can be measured and D is known for the gases of interest, z is the factor that will vary from one water body to another. In the many systems where it has been measured, z can range from a few tens to several hundreds of micrometres in apparent thickness, depending on the size and surface disturbance of a water body (Fig. 5.4).

D/z is often referred to as the piston velocity, as it has units of distance/time. It can be visualized as an enormous piston of area equal to the surface of a water body, sweeping all gases before it in a direction from high to low gas concentration, moving at a rate of D/z, in distance per unit time. Typical rates for D/z range from a few tenths of a metre per day for small lakes, to several metres per day in large

Table 5.1 Rates of molecular diffusion of various gases in water. From Broecker (1974)

Gas	Molecular weight, (g mol^{-1})	Diffusion coefficient ($\times 10^{-5}$ cm^2 s^{-1})	
		0°C	24°C
H_2	2	2.0	4.9
He	4	3.0	5.8
Ne	20	1.4	2.8
N_2	28	1.1	2.1
O_2	32	1.2	2.3
Ar	40	0.8	1.5
CO_2	44	1.0	1.9
Rn	222	0.7	1.4

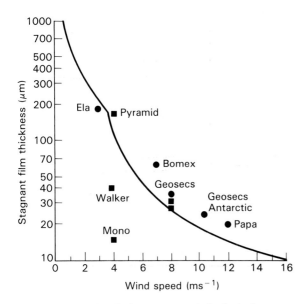

Fig. 5.4 Plot of film thickness against wind velocity for various natural systems. Values were estimated using the radiocarbon method (■), or by the radon method (●). The curve was obtained in a wind tunnel. The break in the slope corresponds to the initiation of capillary waves. From Broeker *et al.* (1980).

water bodies under windy conditions. This value, and the depth of the mixed layer, determine how rapidly the concentration of gas in the surface layers of lakes and oceans is altered. For example, in Lake 227 in the Experimental Lakes Area of northwestern Ontario, D/z is typically 0.5 m day^{-1}. The epilimnion is usually about 3 m deep, so that gases are renewed every 6 days by gas exchange alone.

In contrast, oceans typically have thermocline depths of 100 m. Gas exchange rates are much higher in oceans than in Lake 227, averaging around 4 m day^{-1}. The time required to totally exchange gases is therefore about 25 days. As a rule of thumb, if the concentration of a gas in either the atmosphere or the surface water is altered, three exchange times will be required to approach a new equilibrium, i.e. 18 days would be required in Lake 227, and 75 days in the ocean. This restricted rate of exchange has contributed to many of the problems imposed on water bodies by man, as we shall see in Chapter 6.

5.3.1 Spatial and seasonal differences in the magnitude of gas exchange

When measurements of the type described above are made over broad areas of the ocean, an interesting and complex pattern emerges. In some areas, surface water of the ocean is a sink for atmospheric CO_2, while in others it is a source of CO_2 to the atmosphere. In all areas, oceanic deep water is oversaturated with CO_2. It is prevented from degassing by the thin 'skin' of surface water above the thermocline, where organisms capture CO_2 by photosynthesis and send the carbon back to the depths as sedimenting organic matter, preventing it from reaching the surface.

Several thousand measurements of CO_2 concentrations made over a 10 year period show, surprisingly, that most northern oceans are net sources of CO_2 to the atmosphere, indicated by the positive partial pressures of CO_2. Most of the area of the ocean that is a sink for CO_2 occurs in the southern hemisphere, particularly in the sub-Antarctic, which accounts for 35% of the CO_2 transferred to the oceans. The strongest source area is in the tropics, which emits 60% of the CO_2 transferred from the oceans to the atmosphere. However, there are considerable year-to-year differences in flux patterns. For example, studies of the 1982–1983 El Niño show that the tropical oceans do not emit CO_2 during such periods, because of decreased transport of nutrients from the deep ocean to near-surface water. El Niño is discussed in more detail below.

The net CO_2 removed from the atmosphere by all the oceans is thus the difference between large fluxes into and out of the oceans in different regions. In general, measurements indicate that an average of about 1.1 gigatonnes of CO_2 are emitted by the oceans each year, while 3.4 gigatonnes are absorbed. In Chapter 6, we shall discuss how these fluxes combine with man's emissions from burning of fossil fuels and biomass, to allow estimates to be made of the magnitude and timing of greenhouse warming.

As for nutrient concentrations, seasonal changes in the CO_2 content of lake water are analogous to spatial differences in the oceans. During summer, when phytoplankton are photosynthesizing

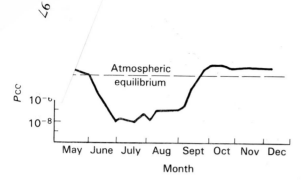

Fig. 5.5 Diagram to show the partial pressure of CO_2 at a depth of 1 m in Lake 227, Experimental Lakes Area. From Schindler *et al.* (1972).

actively, CO_2 in surface water is below atmospheric saturation, so that the net carbon flux is from the atmosphere to the lake. A high concentration of CO_2 occurs in the hypolimnion, due to the build-up of decomposition products. At overturn, surface and deep water are mixed. Surface water concentration of CO_2 tends to be higher than the atmospheric value, and the net flux of carbon is to the atmosphere (Fig. 5.5). The annual net flux is therefore the balance between differences in seasonal input and outputs, as the oceans flux is the balance between spatial differences. Of course, there are seasonal differences in the carbon fluxes of various parts of the oceans too, but they are not yet well-defined.

5.4 EFFECT OF WATER RENEWAL ON CHEMICAL PROPERTIES

Another physical property that defines the chemistry and biology of lakes is their rate of water renewal. Water renewal time is usually expressed in years, and is simply the volume of a water body divided by the volume of outflow (using the volume of outflow, rather than inflow, automatically corrects for the effects of evaporation). Lakes in high rainfall areas tend to have faster water renewal than lakes in more arid climates. The area of a lake's catchment in proportion to its volume and the magnitude of evaporation relative to precipitation, are other factors that affect the water renewal time. The water renewal time for lakes ranges from a few hours in small lakes on large rivers, to hundreds of

years in the St Lawrence Great Lakes, to even thousands of years in closed basins like the ocean and the saline lakes of arid areas, where evaporation is the dominant mechanism by which water is lost.

In lakes with rapid water renewal and in geological settings that have low rates of weathering and ion exchange, the concentration of solutes may be only two or three times higher than in precipitation, while in lakes with little or no outflow, where water is removed only by evaporation (which of course would concentrate chemical substances), chemical concentrations are determined by the solubilities of various combinations of cations and anions. Even moderately concentrated and productive lakes can precipitate calcium carbonate. Some lakes, particularly those with closed basins like Mono Lake, California or lakes in south-central Saskatchewan and Alberta, become so concentrated that they precipitate calcium sulphate, magnesium sulphate, sodium chloride and other salts of high solubility. In some cases, salinities much higher than sea water are reached. Needless to say, few organisms can survive in such concentrated chemical solutions and biotic communities are of very low complexity. In arid areas of North America, like the Great Basin or the Canadian prairies, almost all stages of evaporative concentrations can be found, resulting in lakes that vary widely in chemical composition. Needless to say, 'fresh water' is a misnomer for such systems!

Water renewal is one of the most important factors governing the degree and the rapidity of a lake's response to incoming pollutants. This has been demonstrated using models to predict the effect of nutrient inputs on the eutrophication of lakes. More recently, models based on chemical inputs and water renewal have been shown to allow prediction of the amount of sulphuric and nitric acid that will be retained and neutralized by lakes (Fig. 5.6). The concentration of biologically-conservative elements in lakes will be determined entirely by rates of chemical input, evaporation and of water renewal by outflow, unless concentrations are high enough to allow chemical precipitation. When the input of a chemical is changed, but the water input remains constant, the chemical concentration will change at a logarithmic rate dictated by the rate of water renewal (Fig. 5.7). As a rule of

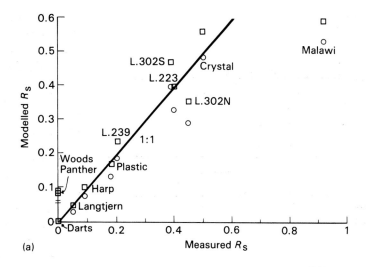

(a)

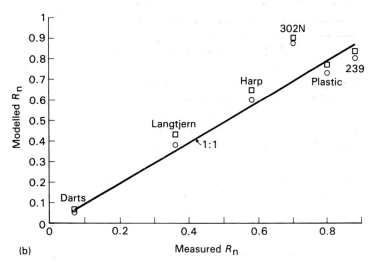

(b)

Fig. 5.6 A comparison of actual coefficients of removal for (a) sulphate (R_s) and (b) nitrate (R_n) with modelled removal coefficients. The latter were calculated using either mass transfer coefficients obtained from the chemical budgets of lakes ($\square$) or mass transfer coefficients obtained from measurements of rates of sulphate reduction or denitrification ($+$). Lakes shown are from the USA, Canada, Norway and Africa (Lake Malawi). Modelled R is given by $R = I/[(z/\tau_\omega) + I]$ where I is input of sulphur or nitrogen, z is the mean depth of the lake, and τ_ω is the water renewal time in years. From Kelly *et al.* (1987).

thumb, it will take three water renewal times for a lake to reach 95% of a new steady-state, after a change in chemical input. This time can range from a few weeks, in small lakes with large catchments in areas of high rainfall, to hundreds of years, in lakes where water renewal times are from decades to centuries. Long water renewal time confers both good and bad characteristics with respect to managing lakes. For example, in Lake Superior, where the water renewal time is over 100 years, the concentration of sulphate is only about one-third of that in acid rain falling on the basin; the current high concentrations of sulphate in rain were not

reached until the 20th century. In contrast, many of the small lakes in the Adirondacks contain concentrations of sulphate as high as those in acidified precipitation, because their water renewal times are only a few months. However, long water renewal times become a disadvantage when pollutant inputs are decreased, as the time to a lower steady-state is as long as that to higher one. As a result, lakes with a short water renewal time both become polluted sooner and recover more quickly, at least in chemical properties.

Of course, water renewal times for the ocean are extremely long; oceans only lose water through

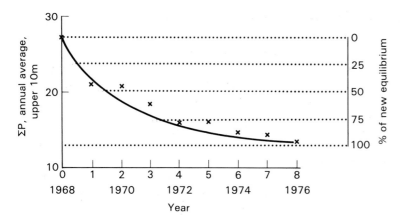

Fig. 5.7 The approach of total phosphorus concentration to a new steady-state following diversion of sewage from Lake Washington. Line shows prediction calculated from average water renewal time. Data from W.T. Edmondson (unpublished).

evaporation. As in closed-basin lakes, concentrations of incoming chemicals increase until they are removed by precipitation. Luckily, even after thousands of millenia, oceans have not reached saturation with more soluble combinations of elements, as the more saline lakes have. They have, however, reached saturation with calcium carbonate in most regions; and carbon in marine sediments, present chiefly as calcium carbonate, accounts for over 99% of the carbon on earth.

5.5 BIOLOGICALLY CONSERVATIVE AND NON-CONSERVATIVE CHEMICALS

The properties of conservative chemicals are rather easy to predict, as their removal is governed by the degree of saturation of various combinations of cations and anions. The solubility products for combinations of common cations and anions in solution are well known. However, many of the chemicals that stimulate or depress the productivity of aquatic systems are not conservative. They are taken up by organisms to varying degrees and are sedimented when the organisms die, or they are precipitated when physical or chemical conditions are biologically altered. As a result, the concentration of many chemicals in lakes may be only a few per cent of that expected from inputs and water renewal alone.

Perhaps the best known examples are the primary nutrients for plant growth: phosphorus and nitrogen. In lakes with a long water renewal

time, 90% or more of the phosphorus and nitrogen entering the system may be biologically removed, so the concentration is a small fraction of what it would be if these two elements were conservative. Another example of a lesser, but still important, degree of non-conservative behaviour is exhibited by sulphur in softwater lakes. Sulphur is used in small quantities by organisms, in roughly the same relative amount as phosphorus. Its supply in the form of sulphate is much larger, so that biological demand alone does not usually cause a detectable decline in conservatism of sulphur. However, sulphate-reducing bacteria in anoxic regions, like sediments and the hypolimnions of productive lakes, remove the oxygen from sulphate in order to metabolize organic matter. The resulting reduced sulphur either remains in the water column as hydrogen sulphide or precipitates as iron sulphide. The latter reaction predominates in lakes with high concentrations of iron in anoxic regions.

Once precipitated, the iron sulphide may undergo further transformation to pyrite, or to organic forms of reduced sulphur. Unless these compounds are reoxidized, the result is significantly non-conservative behaviour, particularly in lakes with a long water renewal time. The retention of sulphur also generates significant alkalinity, affording some (but not complete) protection to softwater lakes from acidification. In contrast to the view once held, that acidification irreversibly reduces the buffering ability of lakes, the reduction of sulphate actually increases with increasing rainfall acidity, as it is determined by the rate of diffusion of sulphur into sediments. Such losses are proportional to the concentration of sulphate in the water overlying the sediments, because sulphate concentration in the sediment is near zero, due to bacterial reduction. Despite its non-conservative nature, the retention of sulphur is a predictable function of water renewal in most lakes, as are phosphorus and nitrogen retention (Fig. 5.6). While the rate of sulphate reduction is not sufficient to prevent lakes from acidifying, it does prevent lakes from becoming as acidic as they would without such processes. Sulphate reduction also allows lakes to recover to some degree after acidic precipitation is reduced, as we shall discuss later.

Similar sulphate-reducing reactions occur in coastal salt marshes, where the bacteria involved supply an important part of the energy requirements of salt marsh food chains, and in deep, anoxic regions of the ocean, which are anoxic due to limited mixing with other water masses.

Obviously, the above reactions with sulphur dictate that the iron cycle is also non-conservative, even though only small amounts of iron are involved in direct metabolic activities. In turn, iron regulates the concentration of many trace elements through its flocculation reactions, causing the behaviour of many other elements to be non-conservative, even though they may not be directly important to biological activity. In some softwater systems, iron seems to be becoming exhausted, as the result of decades of increased precipitation of iron sulphides caused by acid rain, with its much higher than natural concentration of sulphate. As a result, the cycles of many elements may be affected.

The precipitation of calcium carbonate (calcite) occurs in marine systems due to the formation of the compound by organisms in producing tests or shells. In calcium-rich fresh waters, spring warming of surface water reduces the solubility product of calcium and bicarbonate, allowing the compound to precipitate. The precipitation of calcite performs a function analogous to that of iron, as described above, co-precipitating many trace substances which would otherwise remain in solution. In clear waters, such reactions may cause plankton to be deprived of trace elements, limiting their production. However, in most waters, particularly those that are mesotrophic to eutrophic, or those receiving runoff from forests or wetlands, some trace elements are kept in solution by chelation with dissolved organic compounds.

It is clear from the above that we cannot treat the physical, chemical and biological processes in lakes and oceans as isolated entities, as is the common practice in beginning textbooks. The remarkable interplay between physical, chemical and biological factors determine the mechanisms and the degree to which a lake or ocean can synchronize its functions to coordinate various chemical and biological processes. The degree to which such processes can adapt to allow the system as a whole to continue functioning is truly amazing. For example, in the early days of eutrophication research, it was believed that excessive algal growth in lakes could be controlled by limiting human input of phosphorus, nitrogen or carbon. However, when we added various combinations of these elements to a series of experimental lakes, it became obvious that the only effective element to control was phosphorus. If carbon was in short supply, as it was in softwater lakes overfertilized with phosphorus, photosynthesis was limited, but the exchange of CO_2 with the atmosphere slowly corrected the carbon deficiency. Likewise, if nitrogen were limiting, dominance among algae shifted to favour cyanobacteria that were capable of fixing atmospheric nitrogen, again allowing the atmosphere to make up the critical shortage of this element. These corrective processes were not very efficient, as the rates are limited by rates of gas exchange, which can be quite slow, as described above. In addition, nitrogen-fixing cyanobacteria in fresh waters are plentiful only during the warmer part of the summer. In marine systems and saline lakes, the scarcity of trace metals like molybdenum and iron impede the fixation of nitrogen, and atmospheric input of N_2 appears to be of minor importance. The rate at which entering carbon and nitrogen are removed to sediments will, of course, also slow the accumulation of nitrogen and carbon in the water column. In the case of one long-term lake experiment in Lake 227 at the Experimental Lakes Area in northwestern Ontario, full steady-state for the nitrogen cycle had not occurred, even after 15 years of enhanced phosphorus and nitrogen inputs at constant rates (Fig. 5.8).

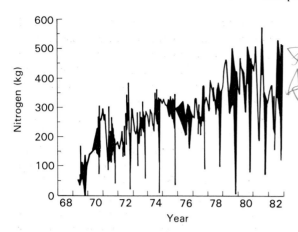

Fig. 5.8 The mass of nitrogen in Lake 227, Experimental Lakes Area, 1968–1982. The upper line is the whole-lake mass, the shaded area is the mass in the epilimnion. From Schindler *et al.* (1987).

Phosphorus has proved to be an exception to the general biogeochemical adaptability of elemental cycles. In lakes experimentally fertilized with nitrogen and carbon, but not phosphorus, or where phosphorus input has been reduced without reducing input of carbon and nitrogen, there has been no tendency for phosphorus to increase in order to balance the three nutrient cycles. The reason for this is, of course, that there is no gaseous atmospheric reservoir of phosphorus that can be drawn upon to balance the proportions of these three critical elements. Phosphine, the only naturally-known gaseous form of phosphorus, has been detected in significant quantities only in sewage lagoons.

Some aquatic ecosystems are regarded as nitrogen-limited, in that the addition of nitrogen will elicit algal growth, while the addition of phosphorus will not. Usually, a high concentration of cyanobacteria is found in such ecosystems, as well as a high concentration of phosphorus. Typically, the response of such systems to addition of nitrogen is quite small, with either phosphorus or light limitation coming into play rather quickly. There is some reason to expect that over long time periods, atmospheric exchanges and recycling from sediments will balance the cycles of these three most critical nutrients for algal growth. Alfred Redfield noticed many years ago that the ratios of phosphorus, nitrogen and carbon in the plankton of various parts of the oceans were remarkably similar, a phenomenon now referred to as 'Redfield ratios'. Similar proportions are maintained in the plankton communities of fresh waters. Obviously, these remarkable similarities must be caused by strong interplay between biological and other sorts of factors, or there would not be such remarkable similarities in the plankton of aquatic ecosystems of different types.

Nearshore marine environments are commonly regarded as nitrogen-limited. However, all measurements have been made in the years after man has enveloped these areas in millions of tonnes of sewage, which has a much higher ratio of phosphorus to nitrogen than Redfield ratios. Consequently, it is difficult to say what the nutrient status of original nearshore phytoplankton communities might have been, before humans altered them. Surely, they were less nitrogen-limited than at present, but it is impossible to say whether they were in as perfect a nutrient balance as in areas of the open ocean where chemical cycles and biological tolerances have been evolving for hundreds of millenia. Long-term experiments, of the sort that have made it possible to explore temporal aspects of nutrient limitation in fresh water, have never been done in marine systems; such experiments might cast nutrient cycles in marine systems in a totally different light.

In oceans, rates of exchange between surface and deep water play a role analogous to that of water renewal in fresh water in determining productivity. We have already mentioned that areas where deep water upwells tend to be the most productive parts of the oceans. Typically, biologically-reactive elements like phosphorus, nitrogen, silica and calcium are depleted in near-surface waters, due to uptake and sinking of photosynthesizing algae. Once such elements leave the upper, mixed zone, they are lost to deep water and sediments for hundreds of millenia, during which time much of the deep water moves horizontally over vast distances before reaching the surface again (Table 5.2). In contrast, elements like sodium and sulphur which are not extensively used by organisms, relative to quantities present in surface water, have

Table 5.2 Time to return to surface waters of the ocean, t, calculated for the elements phosphorus, silicon, barium, calcium, sulphur and sodium. From Broecker (1974)

	t (years)
Biolimiting	
Phosphorus	2×10^5
Silicon	6×10^5
Biointermediate	
Barium	3×10^4
Calcium	1×10^6
Biounlimited	
Sulphur	2×10^7
Sodium	2×10^8

surface water concentrations very similar to those in deep water. Molecules of such elements in deep water tend to turn over even more slowly than nutrients, and may on average reach surface water only every 10–100 million years. Thus, similar cycles of elements between surface and deep waters occur in lakes and oceans, but the rates are vastly different. Internal recycling times for elements of from months to years in lakes, may require from hundreds to thousands of millenia in oceans.

5.6 EL NIÑO

Upwelling areas of marine systems also support high productivity. Typically, such areas occur where offshore winds drive surface waters away from shore, decreasing thermocline depths so that nutrient-rich deep waters can reach the surface. High light level in surface waters allows nutrients to promote the photosynthetic production of phytoplankton, and rates of production can be 10–20 times greater than in the open ocean.

Large upwelling areas occur off the western coasts of North and South America, as well as Africa. In typical years, these productive regions support lucrative fisheries which form an important part of the economies of adjacent countries. Despite their small area, relative to the entire oceans, upwelling areas produce nearly a quarter of the global fisheries harvest. However, these normally productive areas can change dramatically during events that natives of the west coast of South

America have known for centuries as El Niño. (El Niño means 'the child'; a reference to the fact that such events normally begin around Christmas, i.e. the birth of the Christ child.) Until a few decades ago, El Niño was regarded as a rather localized phenomenon, originating in the eastern tropical Pacific and affecting primarily the western coast of South America. Only when modern oceanography and meteorology advanced enough to obtain synoptic data for wide expanses of the ocean and atmosphere, did we begin to realize that El Niños were determined by atmospheric and oceanic events on the other side of the world, and that they could affect current and weather patterns throughout much of the globe. They provide an excellent example of how oceans, the atmosphere and physical, chemical and biological processes in oceans interrelate. The meteorological events connected with El Niño are referred to as the Southern Oscillation, and due to the tight coupling between oceanic and meteorological phenomena, the entire 'package' has recently been referred to as ENSO (an acronym for El Niño–Southern Oscillation).

As mentioned earlier, the combination of differences in barometric pressure and earth's rotation normally produce offshore winds along the western coast of South America. These winds blow warm surface waters westward, causing a thinning of the surface layer in the east. As a result, the thermocline becomes shallower. Overall, the phenomenon is similar to the effects of wind on a lake's thermocline, as shown in Fig. 5.1. As a result, nutrient-rich deep water is normally able to upwell off the coasts of Peru and Chile, causing them to be extremely productive (see Fig. 2.7). In El Niño years, higher atmospheric pressures in the western Pacific cause offshore winds to be weaker (Fig. 5.9). As a result, thermoclines are deeper, upwelling of cold, nutrient-rich deep water is reduced, and surface waters are much warmer. The change in conditions has devastating consequences for both fisheries and agriculture. The normal fisheries of coastal communities are reduced; sea birds, which rely on the same fish stocks, starve or move away, reducing production of guano for the fertilizer trade; torrential rains sometimes occur, which erode soils and cause explosive increases in weeds and insect pests. Sea levels are slightly higher as the

Normal conditions

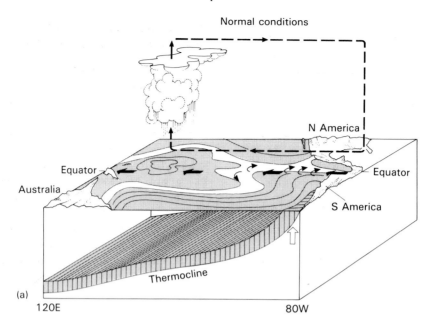

(a)

El Niño conditions

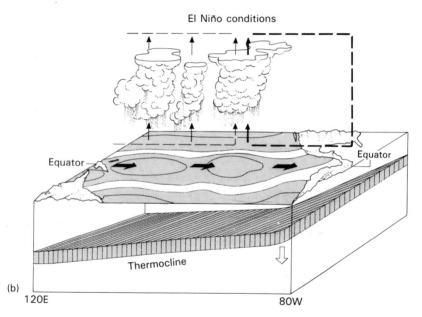

(b)

Fig. 5.9 El Niño theory. (a) During normal conditions the thermocline shallows as it nears the South American coast to bring cold, nutrient-rich water to the surface, while warm water in the western Pacific heats the atmosphere to produce rain. (b) During El Niño conditions, the thermocline slides eastwards, cutting off upwelling and displacing the rainfall zone. From Cromie (1988).

result of the lower migration of surface waters, and storms can cause great damage to coastal communities. Enfield (1989) gives a detailed review of ENSO phenomena. In addition, ENSO events can have dramatic effects on lakes, particularly in coastal areas. Patterns of nutrient circulation, production and water renewal result from the extreme weather patterns that usually accompany ENSOs.

El Niños typically occur every 3–7 years, with extremely strong events every 20 years or so. While the phenomena involved are so complex that it is currently impossible to predict El Niños years in advance, recent models allow warnings several months ahead. Many advances in understanding resulted from the 1982–1983 El Niño, the strongest this century. The powerful linkages between ocean and atmosphere, which El Niño studies have revealed, cause great concern about how greenhouse warming may affect the ocean–atmosphere system, causing secondary effects of even greater magnitude than those due to greenhouse warming alone. Some examples will be discussed in Chapter 6.

5.7 FUNCTIONING OF BIOLOGICAL COMMUNITIES IN LAKES AND OCEANS

The biological communities of lakes and oceans in different climates and geological settings, differ greatly from one another in species composition and diversity. Nevertheless, communities ranging in complexity from a few hundred species to those containing thousands of species, all appear to be capable of the same basic functions: transforming light and nutrients into organic matter, which supports herbivorous and carnivorous animals of various sorts, followed by decomposition of organic matter and regeneration of the nutrients. There seems to be little correlation between the number of species in an ecosystem and its ability to perform these basic metabolic and nutrient cycling functions. A large number of factors, other than productivity, appear to control biological diversity in aquatic systems. A brief discussion of some of these factors will provide some insights into why ecosystems differ in complexity and productivity.

5.7.1 Factors controlling the diversity and productivity of aquatic communities

The number of species in an aquatic community correlates with a wide variety of factors, including age, latitude, temperature and diversity of habitat. In very old lakes, such as Lake Baikal, isolation from similar habitats in other lakes has resulted in diverse endemic faunas of several groups. For example, 240 species of Gammaridae and 53 species of Gastropoda occur in the depths of Baikal, but nowhere else on the planet. Overall, the lake contains over 700 endemic species of animals, including 23 species of fish and one of seal. Similar diversity has been recorded for Lake Tanganyika and other ancient lakes in the Rift Valley of Africa. However, there are some exceptions. The diverse fauna of cichlid fish in Lake Victoria must have developed largely in the last 13 500 years; the lake was almost dry at that time. Some groups of organisms simply seem capable of speciating very quickly. In general, however, lakes of greater age have greater biological complexity than younger lakes. In oceans, even greater ages have allowed the evolution of large numbers of abyssal creatures, many of which have developed unique features to adapt to the absence of light, cold temperature, shortage of food and great pressure (see Chapter 4). At the other extreme of complexity are lakes at high latitudes or altitudes, which contain only a few tens or hundreds of species. These systems have usually evolved since glaciers covered the areas, only 6000–10 000 years ago. For example, many lakes of the Canadian Shield contain one to five species of benthic crustaceans only, instead of the several hundred species in Lake Baikal and several thousand species in the oceans. Lakes at high latitudes contain still fewer. Char Lake, in the high Canadian Arctic, contained no benthic crustaceans, and only two planktonic species. Instead of the hundreds of fish species found in most tropical lakes, or the 6–10 species found in typical lakes of the Canadian Shield, Char Lake contained only one: Arctic char. The same tendencies for lower diversity with increasing latitude and altitude are true of streams and rivers (Schindler, 1990a).

Larger freshwater ecosystems also contain more species than smaller ones, at the same latitude. This is true of rivers as well as lakes (Schindler, 1990a). Other factors known to affect the diversity of freshwater communities are ionic strength and certain chemicals. For example, low calcium waters generally have very few species of mollusc, while those with low silica support few species of diatoms.

Regardless of the complexity of their communities, lakes and marine systems appear to function as viable ecosystems at all latitudes and altitudes, and under most chemical conditions. Fully functional food chains and chemical cycles have been documented in: (i) lakes which are only ice-free for a few weeks a year; (ii) lakes that have salinities greater than ocean water; and (iii) lakes that have pH values as low as 2. Of course, the organisms in these specialized communities are not necessarily those which supply humans with either food or pleasure, but they are capable of performing vital biogeochemical activities. In all but very exceptional cases, ecosystems are able to maintain production and respiration at values controlled by nutrients and light, no matter how few or how many species are present under natural conditions. Even when we artificially eliminate species from such systems, they are able to maintain vital ecosystem functions such as production and respiration, until numbers of species have been extremely depleted (Schindler, 1990b). Given the large number of insults imposed by man on the world's waters, we are extremely fortunate that this is the case. This remarkable resiliency is discussed in greater detail in Chapter 6.

5.7.2 Littoral communities in the functioning of aquatic ecosystems

Much of the vital activity in lakes and oceans occurs in shallow waters, less than a few metres to a few tens of metres in depth. Often, the food webs of such communities support higher fisheries production than the entire pelagic regions of lakes or oceans. For example, almost 50% of fish harvested in the oceans are taken within 3 miles of shore. Such communities are often remarkably complex. In oceans, several specialized classes, or even phyla, exist in the littoral zone, with few representatives in other parts. Many species which are pelagic as adults use littoral areas for spawning, and to rear

their young. For example, the lake trout, which as an adult may be found in waters tens of metres deep, usually lays its eggs in rocky shoals less than 3 m deep. Other species choose weedy areas, where young can feed on zooplankton, which gather to feed on bacteria that decompose plant remains.

One shallow-water marine community of this type is the salt marsh. Salt marshes occupy coastal regions of much of temperate North America (or at least they did before man began to fill, pollute and drain them). Many species of marine fish and invertebrates are supported by salt marshes, which are low-diversity communities dominated by marsh grasses of the genus *Spartina*. Few aquatic herbivores feed directly on *Spartina*; instead, microbially-decomposed matter is washed into creeks, estuaries and other nearshore areas, forming the food for invertebrates and fish larvae. Salt marsh soils are water-saturated and anoxic for a high proportion of the time, and anaerobic bacteria, such as sulphate reducers, can represent a high proportion of the decomposer community, and hence of the food available for small life-stages of marine animals. Among the species which rely heavily on salt marsh communities are mullet, tarpon, shrimp, crabs and oysters.

Other littoral marine areas that support high productivity are off the mouths of rivers, particularly those that carry high concentrations of nutrients from rich soils or urban areas. Typically, such rivers are very turbid, so that productivity is quite low until waters are discharged to continental shelves, where suspended particles settle. Frontal areas, where productivity is quite high, are typical of such coastal areas.

5.7.3 Transfer of energy and nutrients from surface water to deep water

As mentioned previously, partial pressures of CO_2 in the surface ocean are kept low by photosynthesizing plankton, which then carry carbon and other nutrients to the deep ocean when they sink. This particle flux is of considerable interest, for not only is it the only important pathway for transferring carbon from the surface to the deep ocean, but it is the energy source for most deep-ocean dwellers. (Bacterial production in the vicinity of ocean vents supply a considerable amount of energy, but only to small areas of the ocean floor.) The amount of organic carbon reaching the deep ocean is small, but due to low temperatures, long periods of very stable conditions and great specialization among some organisms, food chains of three or more steps occur. Most specialization of deep-sea organisms seems to occur among predatory fish. For example, some species have lures to attract prey within ambush range; others have males of tiny size, which attach themselves to females, and function only to fertilize eggs; still others are little more than a sack with a mouth and eyes, conserving energy by having little body mass, while still being large enough to consume large prey. The rain of energy from above must similarly support the profundal and abyssal communities of lakes, as is dramatically demonstrated by the diverse deep-water fauna of lakes like Baikal. In general, the bizarre specializations found in abyssal marine communities have not occurred in lakes, probably because of the much shorter periods that lacustrine communities, even in such ancient lakes, have had to develop.

In summary, it is clear that both lakes and oceans function efficiently because of complex interactions between physical, chemical and biological features, and with the meteorological and hydrological cycles of the planet. Modern scientific developments are only beginning to allow us to understand such phenomena, and our knowledge of the functioning of these systems has probably doubled in the past 10 years alone. Prospects for future research are exciting and challenging; they involve not only our understanding of how these important ecosystems coordinate and synchronize their functions: under-standing the functioning of lakes and oceans also holds the key to whether man will be able to prevent irreversible changes in climate and fresh waters as the result of widespread alterations to global systems, which he has so far undertaken with only the vaguest notion of what the consequences might be.

FURTHER READING

Broecker, W.S. (1974) *Chemical Oceanography*. Harcourt Brace Jovanovich, New York.

Broecker W.S., Peng, T-H. & Engh, R. (1980) Modelling the carbon system. *Radiocarbon* 22: 565–98.

Cromie, W.J. (1988) Grappling with coupled systems. *Mosaic* 19: 12–23.

Enfield, D.B. (1989) El Niño, past and present. *Rev. Geophys.* 27: 159–87.

Hutchinson, G.E. (1957) *A Treatise on Limnology*. John Wiley & Sons, New York.

Kelly, C.A., Rudd, J.W.M., Hesslein, R.H., Schindler, D.W., Dillon, P.J., Driscoll, C., Gherini, S.A. & Hecky, R.E. (1987) Prediction of biological acid neutralization in acid-sensitive lakes. *Biogeochemistry* 3: 129–40.

Mortimer, C.H. (1954) Models of flow patterns in lakes. *Weather* 9: 177–84.

Schindler, D.W. (1990a) Natural and anthropogenically imposed limitations to biotic richness in fresh waters. In: Woodwell, G.M. (ed) *The Earth In Transition: Patterns and Processes of Biotic Impoverishment*. Cambridge University Press, Cambridge.

Schindler, D.W. (1990b) Experimental perturbations of whole lakes as tests of ecosystem theory. *Oikos* 57: 25–41.

Schindler, D.W., Brunskill, G.J., Emerson, S., Broecker, W.S. & Peng, T-H. (1972) Atmospheric carbon dioxide: its role in maintaining phytoplankton standing crops. *Science* 177: 1192–4.

Schindler, D.W., Hesslein, R.H. & Turner, M.A. (1987) Exchange of nutrients between sediments and water after 15 years of experimental eutrophication. *Can. J. Fish. Aquat. Sci.* 44 (Suppl. 1): 26–33.

6
Aquatic Ecosystems and Global Ecology

D.W. SCHINDLER

6.1 INTRODUCTION

We now recognize that aquatic ecosystems play a vital role in the biogeochemical cycles of the entire planet, affecting the composition of the atmosphere, the climate and the hydrological cycles that allow the maintainence of terrestrial and wetland ecosystems. Conversely, climate determines many of the critical characteristics of aquatic ecosystems; for example, by defining temperatures in which various species and communities must survive, and controlling rates of chemical weathering and evaporation, which in turn determine chemical concentrations.

The most important aquatic ecosystems in global biogeochemical interactions are the oceans, due to their tremendous size. However, the relatively minute amount of fresh water on the planet cannot be overlooked, for without it the earth would be devoid of all terrestrial and freshwater species, including man!

Until the mid 20th century, the earth appeared to be so vast that man did not believe that he could alter it significantly. In particular, the oceans, great lakes and the atmosphere seemed to be practically infinite, and inexhaustible suppliers of resources and sinks for man-made pollutants. It is only in the last decade that we have realized that not only are these vast reservoirs contaminated by human effluent, but in some cases they are fouled to a degree that threatens the biogeochemistry, and thus the ecology, of the entire planet.

Perhaps the first indications that man was capable of contaminating the entire globe, and disrupting the vital linkages between the atmosphere and the ecosystems of the earth, came with nuclear testing in the 1950s and 1960s, when scientists found that carbon-14 and other radioactive substances were carried around the globe for months or years, becoming so well-mixed with the atmosphere that when they did return to the biosphere, they were deposited in oceans, lakes and other ecosystems at all latitudes, even in the ice-caps of Greenland and the Antarctic! A second important discovery was the relatively slow rate at which these substances reached equilibrium in the atmosphere, and in oceans and large lakes. In many cases, these equilibria were found to take decades, centuries, or even millenia to reach. For the first time, it was obvious that man was a significant force in shaping the biogeochemistry of the earth, and that the alterations caused could be very long-lasting.

However, not all aspects of the contamination of the world's oceans with radioactive materials were bad. Scientists quickly realized that the pulse inputs of both biologically-active and inert radioactive materials provided a unique opportunity to measure key mixing processes and biogeochemical reactions in the oceans. The result has been to greatly enhance our knowledge of how the oceans operate. In fact, the contamination of the oceans with carbon-14 probably led to the realization that the oceans would not be able to absorb man's emissions of CO_2 quickly enough to prevent greenhouse warming.

6.2 THE ROLE OF THE OCEANS IN AMELIORATING GLOBAL WARMING

It is now widely realized that the CO_2 content of the atmosphere is increasing rapidly, due to the burning of fossil fuels and forests, and that this increase has the potential for warming the earth

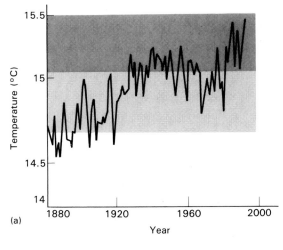

(a)

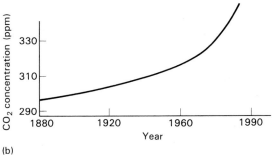

(b)

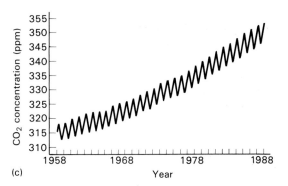

(c)

Fig. 6.1 Plots of (a) global temperature and (b) CO_2 in the atmosphere 1880–1988. (c) Details of the atmospheric CO_2 record from Mauna Loa, Hawaii, 1958–1988. From Udell (1989) and Patrusky (1988).

by several degrees, due to its effect on blocking the earth's re-emission of solar radiation — the so-called 'greenhouse effect' (Fig. 6.1).

Less widely known, is the fact that the atmosphere contains only a tiny proportion of carbon

on earth — only 5×10^{16} of 3×10^{21} g, or less than 0.002%! By far the largest carbon stores are in the ocean, where most of it is present in the sediments as calcium carbonate, magnesium carbonate and kerogen. The amount in the water column, present largely as carbonate and bicarbonate, also exceeds atmospheric CO_2, by many times (Table 6.1). Carbon dioxide is removed from the atmosphere to the surface ocean by gas exchange, as we discussed in Chapter 5, after which it is taken up by organisms as organic carbon or calcium carbonate, and sedimented to deeper layers. Why, then, doesn't the excess CO_2 that we introduce to the atmosphere simply disappear into the oceans or their sediments? The answer is not one of magnitude; this will certainly happen. Instead, it is one of rate, as complex investigations of physical, chemical and biological processes in the oceans have revealed. The oceans have already absorbed about 50% of the CO_2 released by man since the Industrial Revolution, and the increases that we have seen are much less than they would have been if the oceans were not a major component in the global carbon cycle.

In fact, the rate of exchange between the atmosphere and the ocean's surface is quite rapid, as we discussed earlier. At the measured values for mass

Table 6.1 Global carbon reservoirs and fluxes. From la Breque (1988)

Sinks	Carbon (gigatonnes)
Atmosphere	700
Oceans	
Total inorganic	35 000
Particulate organic	3
Land biota	600
Soil humus	3000
Marine sediments	
Organic	10 000 000
Calcium carbonate	50 000 000
Fossil fuels	5000

Sources	Carbon (gigatonnes year^{-1})
Atmosphere–marine biota	45
Atmosphere–land biota	70
Deposition in oceans	1–10
Fossil fuel combustion	5

transport (piston velocity) of 1700 m year^{-1} given in Chapter 5, about 17 mol of CO_2 should enter each square metre of sea surface per year (Broecker, 1974), and the ocean's surface and the atmosphere should reach equilibrium with a lag time of only a few years. However, the well-mixed surface layer is but a small part of the oceans, and transport of the carbon to greater depths, which constitute the main oceanic carbon reservoirs, is the process that limits the rate at which the oceans can absorb atmospheric CO_2.

The surface and deeper ocean mix quite slowly, as shown by the fact that radiocarbon from atomic bombs, released over 30 years ago, has still not reached steady-state with the deep ocean. It will take about 1600 years for carbon reservoirs in the surface and deep ocean to reach a new steady-state.

The major mechanism transporting carbon from the surface to the deep ocean is biological uptake and gravitational settling. However, organisms can only convert CO_2 to particulate carbon (organic particles or calcium carbonate), if they can obtain enough of other nutrients to photosynthesize and grow. The most limiting of these nutrients on an ocean-wide basis are phosphorus and nitrogen. Although we have greatly polluted many coastal waters with these two elements, primary production in pelagic regions of the oceans is almost always limited by one or the other. Even if all of our excess nutrients were sprayed evenly over the oceans, the amounts would not be large enough to significantly affect the rate of production, and hence of carbon transport to the deep ocean. Obviously, fertilizing the oceans or increasing the rates of mixing processes are too vast to be solutions to the greenhouse problem. Reduced burning of fossil fuels, decreased deforestation and perhaps major reforestation efforts appear to be more tractable short-term solutions, but even if these measures could be implemented instantaneously, we could not eliminate the threat of a substantial climatic warming.

6.3 THE ROLE OF OTHER GASES IN GLOBAL WARMING

The amounts of several other gases in the atmosphere are also increasing, contributing to the green-house effect. Recent calculations indicate that together, methane (CH_4), chlorofluorocarbons (CFCs), tropospheric ozone and nitrous oxides double the warming expected from CO_2 alone (Table 6.2). All of these gases are increasing at very rapid rates (Fig. 6.2).

Methane suddenly began to increase, at over 1% year^{-1}, about 200 years ago. While the amount is far lower than CO_2 in the atmosphere, a methane molecule is about 50 times more effective as a greenhouse gas than a CO_2 molecule. At this time, we do not even know the reason for the increase. Some of the possible reasons for the increase involve aquatic mechanisms: for example, increasing culture of rice in paludified soils, and increasing decomposition in warming wetlands are major possibilities. In addition, increased production of cattle and other ruminants, increased termite activity and hydrocarbon exploitation may be contributors to the observed increase in methane. Recent studies of the radioactive carbon isotopes in atmospheric methane indicate that most of it originated from recent biological activity, for it contains considerable amounts of carbon-14, while fossil methane contains only tiny amounts of this isotope, because it has radiodecayed, with a half-life of 5700 years. Recent interpretations indicate that 70–90% of atmospheric methane results from contemporary biological activity. The exploitation of fossil natural gas deposits, therefore, seems to be a minor contributor.

While it is presently impossible to distinguish among the different contemporary sources, investigations employing stable carbon isotopes are underway, which may assist in differentiation, as different metabolic pathways exclude the stable isotope carbon-13, with respect to ordinary carbon-12, to different degrees.

One study gives us a clue that man's activities alone may not be responsible for the increase in atmospheric methane. Ice-cores of several thousand metres in length were recently taken from the Antarctic ice-sheets, in order to investigate long-term changes in climate. Gas samples from these cores were analysed, and changes with time plotted. It was discovered that the patterns for change in CO_2, methane and oxygen-18 (the abundance of which is known to increase relative to

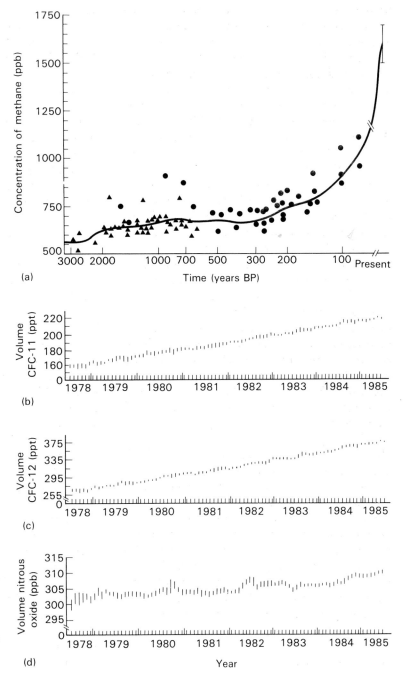

Fig. 6.2 (a) Data from a polar ice-core showing that, beginning about 200 years ago, the concentration of methane in the atmosphere has doubled. (b), (c) and (d) Records showing the rise of concentrations of chlorofluorocarbons and nitrous oxide in the atmosphere, 1978–1985. From Patrusky (1988).

oxygen-16 with increasing temperature), were remarkably similar — so similar that it is impossible to tell whether changes in CO_2 and methane preceded or followed climatic change. One thing is

certain — 150 000 years ago, the changes were not caused by man's activities!

There is some concern that the atmospheric increases in methane may also set off a cycle of

Table 6.2 Current concentrations and greenhouse heating due to trace gases. From Mitchell (1989)

Gas	Concentration (ppm)	Principal absorption bands		Greenhouse heating (W m^{-2})
		Position (cm^{-1})	Strength (cm^{-1} atm^{-1} cm^{-1} STP)	
Water vapour	~3000	—	—	~100
Carbon dioxide	345	677	(Many bands)	~50
Methane	1.7	1306	185	1.7
Nitrous oxide	0.30	1285	235	1.3
Ozone	$10-100 \times 10^{-3}$	1041	376	1.3
CFC-11	0.22×10^{-3}	846	1965	0.06
		1085	736	
CFC-12	0.38×10^{-3}	915	1568	0.12
		1095	1239	
		1152	836	

The main absorption bands and band strengths (a measure of the probability of a molecule absorbing a photon at the band wavelength) are shown for the less abundant gases.

positive feedback, which will enhance the greenhouse effect. For example, if the northern latitudes become warmer and drier as most current models predict, declining water levels in freshwater wetlands could expose waterlogged peat to atmospheric oxygen, promoting oxic decomposition and releasing CO_2.

Nitrous oxides are produced by microbial activities in soil and by combustion of fuels which contain nitrogen. The oceans and freshwater systems are not presently implicated as major sources, though they may be important sinks. Flux rates are still poorly known. At present, atmospheric concentrations of the gas are increasing by 0.25% year^{-1}.

Chlorofluorocarbon (CFC) quantity is presently increasing faster than any other greenhouse gas in the atmosphere. Production is entirely by human industries, and the compounds are so resistant to biological and chemical degradation that they have residence times in the atmosphere of several decades. While attempts have been made to control production of CFCs, they are still on the increase. Large reserves are locked in landfills, in the form of abandoned, decaying refrigeration equipment and plastic foam, and releases can be expected from such sources for decades to come, even if we stop using these substances now. Other synthetic halocarbons show similar properties and increases, including carbon tetrachloride, methyl chloroform and various exotic solvents used as cleaners in the electronics industry and as brominated gases in fire extinguishers. The oceans are again implicated as a sink for such gases, but rates of equilibration will be limited by physical and chemical constraints, due to the biological unreactivity of these substances. They will therefore be much slower to equilibriate than CO_2.

6.4 EFFECTS OF CLIMATE CHANGE ON THE WORLD'S OCEANS AND FRESH WATERS

If the greenhouse effect warms the climate, one major predicted effect is a rise in sea level, due to: (i) the increased flow and melting of glaciers from Greenland, the Antarctic ice-cap and alpine glaciers; and (ii) the thermal expansion of water as it warms. The largest increases in the earth's temperature are expected in polar regions, where major ice-sheets occur. While the melting of ice-sheets and rises in sea level would lag somewhat behind climate change, sea-level rises of up to several metres are predicted to occur eventually (USEPA, 1989). Such rises would inundate large parts of most coastal cities — for example, a rise of

7.6 m would decrease the land area of Florida by almost 50%, and would flood much of Washington DC (Houghton & Woodwell, 1989). Consequences for cities like Venice defy imagination.

Despite detailed tide-gauge records for many areas of the earth, postglacial isostatic adjustments (commonly known as glacial rebound) make it difficult to detect average changes in sea level. Recent models that filter isostatic adjustments from tide-gauge records, indicate that the sea level is rising at the rate of about 2.4 mm year^{-1}, but it is currently impossible to tell whether this rate is increasing, or how long it has been occurring.

Effects of global warming on fresh waters will be profound. While some areas will experience increased rainfall, in general the increased evaporation and transpiration caused by higher temperatures will cause lake levels, stream flows and the size of groundwater aquifers to decrease. In the hypsithermal of 4000–6000 years ago, when temperatures averaged a mere 1°C warmer than at present, Lake Manitoba, one of the largest lakes in southern Canada, was dry; as shown by grasses in bottom sediments deposited at that time. Few, if any, wetlands occurred on the Canadian prairies south of latitude 53°N in the hypsithermal period, and it is doubtful that those present now would survive greenhouse warming.

A 20-year study at the Experimental Lakes Area in northwestern Ontario gives some hints of what global warming may do to lakes. Both air and water temperatures increased about 2°C between the late 1960s and late 1980s. While it cannot be said that the warming is the result of the greenhouse effect, the trend is established enough to allow a preview of what greenhouse warming might do. Stream flows decreased, due to lower precipitation and higher evaporation, causing decreased renewal rates of lake waters. As a result, chemical concentrations in lakes increased. Due to warmer, drier conditions, the area of forest burned by forest fires increased; after the forests burned, increased wind velocities at lakes surfaces caused thermoclines to deepen. This decreased the volume of cold water in the hypolimnions of lakes — the only refuge in summer for cold-temperature species like lake trout, several large crustacean species and other

'glacial relicts' which have evolved in cold environments. Nutrient concentrations increased as the result of higher runoff from land, and reduced water renewal of lakes. As a result, the growth of algae increased. The length of the ice-free season increased by over 2 weeks (Fig. 6.3).

Lakes farther north may be even more affected. In large, shallow lakes of northern Ontario and the northern Praries, lake trout and other cold stenothermic organisms are currently able to live, even though the lakes do not stratify thermally, because surface water temperatures do not exceed their maximum temperature tolerances, which are typically about 16°C. Even a slight climatic warming would render such lakes too warm for many of the species that currently dominate their biological communities. Many of the lakes are located in permafrost, the melting of which will destabilize shorelines, causing bank slumping and siltation of spawning areas, as has commonly been observed when water levels in lakes and rivers have been changed by dams.

Decreased stream flows, as well as warmer waters, have adverse implications for organisms. Lower water levels and lower volumes of flow under winter ice can cause lower winter oxygen concentrations under ice. In many northern rivers, such as the Athabasca River of northern Alberta, winter oxygen concentrations are already low enough to cause concern about sensitive larval fish, due to a combination of natural and man-made phenomena. (The latter include the discharge of sewage and pulp-mill effluents, which contain large amounts of biodegradable organic matter, into the rivers.) Of course, lower stream flows also have adverse implications for human uses, like navigation, waste disposal and hydro-electric generation. In the American Southwest, water has been in short supply for many years. In 1990, the Colorado River had its lowest spring flow of all time. As a result, water for the city of Los Angeles and for irrigating crops is to be severely restricted.

Finally, drier conditions and forest fires promote the oxidation of sulphur in terrestrial soils and wetlands. This can cause pulses of sulphuric acid to reach streams and lakes following rainstorms. This

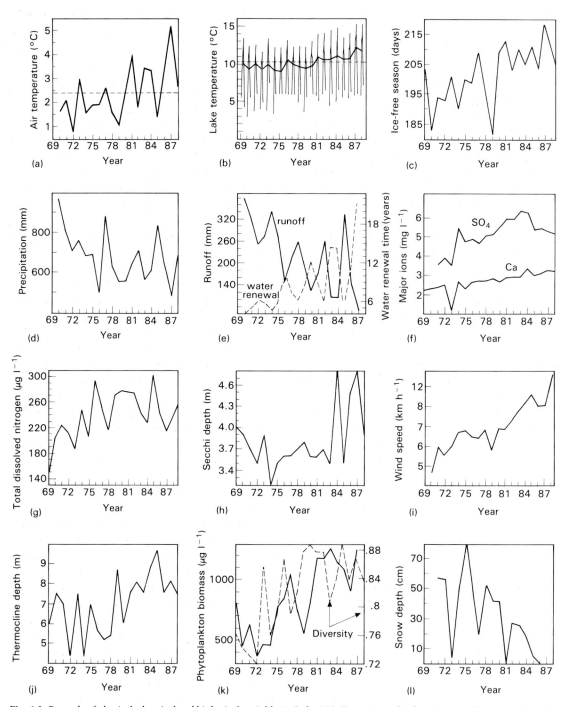

Fig. 6.3 Records of physical, chemical and biological variables in Lake 239, Experimental Lakes Area, northwestern Ontario 1969–1988. In (b), each 'spike' is a plot of monthly water temperature in the ice-free season and the jagged line connects seasonal means. In (h), the secchi depth is a measure of transparency of the water. In (k), the broken line gives Simpson's index of species diversity in the phytoplankton. From Schindler (1990).

problem is probably enhanced by acidic precipitation, which increases the amount of sulphur in ecosystems upon which it falls.

6.5 ACID RAIN AND FRESHWATER ECOSYSTEMS

Another change in aquatic ecosystems, occurring on an almost global scale, is the acidification of softwater lakes and streams throughout much of the industrialized world. The oceans and lakes in calcareous geological regions are not acidified, due to their high buffering capacity, but much of the world's fresh water occurs in areas of igneous or metamorphic rock, where geological buffering is very poor. The number of such lakes in eastern Canada is over one million, and an equal number may be present in western provinces plus the Northwest Territories. Large areas of northern Europe, the Soviet Union and other continents are also acid-sensitive. The release of sulphur and nitrous oxides from the burning of fossil fuels and the smelting of sulphur-containing ores are the major contributors to the acidification of soft waters, since when these are discharged to the atmosphere, they are transformed chemically into sulphuric and nitric acids as they are carried by air masses. These strong acids are then deposited as rain, snow, fog, or as dry fallout, hundreds or even thousands of kilometres from their sources. Not surprisingly, from what we know about carbon-14 from atomic bombs, the sulphate concentration in polar ice-caps has increased in the past 150 years. Many waters, including the oceans, are too well buffered to be affected by the increase in input of strong acids. However, when the fallout occurs over areas where waters and soils are poorly buffered, the chemistry and biology of fresh waters can be changed dramatically. While we do not have good historical data for a large number of lakes, fundamental geochemical principles allow us to estimate what chemical changes have taken place.

In waters that do not receive acid rain, the alkalinity (or acid neutralizing capacity — ANC as it has come to be known) is roughly equivalent to the sum of calcium and magnesium (where all are expressed in equivalents). This is expected, for all calcareous and many igneous and metamorphic rocks yield these proportions when weathered. When strong acids are added to such systems, they can deplete ANC by reaction with bicarbonate

$$H^+ + HCO_3^- = H_2CO_3 = H_2O + CO_2 \quad (6.1)$$

destroying both ANC and hydrogen ion. The CO_2 formed is lost to the atmosphere, if air–water equilibria are exceeded. The second reaction possible is the exchange of H^+ for Ca^{2+} or Mg^{2+} in soils or lake sediments. The mobilized calcium and magnesium is then carried into streams and lakes. In brief, the incoming acid can have two effects on surface waters: (i) it can deplete ANC, or (ii) it can increase calcium and magnesium quantity. While the proportion of H^+ participating in one reaction or the other varies between watersheds, due to differences in soils, sediments, water balances and the history of acidification, the ratio ANC/(Ca+Mg) will decline under the influence of acid precipitation. Instead of the normal values near 1, ratios of 0.6 or less are common in areas of highly acidic deposition, such as eastern North America (Fig. 6.4) (Schindler, 1988).

In extreme cases, the value of ANC reaches 0, indicating that all buffering capacity is gone. The areas in North America where such conditions prevail contain hundreds of thousands of lakes.

Of course, changes in the above ratio do not tell us whether lakes and streams have become more acidic, a key piece of information, for it is H^+ rather than concentrations of other ions that has the major effect on distribution of most organisms in soft waters (a partial exception to this statement is for calcium, which limits the distribution of some molluscs and crustaceans, as well as rendering organisms in these groups less sensitive to acidification). However, a few studies in different parts of the world have now measured the relative changes in ANC and calcium and magnesium, resulting from changing acidity of deposition, allowing us to make some reasonable guesses about the change in pH and therefore the likely degree of biological damage caused by acid precipitation.

Most estimates indicate that the proportion of H^+ exchanging for calcium or magnesium is about 30–50%. In other words, the proportion of incoming acid that either acidifies freshwater eco-

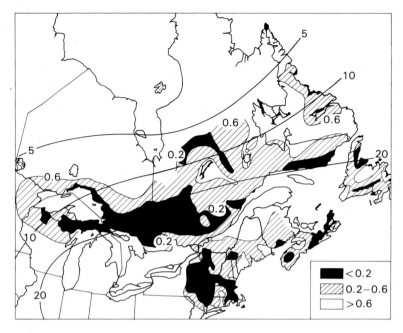

Fig. 6.4 The average ratio of alkalinity (acid neutralizing capacity) to calcium and magnesium ions in freshwater lakes of northeastern North America. Only lakes with Ca^{2+} and $Mg^{2+} < 200\,\mu mol\,l^{-1}$ are included. Values $< 0.2\,\mu mol\,l^{-1}$ indicate sensitive lakes with highly acidic deposition. Lines numbered 5, 10 and 20 indicate sulphate deposition in $kg\,ha^{-1}\,year^{-1}$. From Schindler (1988).

Table 6.3 Median values for original pH (pH_0) and median changes in pH (ΔpH) in Canadian lakes with $Ca + Mg \leqslant 200\,\mu eq\,l^{-1*}$. From Schindler *et al.* (1989)

Subregion	pH_0			ΔpH		
	Median	Q_1	Q_4	Median	Q_1	Q_4
Adirondacks	7.0	6.8	7.3	−0.63	−1.7	−0.30
Poconos/Catskills	7.0	6.9	7.3	−0.73	−1.6	−0.31
Central New England	7.2	7.0	7.4	−0.26	−0.51	−0.14
South New England	7.1	6.9	7.3	−0.55	−1.6	−0.25
Maine	7.2	7.0	7.4	−0.18	−0.27	−0.07
Northeastern Minnesota	7.2	7.0	7.3	−0.09	−0.23	0.07
Upper Michigan	6.7	5.2	7.1	−0.11	−0.35	0.12
North Central Wisconsin	6.7	6.0	7.1	−0.16	−0.64	0.09
Upper Great Lakes	7.1	6.6	7.4	+0.01	−0.13	0.17

* Changes are the differences between pH_0 and currently measured air-equilibrated pH. Q_1 and Q_4 are the first and fourth quantiles.

systems or depletes their buffering is 50–70%. When this assumption is combined with reasonable estimates for pre-industrial inputs of acids, results indicate that many lakes have acidified by one or more pH units as the result of acidic deposition (Table 6.3) (Schindler *et al.*, 1989). The actual degree of acidification varies from one lake to another, depending on the rate of water renewal and the depth and composition of soils.

Other estimates of pre-industrial pH have been made from the changes of diatom flora. The siliceous frustules of diatoms preserve well in the sediments of lakes and oceans. Frustules are unique for each species, so that if sediments can be dated (by radioactive methods, or by counting the annual layers that can be distinguished by seasonally deposited chemicals or by layers of spring pollen), the history of the diatom flora can be reconstructed.

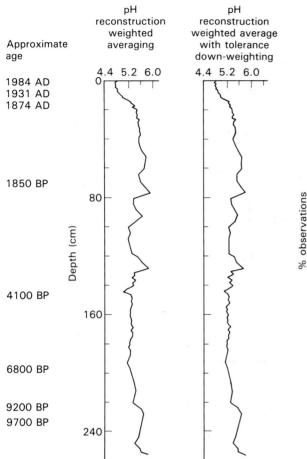

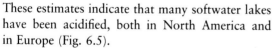

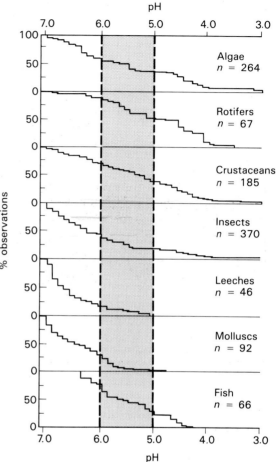

Fig. 6.5 pH values for Round Loch of Glenhead, reconstructed from palaeolimnological analyses of diatoms. In the right-hand plot allowance is made for changes in species composition resulting from different tolerances of the reduced pH. From Birks *et al.* (1990).

Fig. 6.6 Cumulative frequency distribution of minimum environmental pH values, reported for field observations of aquatic taxa. From Mills & Schindler (1986).

These estimates indicate that many softwater lakes have been acidified, both in North America and in Europe (Fig. 6.5).

Monitoring studies, bioassays and whole-lake experiments all indicate a general impoverishment of species numbers in lakes as they become more acidic (Fig. 6.6). Such examples suggest that the biological impoverishment caused by acid rain is not something that occurs suddenly, but little by little as lakes acidify. When models were developed for eastern USA, based on the chemical assumptions mentioned above and the relative changes in

numbers of species in different taxonomic groups from Fig. 6.6, they indicated that many lakes in northeastern USA have lost 30% or more of the species in some taxonomic groups (Fig. 6.7). It seems peculiar that the most technologically-advanced countries on earth are still causing the impoverishment of their ecosystems at a high rate.

In general, ecosystem functions have proven to be more resistant to acidification than species changes, i.e. it is clear that species assemblages are plastic enough to maintain ecosystem functions even under quite extreme and, for them, unusual stresses. There are exceptions: in the experimental

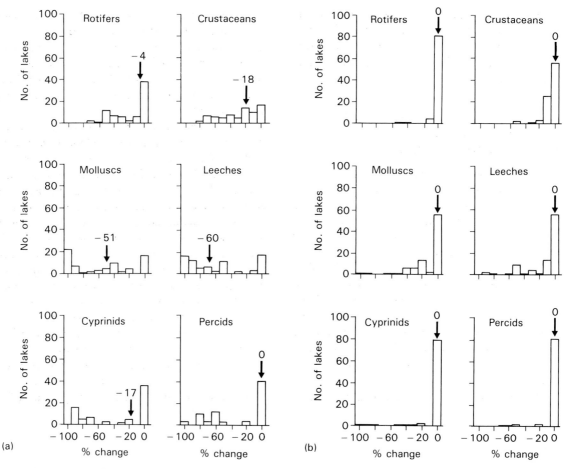

Fig. 6.7 Frequency distribution of modelled declines in relative numbers of species of different taxonomic groups, in (a) the Adirondacks (heavily damaged by acid deposition) and (b) Maine (the least damaged area of northeastern USA). Arrows indicate medians. From Schindler *et al.* (1989).

acidification of two small lakes in the Experimental Lakes Area of northwestern Ontario, net periphyton production decreased to near zero at pH levels of 5.6–5.8, due to increased respiration by the algae. The only species capable of maintaining positive net production were filamentous green algae; thus the structure of the periphyton community was altered dramatically (Schindler, 1990).

The winter respiration of lakes is easily measured under ice, when gas exchange is prevented by ice-cover, and snow-cover prevents photosynthesis. Such measurements showed that at pH values less than 5.1, respiration declined dramatically. Once

pH values were allowed to recover to 5.5, winter respiration rates returned to normal.

At pH values below 5.4–5.7, ammonium began to accumulate in the water columns of both experimental lakes. Further investigations revealed that this was due to the cessation of nitrification at lower pH values. Once the pH was allowed to recover to above 5.7, nitrification resumed, but with a time lag of a year. In summary, even slight acidification appears to cause the loss of species from freshwater lakes, and some ecosystem-level chemical cycles are disrupted at moderate levels of acidification. We wonder how many less-studied pollutants are similarly eroding the diversity of aquatic communities, and perhaps even causing changes in biogeochemical cycling. If small changes in biogeochemical cycles occur over wide areas, consequences for global chemistry and climate can be very great.

6.6 EVIDENCE FOR PAST AND PRESENT COUPLING BETWEEN OCEANS AND CLIMATE

The past history of relationships between the climate of the earth and other factors is still largely a mystery. We know that major climatic changes have been associated with major changes in the wobble of the earth's orbit, known as Milankovich cycles after their discoverer. However, calculations indicate that changes in the earth's orbit alone are insufficient to account for the magnitude of observed change, leading some scientists to hypothesize that orbital changes were simply an initial event that set off a series of reinforcing phenomena known as 'positive feedbacks'. The most important of these feedbacks probably occurred in the oceans.

One example of the enormous influence that the oceans can have on climate is the El Niño, caused

by disruption of the upwelling of cold, nutrient-rich waters along the coast of South America, which was discussed in Chapter 5. El Niños have dramatic effects on fisheries and crop productions in coastal areas (Fig. 6.8). It is not known how global warming will affect El Niños, but changes are likely to be dramatic.

Palaeoecological evidence suggests that in the past, the oceans may have caused much more dramatic shifts in climate than we see in even the most severe El Niños. For example, there is some evidence from studies of foraminiferan fossils in sediments that the circulation of the oceans was reversed in the distant past, controlling the onset and disappearance of ice ages! The ocean, which moves chemicals in patterns that control productivity and gas exchange (as described in Chapter 5), also operates as an enormous heat pump, moving heat from equatorial regions to the poles. If flow patterns were reversed during ice ages, heat would have been moved from, instead of to, the poles, allowing ice-sheets to grow. Changes appear to have occurred very suddenly, and to have been even greater than those that are predicted to occur as the result of greenhouse warming alone (Fig. 6.9) (Broecker & Denton, 1990).

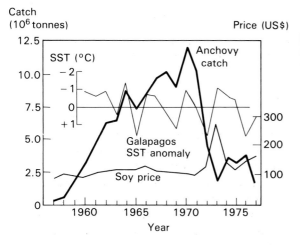

Fig. 6.8 The annual sea surface temperature (SST) anomaly at the Galapagos Islands, the Peru anchovy catch and the cost of soybean meal. Note that the temperature scale is inverted, and warmer temperatures are associated with declining fish catches. Soybean meal is a substitute for fish meal in animal feed. From Enfield (1989).

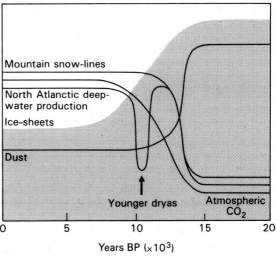

Fig. 6.9 Qualitative summary of changes that began about 14 000 years ago and proceeded at different rates. From Broecker & Denton (1990).

It is hypothesized that the changes occurred as the result of changes in the flow of fresh water into the ocean's surface, resulting from advancements and retreats of the continental ice-sheets, and it is possible that the disruption of freshwater flows in northern regions by damming of large rivers could have similar effects. As a result of the close synchronization between these changes in circulation and ice ages, it is not possible to tell whether climate changes triggered the changes in ocean circulation, or vice versa. It is therefore possible that man-induced climatic warming could trigger similar reversals in oceanic circulation that would cause global warming far more devastating than current models indicate.

In brief, we have learned how to increase the geochemical cycling of elements on an enormous scale, without realizing that we must have similar knowledge of the consequences of these alterations, as they may be capable of setting off a sequence of forces that are still well beyond man's capabilities to comprehend or control.

Some experts even believe that major extinctions of marine species in the past were related to climatic change (Kutzbach, 1989). Presumably, either direct warming or shifts in ocean circulations, similar to those discussed above, could have caused such extinctions. Future research will probably reveal whether the extinctions were caused, either directly or indirectly, by global warming.

Perhaps the most tragic of all manifestations of global change is man's effects on biotic diversity. The effects of deforestation of tropical rain forests are well publicized, and the effects of acid rain are described above. But man's movements of species, including pathogens, and his harvesting practices are even more devastating. There are numerous freshwater examples. The best known aquatic example in North America, is undoubtedly the introduction of the common carp *Cyprinis carpio*. In a period of a few decades, this species invaded most waterways in the USA and southern Canada, displacing other species and destroying their habitats. It is now widespread.

Similar deliberate introductions, usually of trout or char, have been made to many oligotrophic lakes. Most high mountain lakes in Europe and North America were originally fishless. Avid fisher-men, who loved the mountains, have invariably stocked these lakes. In Europe, many lakes had already been stocked in the 15th century. In western North America, only a few per cent of lakes in the Rocky Mountains have escaped stocking with salmonids. In many cases, several species were stocked, with repeated stockings over many decades if necessary. Some of the lakes were also treated with rotenone or even toxaphene to kill species that were judged unsuccessful from previous stockings, prior to introducing a new species! Of course, these chemicals killed many species of invertebrates as well.

The overall effect of such introductions on aquatic communities has not been assessed. It is expected to be profound, as it involves addition of a trophic level to the aquatic community. By the principal known as 'trophic cascading', such predators would be expected to greatly reduce the number of organisms at the next, lower trophic level, by their feeding activity. This would, in turn, allow the trophic level two below the salmonid predator to flourish. It is known from palaeoecological studies of plankton remains in sediments that some large species of planktonic crustaceans were eliminated from alpine lakes in Europe in the 15th century, and have not returned. Periodic sampling of some lakes in the Canadian Rockies indicate similar deletions of prominent species in the 20th century, with no indications of recovery in some cases. It is possible that the lakes in our national parks and wilderness areas, which most people regard as pristine, are among the ecosystems already impoverished by man.

Another recent example is the effect of man's introduction of the Nile Perch into Lake Victoria. The lake was originally occupied by hundreds of species of cichlid (mouthbreeding) fishes, representing a remarkable display of rapid and explosive speciation. Many of the cichlid species were used as food; plus many of them are sought after by hobbyists for aquaria, for their colouration is remarkable. The introduced perch are voracious and catholic piscivores, which are rapidly depleting the native cichlid communities.

In marine systems, similar things have happened. Only two or three of the 20-odd mollusc species in San Francisco Bay are native. Other native species have been eliminated by competition with introduced ones.

There are also botanical examples. The introduction of water hyacinth to the southeastern USA has caused this species to choke most waterways, disrupting boat traffic as well as biological communities. Introduced Eurasian milfoil causes similar problems in lakes of western Canada and Sweden.

Of course, the depletion of whale stocks, of east coast fisheries for several species and the driftnetting of porpoises are among the 'experiments' that man is currently performing on the oceans, which are sure to have massive repercussions throughout marine food webs.

In coastal waters of the North Pacific from California to Alaska, hunting of sea otters removed the natural predators of sea urchins, which became so abundant that they destroyed large areas of kelp forest and reduced productivity of the whole system. Where otter populations have been rebuilt, the sea urchin populations have declined and the kelp beds have returned to their former abundance.

On the east coast of Canada a similar relationship appears to exist between lobsters, sea urchins and kelp. It is less clear that lobsters are important predators of urchins, but over the last 20 years, there has been a period of low lobster abundance, high abundance of sea urchins and massive destruction of kelp beds. The lobster populations are very heavily exploited and many believe it is this removal of predators that has permitted the urchin population to explode. The relationship is made more complicated by the occurrence of a disease organism which decimated the urchin population, permitting the kelp beds to return. The increase in kelp was accompanied by an increase in lobster stocks. This is one of many examples that could be given of the complex changes that occur when man tampers with the natural balance between predators and prey.

Overall, man's addition of biological 'pollutants', destruction of habitats, or overharvesting of species may be the most serious ecological effect in the entire 'global change' package.

In summary, recent research has shown that improved understanding of the interactions be-

tween human activities, the atmosphere and water bodies is critical to predicting and preparing for climatic change, and to the preservation of all ecosystems of the earth. Not only are there numerous exciting and challenging possibilities for future research in both marine and freshwater ecosystems, but such research is essential to the success of global change programmes, and to the maintenance of the integrity of our planet's biology and biogeochemistry.

FURTHER READING

Birks, H.J.B., Line, J.M., Juggins, S., Stevenson, A.C. & ter Braak, C.J.F. (1990) Diatoms and pH reconstruction. *Phil. Trans. R. Soc. Lond.* B327: 263–78.

Broecker, W.S. (1974) *Chemical Oceanography.* Harcourt Brace Jovanovich, New York.

Broecker, W.S. & Denton, G.H. (1990) What drives glacial cycles? *Sci. Am.* 262: 48–56.

Charles, D.F. & Norton, S.A. (1986) Paleolimnological evidence for trends in atmospheric deposition of acids and metals. In: *Acid Deposition: Long-Term Trends*, pp. 335–431. National Academy Press, Washington DC.

Cromie, W.J. (1988) Grappling with coupled systems. *Mosaic* 19: 12–23.

Enfield, D.B. (1989) El Niño, past and present. *Rev. Geophys.* 27: 159–87.

Houghton, R.A. & Woodwell, G.M. (1989) Global climatic change. *Sci. Am.* 260: 36–44.

Kutzbach, J.E. (1989) Historical perspectives: Climatic changes throughout the millenia. In: De Fries, R.S. & Malone, T.F. (eds) *Global Change and Our Common Future*, pp. 50–72. National Academy Press, Washington DC.

la Brecque, M. (1988) A global chemical flux. *Mosaic* 19: 91–101.

Mills, K.H. & Schindler, D.W. (1986) Biological indicators of lake acidification. *Water, Air, Soil Pollut.* 30: 779–89.

Mitchell, J.F.B. (1989) The "Greenhouse" effect and climatic change. *Rev. Geophys.* 27: 115–39.

Patrusky, B. (1988) Dirtying the infrared window. *Mosaic* 19: 24–37.

Schindler, D.W. (1988) Effects of acid rain on freshwater ecosystems. *Science* 239: 149–56.

Schindler, D.W. (1990) Experimental perturbations of whole lakes as tests of hypotheses concerning ecosystem structure and function. *Oikos* 57: 25–41.

Schindler, D.W., Kasian, S.E.M. & Hesslein, R.H. (1989) Biological impoverishment in lakes of the midwestern and northeastern United States from acid rain. *Environ. Sci. Technol.* 23: 573–80.

Tolonen, K., Liukkonen, M., Harjula, R. & Patila, A. (1986) Acidification of small lakes in Finland documented by sedimentary diatom and chrysophycean remains. In: Smol, J.P., Battarbee, R.W., Davis, R.B. & Merilainen, J. (eds) *Diatoms and Lake Acidity*, pp. 169–221. W. Junk, Dordrecht.

Udell, J.R. (1989) Turning down the heat. *Sierra* 74: 29.
USEPA (1983) *Projecting Future Sea Level Rise*. EPA 230–09–007, Office of Policy and Resource Management, Washington DC.

PART 3
AQUATIC INDIVIDUALS AND COMMUNITIES

7

Community Organization in Marine and Freshwater Environments

C.R. TOWNSEND

7.1 INTRODUCTION

This chapter is concerned with factors that determine the composition of communities existing in aquatic ecosystems; we ask what subset of all possible species actually occurs together at a location, and why? Some of the earliest attempts to answer these questions were descriptive and focussed on the occurrence of species in relation to physicochemical features of their environment. Subsequently, attention moved away from abiotic factors, and evidence was found for the importance of biotic interactions in determining which species could coexist. Given the conclusions from simple mathematical models and a multitude of descriptive studies, there developed in the 1960s and 1970s a view that competition was overwhelmingly the most important process determining community composition, sometimes moderated by the influence of predation (Cody & Diamond, 1975). Most recently, community ecologists have shifted from this monolithic view of community organization, to one that gives due weight to stochastic (random) factors, including various kinds of disturbance to the system. Moreover, the focus has shifted in some aquatic systems to the importance of factors that influence recruitment of larvae. The general view to emerge in the 1980s was that no single model can describe all communities (Strong *et al.*, 1984; Diamond & Case, 1986; Gee & Giller, 1987). Some are structured mainly by competition, some by predation, some by factors that affect recruitment and some by unpredictable disturbances. Yet others may be largely unstructured if problems of colonization have led to undersaturation with species and unfilled niches; for example, in landlocked saltwater lagoons (Barnes, 1988).

In this chapter we take it as given that only species whose tolerance limits are encompassed by the range of physicochemical factors prevailing at a location can occur there. Moreover, we ignore tolerant species that are remote from the location and lack the colonizing capacity to reach it. In separate sections, we ask to what extent competition, predation, fluctuating conditions and disturbance play roles in determining which of the available, tolerant species coexist. My contention is that most communities fit into the scheme of so-called patch-dynamics models (Pickett & White, 1985), being patchy in structure and subject to both stochastic disturbances and biotic forces. The comparative importance of competition, predation and disturbance in aquatic communities is discussed, and finally we consider the question of community stability.

Aquatic ecologists have played a very prominent role in the development of ideas about community organization. It is appropriate and instructive to set these ideas in the context of the range of aquatic communities that exist — the benthic communities of streams, ponds, rocky shores, soft sediments and deep seas, the fish communities of streams, lakes and coral reefs and the freshwater and marine plankton, provide a rich variety of examples to compare and contrast.

7.2 COMPETITION, COEXISTENCE AND SUCCESSION

Some of the earliest ideas about species coexistence were based on the competitive exclusion principle, which states that coexisting species must differ in their trophic niches (see, for example, Chapter

7 of Begon *et al.*, 1990). The concept derives from simple mathematical models (Lotka, 1925; Volterra, 1926) that indicate that of two species competing for the same resource in a homogeneous and constant environment, one is certain to become extinct — to be competitively excluded by the other. According to this view, coexisting species partition resources and each is limited by a different resource, or by the same resource but in different microhabitats, or at different times. Communities structured in this way can be described as niche controlled (Yodzis, 1986).

An early and influential study of competition and coexistence was performed by Connell (1961) on two species of barnacle, on a Scottish rocky shore. Adult *Chthamalus stellatus* generally occur higher in the intertidal zone than adult *Semibalanus balanoides*, despite the fact that many young *Chthamalus* settle in the *Semibalanus* zone. The normal cause of mortality in young *Chthamalus* was not the increased submergence times in lower

zones, but competition from *Semibalanus* which, as they grew, smothered, undercut or crushed them. The fundamental niche of *Chthamalus* (where a viable population can persist given the abiotic conditions that prevail), extended down into the *Semibalanus* zone, but competition from *Semibalanus* restricted it to a realized niche higher on the shore. On the other hand, *Semibalanus* was prevented from surviving in the *Chthamalus* zone by its sensitivity to desiccation — its fundamental niche did not extend into the upper zone. Overall, the coexistence of these species was associated with a differentiation of realized niches, involving partitioning of space down the shore.

A similar situation has been described for two species of aquatic plants that coexist in Michigan ponds (Grace & Wetzel, 1981). One species of cat-tail (reedmace), *Typha latifolia*, occurred in shallower water, whereas the other, *Typha angustifolia*, was found in deeper situations (Fig. 7.1). Experiments revealed that *T. angustifolia* is normally

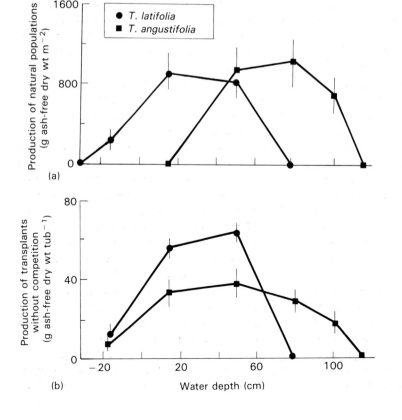

Fig. 7.1 (a) The natural distributions of coexisting populations of *Typha latifolia* (in shallow water) and *T. angustifolia* (in deeper water). (b) When grown alone, *T. angustifolia* grows over a wider range of depths but *T. latifolia* grows in the same range of depths as it did when in competition. From Grace & Wetzel (1981).

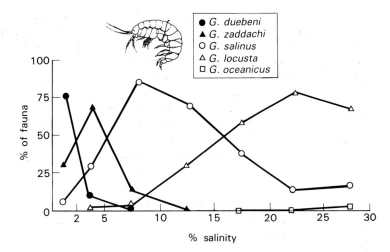

Fig. 7.2 Percentage composition of various species of *Gammarus* from sampling stations in the Limfjord with different salinities. From Fenchel & Kolding (1979).

excluded by *T. latifolia* from shallow water, but the distribution of *T. latifolia* is unaffected by competition with *T. angustifolia*.

Salinity is a factor that shows a marked pattern of variation in estuaries, and in fjords which have a significant freshwater input. Habitat partitioning in relation to such salinity variation, has been described for five species of amphipod crustaceans that inhabit the brackish waters of Denmark's Limfjord (Fenchel & Kolding, 1979; Kolding & Fenchel, 1979). *Gammarus duebeni* was found in the least saline areas, while at the opposite extreme *G. oceanicus* occurred at the highest salinities (Fig. 7.2). The species all have similar diets, and all five species have been reared successfully in the laboratory at a salinity fluctuating between 23‰ and 27‰. It seems that the pattern of habitat use is maintained by interspecific competition, and that the species' realized niches are narrower than their fundamental niches, in relation to salinity. In other regions, such as the Baltic Sea, these species occur at the same ambient salinity, but appear to subdivide the depth gradient between them.

As a final example, six species of net-spinning caddis larvae, in the family Hydropsychidae, show a marked partitioning of space down the length of some European rivers (Fig. 7.3). This distribution pattern was reflected by physiological differences; downstream species such as *Hydropsyche pellucidula* performed best at the higher summer temperatures experienced there, whereas the metabolism of *Diplectrona felix* appeared to be adjusted to the lower temperature regime of the headwaters. Where distributions overlapped, species spun their nets in subtly different microhabitats (Hildrew & Edington, 1979).

These examples, from a variety of aquatic communities, have stressed that, on the large scale, species may coexist in a pond, fjord or river, or on a rocky shore, by virtue of spatial resource partitioning.

On a much smaller scale, it is often true that competitive exclusion is ruled out because of local habitat partitioning (e.g. species that are characteristic either of rock pools or exposed substrate on rocky shores), or because of very different food requirements (e.g. stream invertebrate species that consume either coarse particulate matter or filter fine particles from the flow, or eat other invertebrates — see Chapter 12). In other words, part of the diversity of species in all communities can be accounted for by different ecological requirements, in relation to the range of microhabitats and food types present — i.e. by so called 'niche control'. Thus, Harman (1972), in a study of the species richness of freshwater molluscs at 348 sites, including roadside ditches, swamps, rivers and lakes, revealed a strong pattern — more species coexisted in habitats with more categories of mineral and organic substrate (Fig. 7.4). Similarly, Tonn and Magnuson (1982) noted a significant

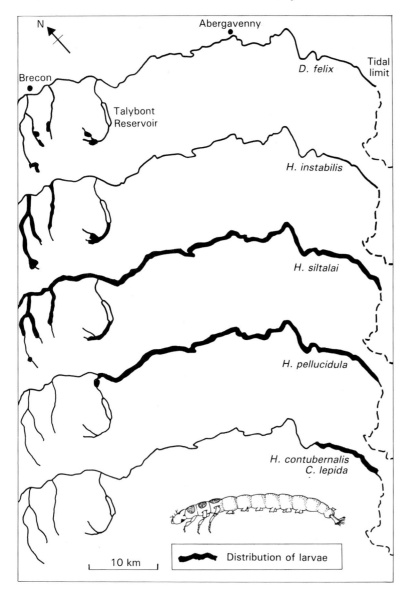

Fig. 7.3 The distribution of hydropsychid caddis larvae in tributaries and lower reaches of the River Usk in Wales. From Hildrew & Edington (1979).

positive correlation between species richness of fish and an index of plant structural diversity, in 18 Wisconsin lakes. Niche control is a mechanism of relevance to the whole range of communities.

The relative abundance of species varies in space, and their different requirements for resources and conditions help to explain the richness of species coexisting in a community. In a similar way, relative abundance often varies with time, as conditions and resource availability change during ecological succession. Succession is defined here as *the non-seasonal, directional and continuous pattern of colonization and extinction on a site by species.*

Sousa (1979a,b) describes a regular sequence of macroalgal species replacement on boulders, in the low intertidal zone of the southern Californian coast. The typical successional pattern can be mimicked using concrete blocks, making it possible for Sousa to obtain a precise description and, through experimental manipulations, to reveal

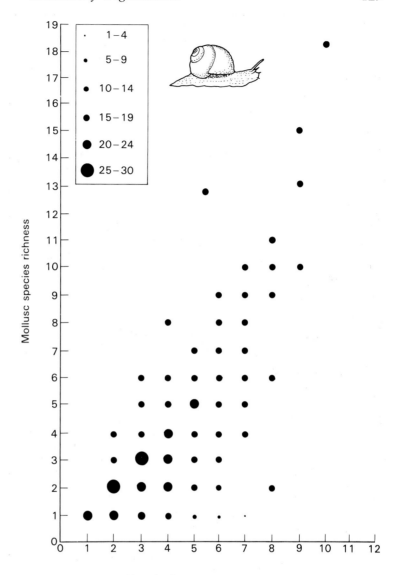

Fig. 7.4 Relationship between species richness of freshwater molluscs and spatial heterogeneity (number of substratum categories present). The size of the points indicates the number of study sites falling into each category. From Harman (1972).

whether competition and predation are mechanisms underlying this succession.

Bare substratum is colonized within a month by a mat of the pioneer green alga *Ulva*; but during the autumn and winter of the first year, various perennial red algae, including *Gelidium coulteri*, *Rhodoglossum affine*, *Gigartina leptorhynchos* and *G. canaliculata*, colonize the substrate surface. Within 2 or 3 years *G. canaliculata* comes to dominate the community. Under benign and calm conditions, each species in the succession was able to persist and the later perennial algae, with their highly seasonal recruitment and slow growth, were inhibited as long as the early colonists remained healthy. The algal succession occurs because individuals of the species that dominate first are more susceptible to the rigours of the physical environment, particularly the harsh conditions of intense sunshine and exposure to air and drying winds. Sousa also discovered that preferential

Table 7.1 Composition of attached algae in riffles and pools before (13 June 1983) and on two occasions after a flood (18 June and 8–10 July). Numbers are proportions of sampled sites where a given algal type dominated. Data from Power & Stewart (1987)

Algae	Riffles			Pools		
	Pre-flood	5 days post-flood	24–26 days post-flood	Pre-flood	5 days post-flood	24–26 days post-flood
Rhizoclonium	0.54	0.09	0.26	0.20	0.09	0.09
Spirogyra	0.22	0.00	0.24	0.68	0.04	0.48
Blue-greens	0.00	0.04	0.00	0.00	0.06	0.02
Diatoms	0.00	0.14	0.10	0.00	0.04	0.11

grazing by crabs on *Ulva* accelerated the succession to the tougher, less palatable perennial species of *Gelidium*.

The situation on intertidal boulders contrasts with the well-known classical successions associated with abandoned fields, or with glacial retreat. In these cases, the later species are generally competitively dominant to the early, rapid colonizers, and this is also true for algal successions in streams. After a severe spate in an Oklahoma stream, microscopic diatoms and blue-greens dominated the benthos. Later, their photosynthetic rates and relative importance declined, as larger and longer-lived filamentous green algae, such as *Rhizoclonium* and *Spirogyra*, overgrew and outcompeted them by shading (Table 7.1) (Power & Stewart, 1987).

The deep sea community, which has been receiving more and more attention since the development of sophisticated underwater craft, provides a stark contrast to boulder beaches and stream beds. Successional patterns in this low productivity/high species richness system, have been studied by introducing organic debris to mimic the natural fall of parcels of organic matter from above. For example, when wood falls to the deep sea floor, it is rapidly colonized by boring bivalves, and may be riddled with holes within 4–15 months. By converting wood to more generally accessible fine particulate organic products, the bivalve activity causes enrichment of local sediment which is colonized in numbers by, for example, capitellid polychaete worms (Smith & Hessler, 1987). This 'degradative' successional sequence comes to an end when the resource has been exhausted.

Successional sequences, such as those described above, enhance the species composition and richness of a region, if different locations are at different successional stages. This will be the case if patches are disturbed to different extents and at different times, producing a successional mosaic with good representation of pioneer, mid-successional and climax species. We will return to this important point in Section 7.5, where disturbances and the patch-dynamics concept are discussed.

7.3 THE INFLUENCE OF PREDATION

The potential foods of consumers vary in many ways, including their size, chemical make-up and defensive features. They are not all equally susceptible to, or preferred by, predators. Thus, predation in a community invariably alters the relative abundance of species and may drive the most vulnerable to extinction. Some particularly dramatic examples of the direct impact of predation have occurred after exotic piscivorous fish have been introduced to lakes. For example, the introduction of largemouth bass (*Micropterus salmoides*) into Lake Atitlan in Panama, caused the extinction of the native fish fauna. Similarly, the invasion of the sea lamprey (*Petromyzon marinus*) into Lake Michigan, North America, resulted in the elimination of two native chub species (*Leuichthys johannae* and *L. nigripennis*). Furthermore, the introduction of the Nile perch (*Lates niloticus*) into Africa's Lake Victoria 'to improve the fisheries' has placed its endemic cichlid fish (more than 200 species) in grave danger of extinction. Indeed, at a stroke, this human intervention could conceivably lead to the single most devastating loss of vertebrate species known (Miller, 1989).

However, rather than the more obvious direct effects of predation, we focus here on the indirect effects that predators have on populations with which their prey normally interact (see Kerfoot & Sih, 1987). In particular, predation can promote the chances of coexistence of species amongst which there would otherwise be competitive exclusion. It does this if the densities of some or all of the prey species are reduced to levels at which competition is relatively unimportant, a phenomenon known as exploiter-mediated coexistence.

The rocky intertidal zone provided the location for pioneering work by Paine (1966) on the influence of the starfish, *Pisaster ochraceus*, a predator of sessile, filter-feeding barnacles and mussels, and of limpets, chitons and whelks. Together with a sponge and four macroscopic algae, these species form predictable communities on rocky shores of the Pacific coast of North America. The main community effect of the starfish, is to make space available for competitively inferior species. It cuts a path free of the dominant mussels that would otherwise outcompete barnacles, other invertebrates and algae for space. This exploiter-mediated coexistence was confirmed by experimental removal of the starfish; the ensuing competitive exclusion led to a reduction in the number of species from fifteen to eight.

Grazing by the periwinkle *Littorina littorea* has a similar effect on algal species richness in tide pools along the rocky Atlantic coast of the USA (Lubchenco, 1978). In pools where the snails are rare or absent, the fast-growing alga *Enteromorpha* outcompetes most other algae, and few species coexist. At moderate periwinkle densities (about $150\,m^{-2}$), the abundance of *Enteromorpha*, the snail's preferred food, is reduced, competitive exclusion is prevented and many algal species coexist. However, at very high periwinkle densities, all palatable algal species are consumed to extinction and prevented from recolonizing, leaving a species-poor stand of tough, unpalatable algal species (Fig. 7.5).

An example of a predator that exerts a profound influence at all levels in an aquatic food web is provided by the sea otter, *Enhydra lutris*. This species, which preys heavily on sea urchins in

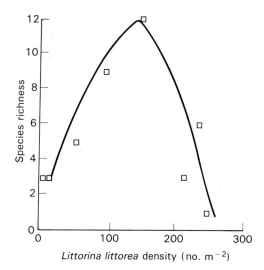

Fig. 7.5 Effects of periwinkle density on species richness of algae in tide pools. From Lubchenco (1978).

kelp forests, was exterminated from large parts of its former range along the Pacific coast of North America by hunters and fur traders. Its demise was followed by large increases in urchin populations, which grazed down kelp forests to almost nothing. The consequent reduction in shelter and food for fish led to reductions in density of the fish and their predators (sea eagles and seals) (Estes *et al.*, 1982). Disease has been shown to have similarly wide-ranging effects on the coastal community of eastern Canada (Scheibling, 1986). An epidemic of pathogenic amoebae caused massive mortality of sea urchins, in years when water temperature was high. The knock-on effects were, as we would predict, an increase in kelp biomass and more favourable conditions for kelp-dwelling species.

The effects of predators on soft-sediment communities have proved more difficult to analyse, because of the third dimension — depth — that is effectively lacking from macrofaunal communities on rocky shores. Some studies (e.g. Reise, 1985), based on the exclusion of mobile, surface-living (epibenthic) predators from experimental areas of mud or sand, have shown that such species can greatly reduce the density, and sometimes the diversity, of their infaunal prey (Table 7.2). Epibenthic predatory fish such as gobies and juvenile flatfish, and crustaceans such as crabs, shrimps and prawns,

Table 7.2 Abundance of macrofauna in muddy substrata in the European Wadden Sea, compared with that in areas enclosed within 1-mm mesh exclusion cages for 4 months during late summer/autumn. From Barnes (1990), data from Reise (1978)

	Abundance (no. per $0.1\,m^2$)	
	Uncaged mud	Caged mud
Hydrobia	2	10
Mya	0	87
Cerastoderma	7	1282
Spisula	0	185
Tubificoides	820	3055
Pygospio	17	350
Spio	2	50
Polydora	0	532
Malacoceros	0	405
Tharyx	7	5322
Capitella	92	140
Heteromastus	240	222
Nephtys	2	5
Eteone	0	117
Corophium	0	490
+ 14 other species	0	87
Total	1 189	12 339

appear to feed indiscriminantly on the young, post-settlement stages of infaunal molluscs, annelids and others. Food webs within the sediment are complex, however (Fig. 7.6), and some of the prey species are themselves predators. If, as argued by Ambrose (1984), the epibenthic consumers *mainly* take prey that are, in their turn, predators, the deposit-feeding infauna may actually increase in the presence of elevated densities of top consumers. Kneib (1988), for example, recorded an increase in density of the burrowing anemone *Nematostella* when numbers of the killifish, *Fundulus*, were increased. The explanation was that the fish preyed on the prawn *Palaemonetes*, itself the main consumer of *Nematostella*. Further complexity is introduced into the picture by refuges in space, such as filamentous algae or seagrass meadows, in which the foraging efficiency of epibenthic predators is greatly reduced. Refuges in time may also occur, if epibenthic species are themselves under attack from larger fish. Small wonder that many studies fail to reveal a large role for epibenthic predators in soft-sediment communities (Thrush, 1986a).

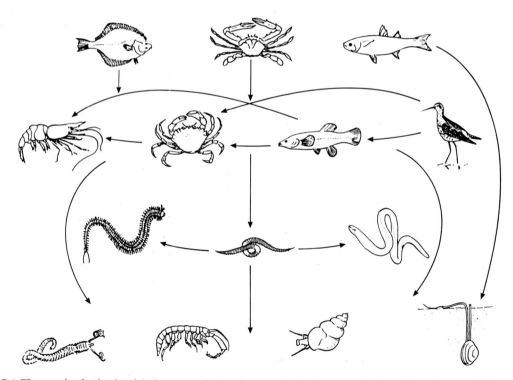

Fig. 7.6 The complex food webs of shallow areas of soft sediment in brackish or marine habitats. From Barnes (1990).

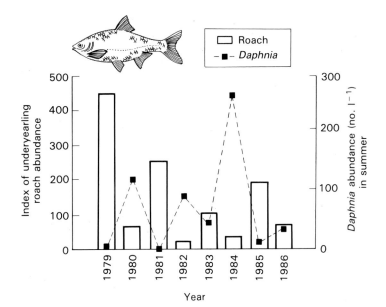

Fig. 7.7 Variation in young-of-the-year zooplanktivorous fish density (*Rutilus rutilus*) in a small lake in England. There is a clear negative relationship between fish density and density of *Daphnia*, testifying to the effect of fish predation on these vulnerable zooplankters. From Townsend (1988b).

In contrast, the impacts of predation, both direct and indirect, are often particularly obvious in lake plankton communities. Thus, a moderate density of planktivorous fish usually suppresses or excludes large-bodied zooplankton, particularly the easy-to-catch cladoceran, *Daphnia* (Fig. 7.7) (Townsend, 1988a). Since competitive interactions among zooplankton can be strong, with large *Daphnia* suppressing smaller-bodied cladocerans, copepods and rotifers (Gilbert, 1988), fish predation can indirectly benefit the smaller species. A further consequence of fish predation may be an increased biomass of phytoplankton, because the small-bodied zooplankton are generally less efficient grazers than the *Daphnia* they replace. A proper understanding of these relationships has implications for the management of culturally eutrophic lakes (those enriched by plant nutrients of human origin). We need to know what action to take to reduce excessive growth of phytoplankton in such lakes. If the biomass of phytoplankton depends mainly on the input of nitrates and phosphates (bottom-up control), the only strategy is to reduce the input of these nutrients. If, on the other hand, biomass is strongly influenced by predation of herbivores (top-down control), regulation may be effected by increasing the biomass of herbivorous zooplankton (by removing zooplanktivorous fish, or by providing zoo-

plankton with physical refuges, such as artificial weeds, which provide protection from fish).

Some intriguing examples have been seen, of the effects that predators can have on species remote from them in the food web. However, it is important to point out that these investigations almost never reveal the whole story. Take, for instance, Paine's rocky shore community. His study was restricted to macrobenthic species such as starfish, molluscs, barnacles and seaweeds. But what of the more cryptic micro- and meiofauna, or the micro-algae that grew, unobserved, on mineral substrate or on the hard outer coverings of animals, and on the epidermis of plants? It is possible that the influence of starfish on large animals and plants was paralleled by equally dramatic, but unobserved, effects on the smaller members of the biota. Similarly, planktivorous lake fish, through their effects on large zooplankton, are likely also to influence the picoplankton (bacteria and minute or small algae in the size range 0.2–2 μm), a group that has generally gone unrecognized or unmeasured. It is only recently, with the advent of the use of epifluorescence microscopy by limnologists, that the very considerable contribution made by algal picoplankton to lake biomass and productivity has been identified (Stockner, 1988). Their principle consumers are probably heterotrophic flagellates and rotifers, themselves eaten by larger zooplankton.

Marine plankton communities do not seem to be so strongly influenced by selective predation; Harris (1987) argues that physical variability and environmental patchiness are more potent forces. The same may be true of stream benthic communities, but the evidence is equivocal. Assessments of predator impacts that compare estimates of consumption rates of stream predators with standing stocks of prey generally indicate heavy use of the invertebrate fauna. However, experimental manipulations of predator densities usually have not shown strong effects on community composition (Hildrew & Townsend, 1987). Overall, the effects of biotic interactions in streams may often be overridden or masked, because of the dynamic and patchy nature of the benthic fauna, in which many species are subject to continuous redistribution and recolonization from outside (Townsend, 1989).

7.4 FLUCTUATING ENVIRONMENTS

Implicit in early considerations of the roles of competition and predation was the assumption that the environment is constant enough for competitive interactions to proceed smoothly and to culminate in exclusion, as depicted in the Lotka–Volterra model (Section 7.2). In reality, however, environmental conditions are rarely constant and often there may simply not be time for a competitively dominant species to exclude a competitively inferior species before their status is altered by a change in conditions. Hutchinson (1961) was the first to highlight this view in his consideration of the 'paradox of the plankton'. He asked how so many species of planktonic algae could coexist in the pelagic zone of a lake or ocean, environments with little apparent scope for niche differentiation, because the algae all have similar requirements for light and nutrients. Hutchinson pointed out that the environment was in a continual state of flux, with radiant energy levels, temperature and nutrient concentrations varying over time scales ranging from hours to years. Thus, competitive interactions may not proceed all the way to equilibrium, and competitive status may be changing continuously, permitting a species-rich community to persist. (In exceptional circumstances, where substantial periods of un-

changing conditions are experienced, for example in ice-covered polar lakes, a species-poor equilibrium may be observed.)

Tilman (1988) has since shown that planktonic algae are able to partition nutrient resources to a greater extent than Hutchinson imagined, and niche control has some role to play in determining species composition and richness of the phytoplankton. However, Hutchinson's view is generally closer to the truth (Harris, 1987; Reynolds, 1987).

7.5 DISTURBANCES AND THE PATCH-DYNAMICS CONCEPT

Disturbances that open up gaps are common in all kinds of aquatic community. On stream beds, for example, the turning of a stone, dislodgement of a leaf pack or scouring of sandy substratum during a spate, create patches available for recolonization by stream biota. On the sea shore, storms tear away *Laminaria* plants, leaving gaps in kelp forest canopies. Agents of disturbance on rocky shores and coral reefs include pollution events, severe wave activity during hurricanes, battering by logs or moored boats and even action by predators such a starfish cutting a swathe through a mussel bed (Section 7.3). Abiotic processes, associated with near-bed currents, and biotic processes, such as the action of burrowing animals, contribute to the environmental patchwork of soft-bottom communities of lake, salt marsh, river and ocean. The formation of gaps is of considerable importance to sedentary species with a requirement for open space on a surface. However, we can also envisage the creation of three-dimensional 'patches' (in the vicinity of a feeding fish for example) that are devoid of zooplankton and ripe for recolonization. The plankton is certainly patchy in its distribution but patches are generally more ephemeral than those on and in the bed.

An important class of models views communities as consisting of a number of patches which can be colonized, at random, by individuals of a number of species (patch-dynamics models). Competition within patches proceeds in accordance with the Lotka–Volterra equations, and results in local cases of competitive exclusion. However, the stochastic elements in the model, the patchy nature of the community and the opportunities presented each generation for dispersal between patches, enables species-rich communities to persist, without the need to invoke niche differentiation (Yodzis, 1986). Early models assumed the community to be a closed system, a single unit within which any extinction was once and for all. This naïve 'Lotka–Volterra in a bottle' view of community ecology has been superceded by the concept of a community as an open system, consisting of a mosaic of patches within which biotic interactions proceed but with a crucial extra feature — migration can occur from patch to patch. An extinction within a patch of an open system does not have to be the end of the story, because of the possibility of reinvasion from other patches.

Implicit in the patch-dynamics view is a critical role for disturbance as a reset mechanism. Disturbance is defined here as *any relatively discrete event in time that removes organisms and opens up space which can be colonized by individuals of the same or different species.*

In his patch models that permit dispersal, Yodzis (1986) distinguishes two quite different categories of community organization, according to the type of competitive relationships that occur. Communities in which some species are strongly competitively dominant are described as dominance-controlled. Those in which all species have similar competitive (and colonizing) abilities are described as founder-controlled.

7.5.1 Dominance-controlled communities

In patch-dynamics models, where some species are competitively superior to others, an initial colonizer of a patch cannot necessarily maintain itself there. If the colonizer is a poor competitor, the later arrival in the patch of a superior species will lead to the competitive exclusion of the first. Despite such local exclusion, the poor competitor may nevertheless persist in the community. This occurs because there are likely to be patches where the poor competitor exists and where its superior has yet to arrive, and by the time the latter colonizes the location, the poor competitor will have established itself in other patches.

In such dominance-controlled communities, dis-

turbances that open up gaps can lead to reasonably predictable successional sequences — early species being efficient colonizers and fast growers, in contrast to later species that can tolerate lower resource levels and mature in the presence of early species, eventually outcompeting them. The effect of the disturbance is to reset the patch to an earlier stage of succession.

In dominance-controlled communities, succession and disturbance go hand in hand. Once the competitively superior, climax species are established, no further succession will be evident unless there is a disturbance to reset the system. If the disturbance is in phase over a large area, the whole community may proceed through a synchronized successional sequence, culminating in a relatively species-poor climax. This applies to a closed, non-patchy community. The essence of dominance-controlled, patch-dynamics models is that communities are open, and disturbances are patchy in space and time, producing a species-rich mosaic of patches at different stages of succession.

The algal succession on boulders on a Californian beach involves early colonization by the ephemeral green alga *Ulva*, later colonization by several mid-successional perennial red algae and eventual domination by the climax species *Gigartina canaliculata* (see Section 7.2). Since boulders of different size are turned over by wave action (disturbed) at different frequencies, the beach as a whole consists of a variety of boulders (patches) at different successional stages. If there were no disturbances, all boulders would eventually be dominated by *G. canaliculata* and the community would have relatively few species. If all boulders were the same size, and disturbance was phased (all turned at the same time), the whole community would go through a synchronous succession, being species-poor shortly after the disturbance (only early colonists present), species-rich some time later (some early colonists persisting along with several mid-successional species, and climax species beginning to appear), and culminating in a species-poor climax (Fig. 7.8). The presence of a range of boulder sizes and the negative relationship between size and frequency of disturbance (smaller boulders generally have a higher monthly probability of movement than larger ones) ensure that the beach consists of a

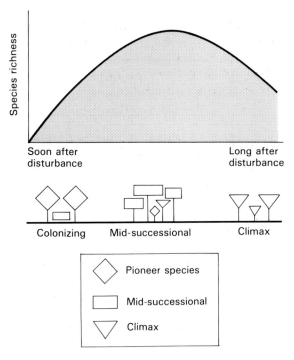

Fig. 7.8 A simplified description of algal succession on boulders on a Californian beach. Pioneer species include *Ulva* spp., mid-successional species include *Rhodoglossum affine* and climax species, *Gigartina canaliculata*. From Begon *et al.* (1990).

within individual patches were monitored to produce a transition matrix. The most common sequence of transitions was 'bare' sand to diatoms to blue-greens to *Cladophora*/blue-green mat. The sequence can probably be described as a dominance-controlled process, but whether the later species outcompeted earlier ones, or whether the 'colonist' diatoms maintained their density after the 'competitive' filamentous forms became established, is unclear. Even long after the flood, patches exposed to the most extreme hydraulic regimes were still dominated by diatoms.

Dominance control may also occur in communities of animals that are relatively immobile. Hemphill and Cooper (1983) studied the interactions between two sessile filter feeders, *Simulium virgatum* and *Hydropsyche oslari*, in a small stream in California. *Simulium* quickly colonized cleared patches on hard stream substratum, but were then gradually replaced by the competitively superior *Hydropsyche*. Natural disturbances, such as those associated with spates and floods, promoted coexistence of the two species by preventing the permanent attainment of a climax state where *Hydropsyche* monopolizes space (Fig. 7.9).

7.5.2 Founder-controlled communities

Dominance-controlled communities exhibit the familiar *r*- and *K*-selection dichotomy, in which colonizing ability and competitive status are negatively related. Conversely, all species in founder-controlled communities are both good colonists

mosaic of 'patches' at different successional stages.

In their detailed study in an Arizonan stream, Fisher *et al.* (1982) described a recovery sequence after an intense, late summer flash flood, which removed a high proportion of algal species. Changes

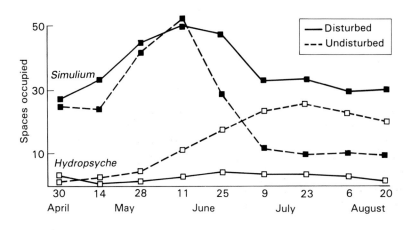

Fig. 7.9 Temporal changes in the amount of space occupied by *Simulium virgatum* and *Hydropsyche oslari* in Refugio Creek, California, in quadrats which were disturbed every 2 weeks or were left undisturbed. From Hemphill & Cooper (1983).

and equal competitors — succession is not expected in such cases. We can imagine that if many species: (i) are approximately equivalent in their ability to invade gaps opened by a disturbance; (ii) are equally tolerant of the prevailing abiotic conditions; and (iii) once established can hold the gaps against all-comers in their lifetime, then the probability of competitive exclusion is reduced in an environment where gaps appear randomly and continually. Every time an organism dies or is killed, the gap is re-opened for invasion. All conceivable replacements are possible in this 'competitive lottery' and species richness remains high.

Certain highly diverse, tropical reef fish communities seem to conform to this model (Sale, 1977; Sale & Douglas, 1984). Most species of reef fish have a bipartite life cycle, with a planktonic larval phase prior to metamorphosis and recruitment to the reef population. As an example of the founder-control mechanism, consider three species of herbivorous pomacentrid fish that co-occur on the upper slope of Heron Reef, part of the Australian Great Barrier Reef. The available space is occupied by a series of non-overlapping territories, each up to $2\,m^2$ in area. Individuals defend and hold territories throughout their juvenile and adult lives, but there is no particular tendency for space held by one of the species to be taken up, after death, by the same species. Nor is any successional sequence evident. The richness of these communities depends, at least partly, on the unpredictable supply of living space. As long as each species wins some of the time and in some places, they will continue to put larvae into the plankton and into the lottery for new sites.

7.5.3 The influence of recruitment

The patch-dynamics models outlined in Sections 7.5.1 and 7.5.2, deal with the processes occurring *within* a community mosaic and tend to focus on factors causing death within patches. However, many benthic species have a life history stage that occurs *outside* the benthic community — most stream insects have a winged, terrestrial adult stage and many marine invertebrates have a planktonic stage during which widespread dispersal and much mortality occur. The numbers surviving to recruit

into the benthic community can be highly variable. Roughgarden *et al.* (1987) refer to this as 'supply side ecology' to emphasize that, in some communities, the episodic supply of recruits has a major influence on community organization.

The causes of episodic recruitment are usually unknown. However, in a study of recruitment of the barnacle, *Balanus glandula*, Gaines and Roughgarden (1987) established that substantial numbers of larvae died as they passed through offshore kelp forests (*Macrocystis pyrifera*), and they showed that yearly variation in the kelp-forest community was correlated with yearly variation in barnacle recruitment. During the four years 1982–1985, the density of barnacles recruiting to a central Californian intertidal zone was inversely related to the areal extent of the surrounding kelp forest (Fig. 7.10). Four hypotheses might account for this pattern: (i) by exerting drag and reducing flow velocities, the kelp plants may reduce the number of barnacle recruits reaching their appropriate settlement zone; (ii) the plants may take up nutrients that are needed for larval survival; (iii) kelp forests may provide a substrate for larval settlement, depleting numbers before they reach their intertidal habitat; (iv) kelp forests may harbour predators of larvae so that the greater the extent of the forest, the more predators there will be and the fewer barnacle larvae manage to 'run the gauntlet'. Detailed study revealed that by far the most important process was

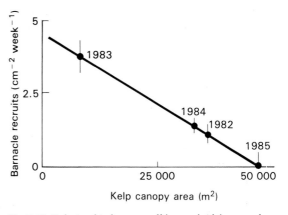

Fig. 7.10 Relationship between offshore subtidal area and recruitment, in the intertidal zone, of barnacles (*Balanus glandula*). Error bars are 95% confidence limits. From Gaines & Roughgarden (1987).

predation by juvenile fish (particularly rockfish, *Sebastes* species) which can be exceedingly abundant in the kelp, often with over 100 individuals m^{-3} near the forest margins. This study showed that the dynamics of two apparently distinct communities are strongly coupled. Subtidal kelp communities, and the processes affecting them, in turn affect intertidal communities through their impact on recruitment to the intertidal zone.

Variations in recruitment appear also to be responsible for periodic devastation of large areas of coral reef around Japan, in the Red Sea and on the Great Barrier Reef of Australia. The predator responsible is *Acanthaster planci*, the crown-of-thorns starfish. Outbreaks become apparent as large numbers of cryptic juvenile starfish mature and start to feed in the open, at about 2 years of age. Adults many always be present in small numbers, but outbreaks, resulting from unusually heavy settlement of larvae, involve an increase in density of several orders of magnitude. The 'parent' populations may be several hundred kilometres distant, and the larvae are swept onto new areas by prevailing currents (Moore, 1989).

7.5.4 The importance of mobility

The sessile communities of plants and sedentary animals on rocky shores and coral reefs are well described by patch-dynamics models (modified to include variation in recruitment where appropriate). However, patch-dynamics models cannot be applied to communities of soft-bottom substrata (in lakes, estuaries, marine mud-flats, etc.) until a further factor is taken into account. The dominant members of the infauna of these habitats generally possess considerable mobility, so that patches may be vacated or colonized at any time. The classic patch-dynamics models assume that vacation of patches is due to mortality (e.g. competitive exclusion) and colonization occurs at discrete intervals (usually as a result of reproduction and dispersal from other patches).

The benthic molluscs and polychaete worms that dominate many soft-bottom communities, retain the ability to move, as adults, laterally through the mud, across the surface or even in the water column (Thrush, 1986b). Coastal salt marshes pro-

vide a region of intertidal mud and sand containing marked spatial heterogeneities, often formed by biological disturbances associated with burrowing and predation. The competitive abilities of species vary between patch types, as does predation pressure (Frid & James, 1988a,b). This infauna possesses considerable mobility, and the continual redistribution of individuals between patches promotes species richness by providing refuges from competition and predation (Frid & Townsend, 1989).

In streams, the fauna is composed mainly of mobile grazers, detritivores and predators, which are continuously redistributed over the benthos, particularly via 'invertebrate drift' (the movement of large numbers of benthic animals, swept off the substrate in the current and deposited a short distance downstream) (Townsend, 1980 and see Chapter 12). This situation is analogous to that described for soft-bottom communities. Both communities are patchy and dynamic, but mobility of the biota is much greater than is assumed in the classic models. The communities may be described as mobility-controlled (Townsend, 1989).

7.6 THE RELATIVE ROLES OF COMPETITION, PREDATION AND DISTURBANCE

It would be wrong to replace the monolithic view of community organization, prevalent in the 1960s and 1970s (the overriding importance of competition and niche differentiation), with another, equally monolithic view (the overriding importance of temporal variation and disturbance). Communities structured by competition are not the general rule, but neither, necessarily, are communities structured by any other single agency (see reviews by Connell, 1983; Schoener, 1983; Sih *et al.*, 1985). Even in the same community, some patches may be dominated by dominance-controlled, and others by founder-controlled or mobility-controlled processes (Townsend, 1989). For example, the community dynamics of sessile species in streams, such as algae, macrophytes and retreat-building invertebrates, depends upon the supply of attachment space and may be described well by dominance-controlled dynamics. On the other hand, organization of the non-sessile component of the stream benthic community is likely to be closer to the mobility-controlled case.

Most communities are probably organized by a mixture of forces — competition, predation and disturbance — but their relative importance depends on a combination of circumstances. Menge and Sutherland (1976; 1987), for example, argue that the relative importance of physical and biotic factors varies along a gradient of environmental harshness (a feature that may be easy to imagine but difficult to define). Thus, in 'stressful' environments, predators have little effect, because they are absent or inactive, and competition for space is prevented because harsh conditions keep densities low. Both the mobile and sessile components of the community are structured by physical factors. In more moderate environments, predators are still considered to be ineffective (predators are assumed to be more strongly affected by physical conditions than species lower in the food web), but sessile organisms occur at high densities, leading to competition for space. Finally, in benign environments, predators keep their prey at low density and prevent competition for space (Fig. 7.11). Menge and Sutherland's model seems close to reality, at least in the case of sessile communities of rocky intertidal zones, for which it was developed. Here the predators are mobile and more susceptible to harsh conditions, associated with storms and wave action, than their sessile prey.

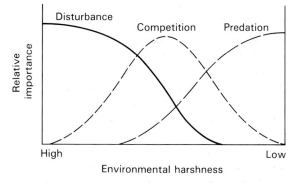

Fig. 7.11 Hypothetical relationship between environmental harshness and the relative importance of disturbance, competition and predation in structuring communities of sessile organisms in the rocky intertidal zone. From Menge & Sutherland (1976; 1987).

Physical disturbance can certainly modify competitive and predator–prey interactions that would otherwise play a dominant role (as will already be evident from several studies cited in Section 7.5). McAuliffe (1984) provides a good example of how physical disturbance in a North American stream influences competitive interactions between shelter-building larvae of the caddisfly *Leucotrichia pictipes*, the moth *Parargyractis confusalis* and the chironomid *Eukiefferiella*. *Leucotrichia* seems capable of establishing a competitive monopoly in the absence of disturbance; it is aggressive in defence of the territory around its silken retreat, and it has the further advantage that its retreats persist from year to year and can be occupied by individuals of subsequent generations. Thus, *Leucotrichia* gets a head start over *Parargyractis*, whose larvae always build their retreats afresh. Disturbances that cause stones to overturn, making previous generations' *Leucotrichia* retreats unavailable, allow the two species to coexist (Fig. 7.12). Another kind of disturbance is responsible for the coexistence of *Eukiefferiella* with the dominant *Leucotrichia*. Low stream discharge, late in the summer, causes the surfaces of some stones to dry out, and *Leucotrichia* larvae perish in retreats that they cannot abandon. This permits the establishment of *Eukiefferiella*, a species with several generations per year, and the ability to exploit the ephemeral resource of stone surfaces in shallow water.

Meffe (1984) studied the predator–prey interaction between an introduced mosquitofish (*Gambusia affinis*) and a native topminnow (*Poeciliopsis occidentalis*) that it has eliminated from most streams in Arizona. Cases of apparent coexistence are all in habitats that experience severe spates, and Meffe showed that coexistence occurs because the prey species is less susceptible to removal by spates than the predator. Unlike *Gambusia*, *Poeciliopsis* responds rapidly at the onset of a spate and its behaviour allows it to maintain station throughout the period of high discharge (Fig. 7.13).

We noted earlier how disturbances associated with spates in streams can set up an algal succession that commences with a thin layer of diatoms, and develops into a thick mat with an overstorey of filamentous forms. Disturbance, colonization and competition between algae all seem to play import-

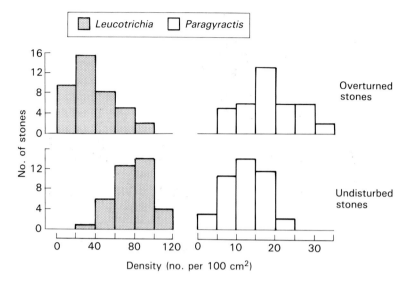

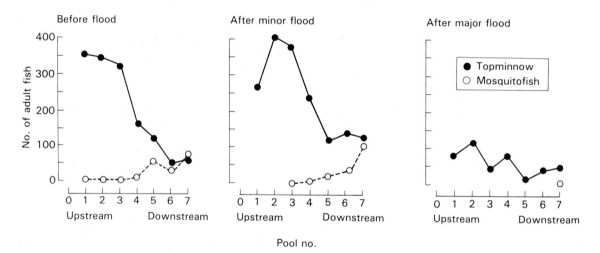

Fig. 7.12 The frequency distributions of densities of *Leucotrichia* and *Parargyractis* on 80 stones selected by McAuliffe (1984), in a stream in Montana. Forty were overturned in October and 40 were undisturbed controls. The densities were measured in the August, after colonization.

Fig. 7.13 Numbers of adults (catch-per-unit-effort) of the topminnow, *Poeciliopsis*, and mosquitofish, *Gambusia*, in each of seven pools in Sharp Spring, Arizona, before and after floods. From Meffe (1984).

ant roles. In studies using replicated stream channels, DeNicola *et al.* (1990) showed how predation (grazing) can modify the process. In the presence of larvae of the caddisfly *Dicosmoecus gilvipes*, which scrapes algae from the substratum using its mandibles and tarsal claws, algal biomass stayed low and the community remained dominated by early successional diatoms. Another herbivore, the snail *Juga silicula*, which rasps the substrate with a radula, had a similar effect on algal biomass and

composition, but tended to graze substrata less uniformly, producing a fine-scale mosaic of patches at different successional stages.

In an elegant series of experiments, Wilbur (1987) investigated the relative roles and interactions amongst competition, predation and disturbance in amphibian communities in North American ponds. Hatchlings of four species of frog and toad (*Rana utricularia*, *Scaphiopus holbrooki*, *Bufo americanus* and *Hyla chrysoscelis*) were introduced to replicate

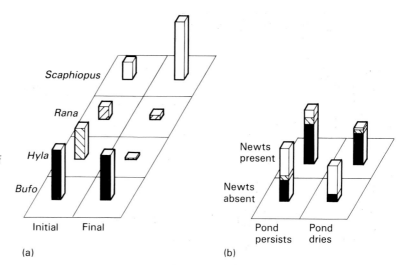

Fig. 7.14 (a) Relative abundance of tadpoles of each of four species introduced at high density into ponds (initial) and the relative abundances of metamorphs (final) at the end of the experiment. (b) Number of metamorphs of four species in the absence and presence of predatory newts and in ponds which persist or which dry up 100 days into the experiment. From Wilbur (1987).

experimental ponds, at high density. The relative abundance through the course of the experiment changed in a consistent manner — *Scaphiopus* showed a 353% increase in relative abundance, *Rana* a 60% decrease and *Hyla* a large reduction of 93%. Thus, *Scaphiopus* tadpoles were normally competitively dominant and *Hyla* had a very low competitive status (Fig. 7.14a). The presence of predatory salamanders, *Notophthalmus viridescens*, did not alter the total number of individuals reaching metamorphosis, but relative abundance was shifted as *Scaphiopus*, the competitive dominant, was selectively eaten (Fig. 7.14b). These results demonstrate elements of predator-mediated coexistence, as discussed in Section 7.3. Finally, Wilbur subjected his high density tadpole communities, in the presence and absence of predators, to water loss, to simulate a natural drying regime (disturbance). The effect of competition was to slow growth, to retard the time of metamorphosis and thus to increase the risk of desiccation in drying ponds. *Scaphiopus* had the shortest period of larval life and constituted a greater proportion of metamorphs at the end of the experiment in which ponds dried. The presence of predators ameliorated the effects of competition, allowing surviving tadpoles of several species to grow rapidly enough to metamorphose before ponds dried up.

Although these experiments were necessarily artificial, they simulate various aspects of amphibian communities in ponds quite faithfully, and it is reasonable to conclude that the interactions shown between disturbance, competition and predation often act in nature to produce the community structure we observe.

7.7 COMMUNITY STABILITY AND THE QUESTION OF SCALE

No community is the homogenous, constant system visualized in basic Lotka–Volterra mathematical models and exemplified by simplistic laboratory microcosms. In most real communities, population dynamics will have a strong spatial element (making patch-dynamics models appropriate) and temporal variation will be present, providing a variety of ways in which the likelihood of coexistence is enhanced and diversity is increased. In the cases presented, there have been many different examples of communities that conform reasonably closely with assumptions built into non-equilibrium models (those incorporating varying conditions, such as envisaged in the paradox of the plankton, or in founder-controlled or dominance-controlled patch dynamics). Such models provide an insight into the reasons that species-rich communities persist in nature and, as Chesson (1986) has pointed out, coexistence under a stochastic, non-equilibrium mechanism can be just as strong and stable as that occurring in an equilibrium, niche-differentiation

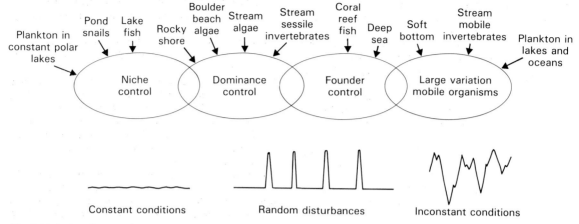

Fig. 7.15 A classification of some aquatic communities discussed in this chapter according to models of community organization and aspects of environmental variability.

model. Although communities in nature cannot readily be slotted into pigeon-holes, defined by factors responsible for their organization, a tentative classification of aquatic communities is presented in Fig. 7.15.

In a closed system, made up of a single patch, extinctions may occur for two very different reasons: (i) as a result of biotic instability associated with competitive exclusion, predator–prey overexploitation and other strongly destabilizing species interactions; and (ii) as a result of environmental instability caused by unpredictable disturbances and changes in conditions. DeAngelis and Waterhouse (1987) have emphasized that by integrating unstable patches of either type into the open system of large landscape (made up of many patches out of phase), persistent species-rich communities can result. This is the important message to emerge from a patch-dynamics perspective (Pickett & White, 1985).

FURTHER READING

Ambrose, W.G. (1984) Role of predatory infauna in structuring marine soft-bottom communities. *Mar. Ecol. Progr. Ser.* **17**: 109–15.

Barnes, R.S.K. (1988) The faunas of land-locked lagoons: chance differences and the problems of dispersal. *Est. Coast. Shelf Sci.* **26**: 309–18.

Barnes, R.S.K. (1991) Macrofaunal community structure and life histories in coastal lagoons. In: Kjerfve, B. (ed) *Coastal Lagoon Processes*. Elsevier, Amsterdam. (In press).

Begon, M., Harper, J.L. & Townsend, C.R. (1990) *Ecology: Individuals, Populations and Communities*. 2nd edn. Blackwell Scientific Publications, Oxford.

Chesson, P.L. (1986) Environmental variation and the co-existence of species. In: Diamond, J.M. & Case, T.J. (eds) *Community Ecology*, pp. 240–56. Harper & Row, New York.

Cody, M.L. & Diamond, J.M (eds) (1975) *Ecology and Evolution of Communities*. Belknap, Cambridge, Massachusetts.

Connell, J.H. (1961) The influence of interspecific competition and other factors on the distribution of the barnacle *Chthamalus stellatus*. *Ecology* **42**: 710–23.

Connell, J.H. (1983) On the prevalence and relative importance of interspecific competition: evidence from field experiments. *Am. Nat.* **122**: 661–96.

DeAngelis, D.L. & Waterhouse, J.C. (1987) Equilibrium and nonequilibrium concepts in ecological models. *Ecol. Monogr.* **57**: 1–21.

DeNicola, D.M., McIntire, C.D., Lamberti, G.A., Gregory, S.V. & Ashkenas, L.R. (1990) Temporal patterns of grazer-periphyton interactions in laboratory streams. *Freshwater Biol.* **23**: 475–90.

Diamond, J.M. & Case, T.J. (eds) (1986) *Community Ecology*. Harper & Row, New York.

Estes, J.A., Jameson, R.J. & Rhode, E.B. (1982) Activity and prey selection in the sea otter: influence of population status on community structure. *Am. Nat.* **120**: 242–58.

Fenchel, T. & Kolding, S. (1979) Habitat selection and distribution patterns of five species of the amphipod genus *Gammarus*. *Oikos* **27**: 367–76.

Fisher, S.G., Gray, L.J., Grimm, N.B. & Busch, D.E. (1982) Temporal succession in a desert stream following flooding. *Ecol. Monogr.* **52**: 93–110.

Frid, C.L.J. & James, R. (1988a) Interactions between two species of saltmarsh gastropod: *Hydrobia ulvae* and *Littorina littorea*. *Mar. Ecol. Progr. Ser.* 43: 173–9.

Frid, C.L.J. & James, R. (1988b) The role of epibenthic predators in structuring the marine invertebrate community of a British coastal salt marsh. *Neth. J. Sea Res.* 22: 307–14.

Frid, C.L.J. & Townsend, C.R. (1989) An appraisal of the patch dynamics concept in stream and marine benthic communities whose members are highly mobile. *Oikos* 53: 137–41.

Gaines, S.D. & Roughgarden, J. (1987) Fish in offshore kelp forests affect recruitment to intertidal barnacle populations. *Science* 235: 479–81.

Gee, J.H.R. & Giller, P.S. (eds) (1987) *Organization of Communities, Past and Present*. Blackwell Scientific Publications, Oxford.

Gilbert, J.J. (1988) Suppression of rotifer populations by *Daphnia*: A review of the evidence, the mechanisms, and the effects on zooplankton community structure. *Limnol. Oceanogr.* 33: 1288–303.

Grace, J.B. & Wetzel, R.G. (1981) Habitat partitioning and competitive displacement in cattails (*Typha*): experimental field studies. *Am. Nat.* 118: 463–74.

Harman, W.N. (1972) Benthic substrates: their effect on freshwater molluscs. *Ecology* 53: 271–2.

Harris, R.P. (1987) Spatial and temporal organization in marine plankton communities. In: Gee, J.H.R. & Giller, P.S. (eds) *Organization of Communities, Past and Present*, pp. 327–46. Blackwell Scientific Publications, Oxford.

Hemphill, N. & Cooper, S.D. (1983) The effect of physical disturbance on the relative abundances of two filter-feeding insects in a small stream. *Oecologia* 58: 378–82.

Hildrew, A.G. & Edington, J.M. (1979) Factors facilitating the coexistence of hydropsychid caddis larvae (*Trichoptera*) in the same river system. *J. Animal Ecol.* 48: 557–76.

Hildrew, A.G. & Townsend, C.R. (1987) Organization in freshwater benthic communities. In: Gee, J.H.R. & Giller, P.S. (eds) *Organization of Communities, Past and Present*, pp. 347–71. Blackwell Scientific Publications, Oxford.

Hutchinson, G.E. (1961) The paradox of the plankton. *Am. Nat.* 95: 137–45.

Kerfoot, W.C. & Sih, A. (1987) *Predation: Direct and Indirect Impacts in Aquatic Communities*. University Press of New England, Dartmouth, New Hampshire.

Kneib, R.T. (1988) Testing for indirect effects of predation in an intertidal soft-bottom community. *Ecology* 69: 1795–805.

Kolding, S. & Fenchel, T. (1979) Coexistence and life-cycle characteristics of five species of the amphipod genus *Gammarus*. *Oikos* 33: 323–7.

Lotka, A.J. (1925) *Elements of Physical Biology*. Williams & Wilkins, Baltimore.

Lubchenco, J. (1978) Plant species diversity in a marine intertidal community: importance of herbivore food preference and algal competitive abilities. *Am. Nat.* 112: 23–39.

McAuliffe, J.R. (1984) Competition for space, disturbance, and the structure of a benthic stream community. *Ecology* 65: 894–908.

Meffe, G.K. (1984) Effects of abiotic disturbance on coexistence of predator–prey fish species. *Ecology* 65: 1525–34.

Menge, B.A. & Sutherland, J.P. (1976) Species diversity gradients: synthesis of the roles of predation, competition, and temporal heterogeneity. *Am. Nat.* 110: 351–69.

Menge, B.A. & Sutherland, J.P. (1987) Community regulation: variation in disturbance, competition, and predation in relation to environmental stress and recruitment. *Am. Nat.* 130: 730–57.

Miller, D.J. (1989) Introductions and extinction of fish in the African Great Lakes. *Trends Ecol. Evol.* 4: 56–9.

Moore, R.J. (1989) A hit-and-run reef vandal: progress in understanding the crown-of-thorns problem. *Trends Ecol. Evol.* 4: 36–7.

Paine, R.T. (1966) Food web complexity and species diversity. *Am. Nat.* 100: 65–75.

Pickett, S.T.A. & White, P.S. (eds) (1985) *The Ecology of Natural Disturbance as Patch Dynamics*. Academic Press, New York.

Power, M.E. & Stewart, A.J. (1987) Disturbance and recovery of an algal assemblage following flooding in an Oklahoma stream. *Am. Nat.* 117: 333–45.

Reise, K. (1978) Experiments on epibenthic predation in the Wadden Sea. *Helgolander Wiss. Meeresunters.* 31: 55–101.

Reise, K. (1985) *Tidal Flat Ecology*. Springer-Verlag, Berlin.

Reynolds, C.S. (1987) Community organization in the freshwater plankton. In: Gee, J.H.R. & Giller, P.S. (eds) *Organization of Communities, Past and Present*, pp. 297–326. Blackwell Scientific Publications, Oxford.

Roughgarden, J., Gaines, S.D. & Pacala, S.W. (1987) Supply-side ecology: the role of physical transport processes. In: Gee, J.H.R. & Giller, P.S. (eds) *Organization of Communities, Past and Present*, pp. 491–518. Blackwell Scientific Publications, Oxford.

Sale, P.F. (1977) Maintenance of high diversity in coral reef fish communities. *Am. Nat.* 111: 337–59.

Sale, P.F. & Douglas, W.A. (1984) Temporal variability in the community structure of fish on coral patch reefs and the relation of community structure to reef structure. *Ecology* 65: 409–22.

Scheibling, R.E. (1986) Increased macroalgal abundance following mass mortalities of sea urchins (*Strongylocentrotus droebachiensis*) along the Atlantic coast of Nova Scotia. *Oecologia* 68: 186–98.

Schoener, T.W. (1983) Field experiments on interspecific competition. *Am. Nat.* 122: 240–85.

Sih, A., Crowley, P., McPeek, M., Petranka, J. & Strohmeier, K. (1985) Predation, competition and prey communities: A review of field experiments. *Ann. Rev. Ecol. Syst.* 16: 269–311.

Smith, C.R. & Hessler, R.R. (1987) Colonization and succession in deep-sea ecosystems. *Trends Ecol. Evol.* **2**: 359–63.

Sousa, W.P. (1979a) Experimental investigation of disturbance and ecological succession in a rocky intertidal algal community. *Ecol. Monogr.* **49**: 227–54.

Sousa, W.P. (1979b) Disturbance in marine intertidal boulder fields: the nonequilibrium maintenance of species diversity. *Ecology* **60**: 1225–39.

Stockner, J.G. (1988) Phototrophic picoplankton: An overview from marine and freshwater ecosystems. *Limnol. Oceanogr.* **33**: 765–75.

Strong, D.R., Simberloff, D., Abele, L.G. & Thistle, A.B. (eds) (1984) *Ecological Communities: Conceptual Issues and the Evidence.* Princeton University Press, New Jersey.

Thrush, S.F. (1986a) Community structure on the floor of a sea-lough: are large epibenthic predators important? *J. Exp. Mar. Biol. Ecol.* **104**: 171–83.

Thrush, S.F. (1986b) Spatial heterogeneity in sub-tidal gravel generated by the pit digging activities of *Cancer pagurus*. *Mar. Ecol. Progr. Ser.* **30**: 221–7.

Tilman, D. (1988) *Plant Strategies and the Dynamics and Structure of Plant Communities.* Princeton University Press, New Jersey.

Tonn, W.M. & Magnuson, J.J. (1982) Patterns in the species composition and richness of fish assemblages in northern Wisconsin lakes. *Ecology* **63**: 1149–266.

Townsend, C.R. (1980) *The Ecology of Streams and Rivers.* Edward Arnold, London.

Townsend, C.R. (1988a). Fish, fleas and phytoplankton. *New Scientist* **118**: 67–70.

Townsend, C.R. (1988b) Population cycles in freshwater fish. *J. Fish Biol.* **35** (suppl. A): 125–31.

Townsend, C.R. (1989) The patch dynamics concept of stream community ecology. *J. Am. Benth. Soc.* **8**: 36–50.

Volterra, V. (1926) Variations and fluctuations of the numbers of individuals in animal species living together. In: Chapman, R.N. *Animal Ecology.* McGraw-Hill, New York. (Reprinted in 1931.)

Wilbur, H.M. (1987) Regulation of structure in complex systems: experimental temporary pond communities. *Ecology* **68**: 1437–52.

Yodzis, P. (1986) Competition, mortality and community structure. In: Diamond, J.M. & Case, T.J. (eds) *Community Ecology*, pp. 480–91. Harper & Row, New York.

8
Reproduction, Life Histories and Dispersal

R.S.K. BARNES

8.1 INTRODUCTION

An almost bewildering variety of reproductive and life-history patterns are displayed by aquatic organisms, and a huge and growing body of literature is devoted to an attempt to explain this diversity in terms of selective advantage under specific ecological circumstances. Thus, for example, poor conditions for juvenile growth are believed to select in favour of the investment of parental resources in the offspring, and the production of few offspring per breeding episode; whilst poor conditions for juvenile survival select in favour of long adult life and repeated reproduction (Sibly & Calow, 1985; and see Begon, 1985).

It is clear, however, that a number of the patterns of life history observed in organisms cannot be explained solely in terms of their ecology, physiology or any other attribute of their current biology. Such features have been constrained by the evolutionary past of the organisms. This is enshrined in the 'principle of historicity': organisms do whatever has contributed to the reproductive success of their ancestors (see Ghiselin, 1987). In some cases, the constraint appears simply as a phylogenetic one. All nereid polychaetes (ragworms) and most coleoid cephalopods (squid, cuttlefish and octopi), for instance, whatever their lifestyles and habitats, die after their one and only breeding episode. The animals may be minute or huge, live long lives or short, be pelagic or benthic, be at the top of the food web or much lower down, occupy habitats of apparent resource abundance or of scarcity; all are programmed to die after having bred and they pass this programme on to their descendents.

Alternatively, constraints may have arisen during evolutionary transitions from one habitat type to another. Several patterns shown by freshwater organisms, for example, would appear to be more a result of adaptation by their ancestors to the terrestrial environment, than of any necessity associated with their present aquatic one.

The interplay of past and present selective forces, of greater or lesser constraints and of evolutionary flexibility or rigidity of response, has therefore resulted in a broad spectrum of strategies, and biologists have responded with as wide a range of attempts at explanation. Although new ideas are appearing faster than they can be assimilated, we are still far from understanding fully the answers to such fundamental questions as: 'why be diploid?'; 'why be sexual?'; 'why reproduce whilst still larval in form (i.e. be paedomorphic)?'; and 'what are the evolutionary consequences of isolating the germ line from the soma?'. In this chapter, we limit our consideration to those aspects of the life cycle that seem to be specifically associated with the aquatic habitat. In particular, we will investigate two lifestyles virtually impossible on land: that of the sessile phagotroph (attached benthic consumers of preformed, particulate organic compounds) and that of organisms or life-history stages that are suspended in the mobile fluid part of the aquatic ecosystem. As emphasized above, however, it is necessary to begin by considering the evolutionary past of such organisms.

8.2 THE EVOLUTIONARY PERSPECTIVE

Eukaryotic life is thought to have originated in the form of single-celled, free-living protists, probably as a result of a series of symbiotic unions of various Precambrian prokaryotes (Margulis, 1970). Judg-

145

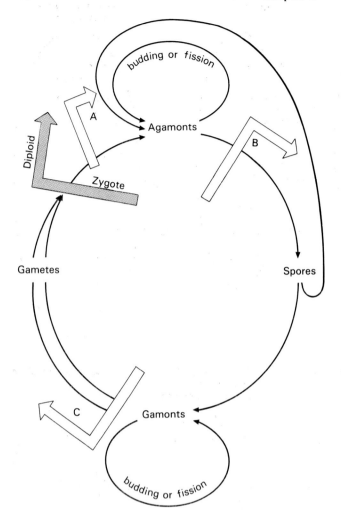

Fig. 8.1 Generalized life cycle of meiotic eukaryotes. A, B and C are alternative positions in the cycle for meiosis and initiation of the haploid phase. Note that gametes are haploid, but that meiosis may restore the haploid state at any point in the cycle, after formation of the zygote and before the production of more gametes — thus, agamonts, spores and gamonts may either be haploid or diploid.

ing from the array of living groups known, the most basic form of life cycle is — and presumably was — growth followed by simple binary fission, without any sexual processes. The large majority of unicellular protist groups have remained haploid and asexual, although a number of others do release gametes that create, at least temporarily, a diploid zygote on fusion. In almost all such forms, the male and female gametes are effectively identical; both being flagellated for example.

The presence of both haploid and diploid phases in the life cycle has been elaborated much further in several phylogenetic lines, especially in those that are characterized by a sessile multicellular stage or stages. The same lines also tend towards differ-

entiation of the gametes, the female one ranging from being larger than the male, although still mobile, to being very large and immobile. A generalized life cycle of one of these multicellular protists (Fig. 8.1) is of the form: (i) haploid individuals mitotically produce gametes (cells or 'spores' capable of fusing with another such cell or 'spore'); (ii) male and female gametes fuse and create a diploid zygote; (iii) each diploid zygote grows (mitotically) into a diploid multicellular individual, which then (iv) meiotically produces haploid spores (cells that do not fuse with other such spores); and (v) each haploid spore grows (mitotically) into a haploid multicellular individual to complete the cycle. There is thus an alternation of asexual haploid and

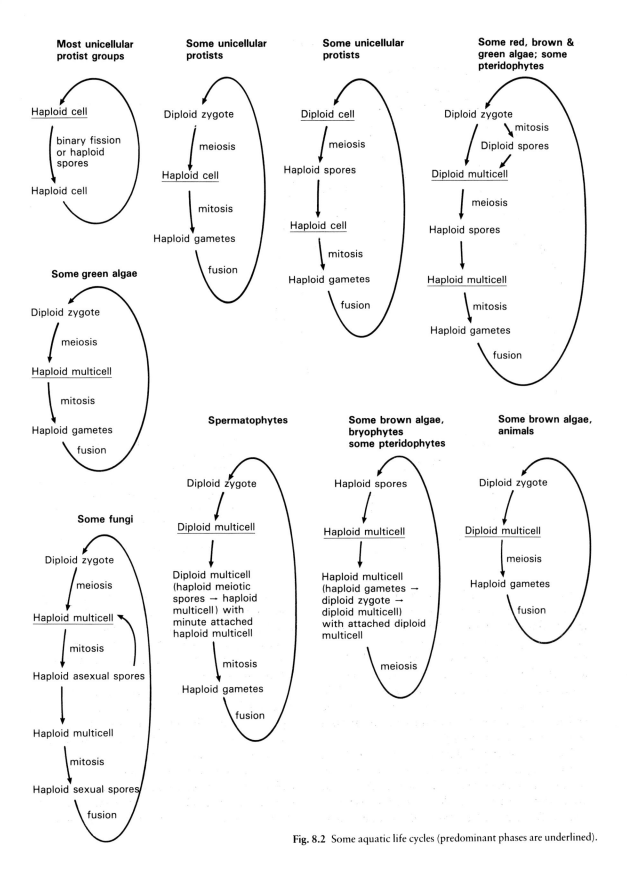

Fig. 8.2 Some aquatic life cycles (predominant phases are underlined).

sexual diploid generations, although additional phases of asexual multiplication by either or both of the haploid and diploid stages may be inserted into the cycle.

A number of variants on this basic pattern involve the reduction in importance of one of the two generations, sometimes down to a small reproductive structure borne on part of the dominant multicellular stage: the haploid form dominates the life history of many fungi and the bryophytes; the diploid form dominates that of the tracheophytes and of other fungi (the dikaryophase of the basidiomycete fungi is effectively diploid). In yet other lines, one of the two generations has been lost or, in the case of the diploid, was never elaborated. In some green algae and many fungi, for example, the gametes fuse to form a diploid zygote in the classic manner, but this zygote immediately undergoes meiosis to give rise to haploid cells (just as in the unicellular, haploid, sexual protists), which then form the multicellular seaweed or fungus, and eventually, by mitosis, the gametes. In the fucoid wracks and in all animals, the haploid phase has been reduced right down to the gametes: the diploid multicellular organism produces haploid gametes by meiosis, and by fusing, these haploid cells then reconstitute the diploid individual (Fig. 8.2).

The ancestral animals, which originated in the sea, are likely to have displayed the following reproductive characteristics: first, they would have possessed lineages of cells from which both gametes and specialized somatic cells could differentiate as required (totipotent cells), or else a system in which somatic cells could dedifferentiate to form the gametes. In other words, there was no separation of a 'germ line' in early ontogeny. Thus, the ancestral animals could multiply asexually by budding or fission: isolated regions of the body could generate complete new individuals (individuals, that is, in the somatic not genetic sense). This pattern is observed in several surviving aquatic phyla (especially in the sponges, cnidarians, lophophorates and in some primitive groups within the flatworms, annelids and deuterostomes) (Table 8.1). Most descendent groups of animals, however, have their gamete-producing cells isolated from the other body cells in discrete gonads, and they are therefore unable to multiply by the ancestral form of asexual

division (Table 8.1). This has marked repercussions on the phyletic distribution of the ability to form multicellular, resistant propagules (overwintering and/or dispersal bodies) and to produce modular clonal colonies (Buss, 1987) — both are processes that rely on asexual multiplication by budding (Section 8.3). (Animals that can bud asexually can also regenerate lost sections of the body and usually show a regulative (= indeterminate) pattern of development.)

Secondly, the ancestral animals were capable of sexual reproduction as well as asexual fragmentation. Most of their descendents have remained sexual; however, some of them, being denied fission when adult because of sequestration of the germ line, have evolved various new, secondary forms of asexual multiplication based on the production of normal or aberrant gametes ('gametogenic asexual reproduction') (Table 8.1). Others undergo fission before the germ line has been isolated ('polyembryony'). A wide variety of forms of gametogenic asexual processes have evolved. In the common 'apomixis', as seen in bdelloid rotifers, tardigrades, digenean flukes, nematodes, cladocerans, etc., diploid eggs are produced, which develop without the need for fertilization into further apomictic females. In the less common 'arrhenotoky', haploid gametes are generated, which on fertilization develop into only (diploid) females; if ova are unfertilized, they develop into haploid males (in the monogonont rotifers the arrhenotokic females then pass through apomictic cycles) (Fig. 8.3). The eggs of apomictic females are often adapted to form resistant stages in a parallel manner to the asexual propagules formed by budding or fission. Polyembryony is known in the cnidarians, monogenean flukes, entoprocts and bryozoans, amongst others (Table 8.1). In the cyclostome bryozoans, for example, the primary blastula buds off secondary embryos, which in turn bud off tertiary ones, so that more than 100 clonal embryos may develop from each zygote.

Thirdly, it appears likely that either the ancestral female animals produced small ova that were discharged from the body into the surrounding water, where fertilization occurred (as in the ancestors of the Bilateria), or else only the sperm were shed freely into the water to enter the body of a nearby

Table 8.1 The distribution, where known, of forms of asexual multiplication through the animal kingdom. From Nieuwkoop & Sutasurya (1981), Hughes (1989) and others

	Via body fragments or multicellular propagules	Gametogenically	Via polyembryony	Sequestration of germ line during early development
Porifera	Common	—	—	—
Placozoa	Universal	—	—	—
Mesozoa	—	—	Universal	√
Cnidaria	Common	—	Rare	—
Ctenophora	Uncommon	—	—	—
Platyhelminthes				
Trematoda	—	Universal	?	√
Others	Common	Rare	Rare	—
Nemertea	Rare	Rare	—	—
Gastrotricha	—	Common	—	√
Gnathostomula	—	—	—	—
Kinorhyncha	—	—	—	—
Priapula	—	—	—	—
Nematomorpha	—	—	—	—
Nematoda	—	Rare	—	√
Rotifera	—	Common	—	√
Acanthocephala	—	—	—	√
Entoprocta	Universal	—	Rare	—
Mollusca	—	Rare	—	In some
Sipuncula	Rare	—	—	?
Echiura	—	—	—	—
Annelida				
Hirudinea	—	—	—	√
Others	Uncommon	Rare	Rare	—
Pogonophora	Rare	—	—	?
Tardigrada	—	Rare	—	√
Pentastoma	—	—	—	?
Onychophora	—	Rare	—	In some
Chelicerata	—	Rare	—	In some
Crustacea	—	Uncommon	—	In some
Uniramia	—	Uncommon	Rare	In some
Phorona	Common	—	—	—
Bryozoa	Common	—	Uncommon	—
Brachiopoda	—	—	—	?
Hemichordata	Common	—	—	—
Echinodermata	Rare	Rare	—	—
Chordata	Rare	Rare	Rare	In some
Chaetognatha	—	—	—	√

√ Occurs; — does not occur; ? not definately known.

individual, where they fertilized the retained ova (as in the sponges). In either event, it is widely believed that the zygotes then developed into small, ciliated larval stages, which spent some time swimming and feeding as part of the plankton. External fertiliz-ation requires the synchronous discharge of relatively large numbers of small gametes, and Olive (1985) has argued that animals with these reproductive characteristics must have bred only when relatively large, because only large individuals

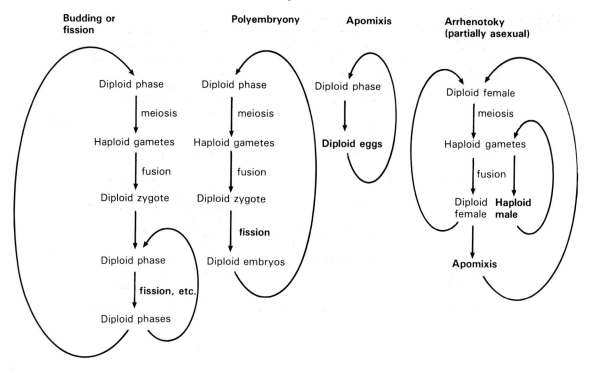

Fig. 8.3 Animal life cycles involving primary or secondary means of asexual multiplication (asexual stages or processes in bold).

would have had the available internal space in which to accumulate sufficient gametes for such mass spawning.

Many descendent lines have modified this ancestral condition quite markedly, especially the paedomorphic ones and those that colonized the land. Small animals, including those originating from larval stages, have a restricted body space in which to store gametes, and species inhabiting the interstitial pore-water spaces within sediments, for example, tend to share a common suite of features, in which internal fertilization via copulation, the production of small numbers of relatively large ova, frequent breeding and the lack of a free-living larva are predominant; they may also possess sacs in which to store sperm, so that a single copulation will suffice to fertilize several successive batches of eggs. Lacking larval stages, development is 'direct', with a small version of the adult hatching from the egg.

Terrestrial species, regardless of their size, show exactly the same set of characteristics, so that even large land animals tend towards the production of

small broods at frequent intervals, even though the size constraint mentioned above cannot be operating. We can either regard the similarity of reproductive features of small aquatic and land animals as coincidental — it just so happens that the land requires the same adaptations as the microhabitats of minute species — or it may be that the terrestrial fauna are the descendents of small interstitial marine species that invaded the land via the soil/groundwater system. The latter certainly appears to be one of the more important routes onto land (see Little, 1983; 1990).

Once lost, it seems that the ancestral ciliated larval stage could not be re-acquired, so that any later selective pressures in favour of juvenile adaptations, markedly different from those of the adult, has led, in the crustaceans and some insects, to the evolution of new, secondary types of larva. It is this evolutionary loss that is likely to be behind the lack of a free-swimming larval stage in many freshwater organisms. Although it is often stated that freshwater aquatic species lack larvae as an adaptation to the one-way flow system typifying most con-

Table 8.2 Ancestral and derived reproductive characters in aquatic animals

Character	Ancestral condition	Derived conditions
Germ line	Not isolated	Isolated
Sexuality	Gonochoristic	Hermaphroditic
Sperm	Round-headed	Other forms, including in spermatophores
Ova	Small	Large
	Without yolk	With yolk
Zygotes	Planktonic	Brooded, in capsule, cocoon, etc.
Gamete discharge	Mass discharge after storage	Many minor discharges without mass accumulation
Fertilization	External	Internal, sperm stored by recipient
Development	Hollow blastula	Solid blastula
	Gastrulation by invagination	Gastrulation by other means
	Via ciliated, free-living planktotrophic larva	Via ciliated, free-living lecithotrophic larva, non-ciliated larva, larval stage passed within egg, or direct development

tinental waters (any larval stage would simply be displaced downstream), or because of the need of their young to hatch with fully developed osmoregulatory capabilities (Lopez, 1988); it is also the case that most types of animal that have invaded fresh water via the entirely aquatic route (e.g. sponges, cnidarians, bryozoans and bivalve molluscs), have retained their ancestral marine life-history system largely unchanged, including, where appropriate, the presence of a free-swimming larva (albeit a precocious and somewhat specialized one). It is only those forms that arrived overland that most obviously lack their ancestral larva (e.g. the pulmonate molluscs and the clitellate annelids). Many of the animals that dominate fresh waters — the insects and pulmonates, for instance — have even remained air-breathers like their terrestrial relatives.

Sequestration of the germ line, direct development, copulation, the production of spermatophores and gametogenic asexual multiplication are therefore all derived, secondary characteristics of animal life cycles (Table 8.2); as, indeed, is the genetically distinct 'individual', because the ancestral forms and many of their descendents were and

are colonies of separate or partially-separated clones. Such modular colonies are characteristic of benthic animals in all types of aquatic habitat.

8.3 MODULAR COLONIES

8.3.1 What is a colonial organism?

Multicellular algae, sponges, bryozoans, and many cnidarians, entoprocts and tunicates are, like terrestrial plants, sessile organisms that rely on the mobile fluid overlying or surrounding them to supply their nutritional requirements. They are also all modular colonies, which form by a process of repeated budding in a similar manner to other clonal organisms (Hughes, 1989). However, whereas the asexual division products of, say, a sea anemone or a hydra move apart and thereafter live separate lives; in modular organisms, the division products remain partly attached to each other. Aggregations of millions of interconnecting 'individuals' can develop by this means. This gives rise to problems of terminology: what is the individual organism? Is it the colony as a whole, or is it one of the modular units of which it is comprised? There is no simple

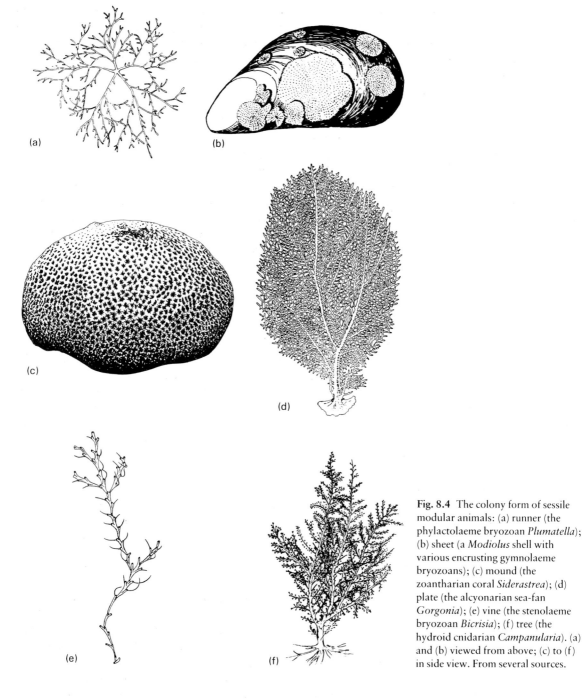

Fig. 8.4 The colony form of sessile modular animals: (a) runner (the phylactolaeme bryozoan *Plumatella*); (b) sheet (a *Modiolus* shell with various encrusting gymnolaeme bryozoans); (c) mound (the zoantharian coral *Siderastrea*); (d) plate (the alcyonarian sea-fan *Gorgonia*); (e) vine (the stenolaeme bryozoan *Bicrisia*); (f) tree (the hydroid cnidarian *Campanularia*). (a) and (b) viewed from above; (c) to (f) in side view. From several sources.

answer, largely because the concept of an 'individual' is inappropriate and vague. If separated, for example by fracture of an interconnecting stolon, the modular units can function independently as 'individuals' (it may well be the case, for example, that all the salt marsh grass *Spartina anglica* in Britain is one single genetic individual) (A.J. Gray, personal communication), but at the same time, in some colonies there is polymorphism of module type and a high level of integration of all the modules into a single functional (and moving) whole, so that the colony could reasonably be regarded as the individual (see below). The concept of individuals arose in unitary organisms and we will largely avoid its use here; we can simply refer to the assemblage of asexually-produced units as the colony, and to the units themselves as the modules.

The lack of any generally accepted terms to apply to modular colonies reflects the fact that it is only very recently that it has been recognized that trees, kelp, sea mats and corals are all expressions of the same basic theme (the concept of modular organization was only introduced by Harper in 1977), and that they differ in quite fundamental ways from unitary organisms such as insects, molluscs and fish.

The diagnostic difference is the colonial form outlined above. At a macroscopic level, colonies grow by repeated branching (although branches may not separate from each other), and thus a series of growth forms based on branching pattern can be distinguished (Jackson, 1979), of which the main variants are: (i) 'runners' — stolon-like systems that spread out in a diffuse, open network across a surface, and give rise to single or groups of modules at intervals; (ii) 'sheets' — two-dimensional encrustations forming a complete cover of closely-packed modules over the substratum; (iii) 'mounds' — massive, three-dimensional encrustations, similar to sheets but growing out from the substratum over the remains of previous 'generations' of modules; (iv) 'plates' — sheets of modules growing perpendicularly away from the substratum like a fence; (v) 'vines' — linear or sparsely-branching, cylindrical forms extending away from a limited number of attachment points; and (vi) 'trees' — erect, densely-branching systems of modules growing from a single anchorage point (Fig. 8.4). Not surprisingly, the more open and diffuse colonies, such as runners, can spread rapidly and these are the forms adopted by opportunistic species, whilst the more massive systems, such as mounds, may take tens or hundreds of years to grow. With the exception of the kelps and some other seaweeds, growth of a colony takes place at the periphery, or along surfaces furthest away from the substratum. This means that much of the mass of a living colony is not living tissue (biomass) at all, but is the accumulated dead remains of previous overgrown modules. It is necromass, the substance of which coral reefs and forests are composed.

8.3.2 Other differences between unitary and colonial organisms

In contrast to a unitary organism, a single, genetic, clonal individual can in theory live to an infinite age, in that there is no senescence. A given module or colony may have a finite lifespan, but: (i) on dividing asexually, the daughter modules are 'rejuvenated', so that the colony survives for much longer than any single module; and (ii) fragments may detach, or be detached, from a colony and be dispersed passively to give rise to new colonies elsewhere, beginning the process anew without the intervention of any sexual meiotic phase. Some freshwater plants are solely dispersed in this way. There is, therefore, no relationship between size of colony and its age. A small colony could be the direct lineal descendent, via asexual fragmentation, of an original colony founded thousands of years earlier.

In marked contrast to unitary organisms, within certain energetic constraints (Sebens, 1987) there are no allometric or scale effects of colony size on metabolic rate. As a result of surface area to volume relationships, unitary organisms display a decreasing metabolic rate per unit bodyweight as they increase in size. In simple terms, it becomes progressively more difficult to fuel a volume-dependent total metabolism across a surface area-dependent interface with the external environment. Modular colonies do not suffer this limitation: their form is such that 100 interconnected modules will have exactly the same combined surface area to volume ratio as would one single module. A single module 100 times larger in volume, however, would have that ratio decreased by 78.5%. Therefore, the growth rate of modular colonies does not necessarily slow down as they get larger, and they have no physiological size limit. They can increase in size until mechanical instability limits further growth, or until they extend into areas of resource shortage or abut onto other colonies.

Sessile colonies cannot move away, down gradients of competition (except through differential growth), nor can they escape from consumers by flight in the way that mobile organisms can. Defence must, therefore, be based on processes fundamentally different from those of most unitary organisms, although some can be shared with other sessile species. Thus colonies can be encased in a protective communal matrix of calcareous, chitinous, gelatinous, collagenous or cellulose material that may bear spines or other defensive structures; and supporting tissues, whether external or internal, woody or spicular, can dilute the nutritive content of the ingestible material and render it an indigestible and nutrient-poor diet. Other systems are particularly characteristic of modular colonies. Many, for example, invest in the production of 'secondary compounds' which by being noxious or toxic exert a deterrent role: compounds such as alkaloids and terpenoids may comprise up to 5% of the dry weight of some species. The same chemicals have been utilized by unrelated organisms. Thus citral and geraniol are as characteristic of some bryozoan animals as they are of a number of angiosperm plants.

Chemical warfare is also involved in intra- and interspecific competition. When, for example, two colonies of a sponge meet as a result of growth, they may fuse, if genetically similar, or react allelochemically, if genetically distinct, by producing substances that interfere with cell adhesion. Indeed, such modular colonies are amongst the very few groups of organisms to have evolved self/non-self recognition systems. Competition for space between different colonies is an ever-present, but highly complex, process. Competitive loops or networks are common. Thus fast-growing trees can overtop sheets or mounds, and in so doing intercept the supply of water-borne food particles, but the trees may be susceptible to chemicals produced by mound-forming species, and their architecture makes them more liable to be broken by wave or storm action. It often appears to be the case that whilst species A can outcompete species B, and species B likewise species C, species C can suppress species A chemically (see Jackson *et al.*, 1985).

Not all sessile organisms are modular, and not all modular organisms are sessile. A few benthic colonies can 'creep' slowly over the substratum (e.g. some pennatulid and coral cnidarians, and one or two bryozoans and compound sea squirts) — the bryozoan *Selenaria* at the rate of $1\,\mathrm{m\,h^{-1}}$ — but the most mobile of colonies are pelagic (Fig. 8.5). In contrast to benthic colonies, in which the modules may have a large degree of local autonomy, pelagic colonial forms are much more tightly integrated systems, with less plasticity of growth form and a more determinate body plan (see Mackie *et al.*, 1987). Modular polymorphism (see below) reaches its greatest development in the pelagic siphonophores, with the numbers of the different morphs being under rigid control, and the colonies of these animals, which are coordinated by a well-developed nervous system, do have a real claim to be considered as individual organisms.

8.3.3 What are the advantages of modular construction?

The escape from allometric constraints is probably one of the main advantages of modular construction. To increase in size carries with it several advantages: for example, larger organisms are relatively immune from casual predation; they are usually the victors in competitive situations; and they can produce a larger total number of reproductive propagules. Yet, there are also advantages attendant on small size. Small organisms do not require circulatory systems; small animals need not invest resources in excretory organs; and in filter feeders, for example, crowns of ciliated tentacles need to be below a certain size to function effectively because of hydrographical constraints. (Ryland and Warner, 1986, have shown that tree-like colonies of larger modules are suited to slow, multi-directional flow conditions, whilst plate-like colonies of smaller modules are favoured in faster, more uni- or bidirectional flows.)

The size of unitary organisms can be regarded as being a compromise between these conflicting pressures, but modular organisms can 'get the best of both worlds'. The module possesses the advantages consequent on small size, and the colony those associated with large size. In encrusting species, for example, the small modules can maximize the area of filtering surface relative to the

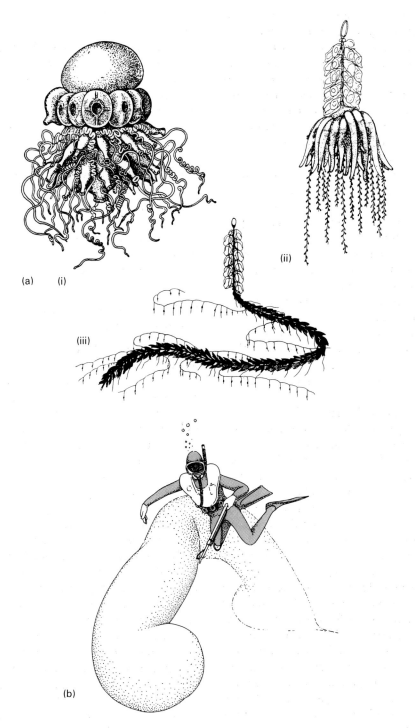

(a)   (i)

(ii)

(iii)

(b)

Fig. 8.5 Mobile pelagic modular colonies: (a) siphonophore cnidarians with polymorphic modules ((i) *Nectalia*, (ii) *Physophora*, (iii) *Nanomia*); (b) the pyrosome tunicate, *Pyrosoma*. From several sources.

colony mass, whilst the colony can rapidly and asexually bud off new modules to preempt local space on the substratum (the high growth rate effect). In this context it is notable that the size of a given module is almost always much smaller than that of its unitary counterpart (e.g. coral polyps and sea anemones; compound ascidian zooids and solitary sea squirts; bryozoan zooids and their phoronan and brachiopod allies). The independence of the modules also means: (i) that those portions of the colony lost to grazing consumers do not adversely affect colony survival, and lost modules can be replaced by compensatory budding; and (ii) that the growth form of sessile colonies can be flexible and can take up the shape most suited to the local conditions.

A second category of advantage is polymorphism of module form. Since the modules are genetically identical (except for 'somatic' mutation) and since they interconnect, some can be specialized for the performance of certain roles — defence, feeding, gamete production, brooding of embryos, propulsion, anti-fouling, packing, anchorage, support, communication, etc. — and suites of modules can cooperate in the performance of common tasks. In organisms that do not possess organ systems, suites of polymorphic modules can carry out equivalent functions. It is possible that the pelagic colonies derive yet another advantage — it has been argued that the serial arrangement of coordinated modules provides a smoother and more economical form of jet propulsion through the water, than could be achieved by a single large unitary organism (e.g. a squid), without sacrificing any manoeuvrability or rapidity of reaction.

Given these advantages of modular construction, it is not unreasonable to question why more organisms are not colonial. The answer seems to be the incompatibility of segregation of the germ line (and, more generally, of mosaic development) with modular construction; most 'higher' organisms have adopted either or both of the two former, for whatever reason, and are therefore denied the latter.

8.3.4 Internally-budded resistant modules

Beside the production of external buds that leads to the colonial form, modular aquatic organisms may also produce internal buds, or equivalent aggregations of totipotent cells. These can be provided with yolk-rich 'nurse cells' and are invariably surrounded by a resistant coating of one or more layers. Such 'statoblasts' (in phylactolaeme bryozoans), 'hibernaculae' (in gymnolaeme bryozoans) or 'gemmules' (in sponges) are then either discharged from the living module, or remain within its body and are liberated only after its death and disintegration. Discharge and, in many cases, liberation results in release into the free water mass, but in some, the resistant ball of cells is anchored to the substratum by the tissues of its founding module, so that liberation only involves exposure to the external environment.

Characteristically, these resistant modules are produced at the onset of adverse environmental conditions — seasonal temperature extremes (both high and low), falling water levels and so on — and they are therefore especially associated with freshwater species, because of the greater lability and lesser persistence of most bodies of fresh water in comparison to the larger, more stable sea. Nevertheless, some marine and brackish-water species display the phenomenon as well (e.g. the sponges *Haliclona* and *Suberites* and the bryozoans *Victorella* and *Bowerbankia*), producing resistant modules as a reaction both to supposedly unfavourable temperatures and to decreased food availability. Although this relationship between seasonal environmental adversity and resistant bud formation seems clear, the precise stimuli involved are still largely unknown; indeed, it is likely that these vary from species to species: some modules enter true diapause, for example, whilst others are capable of 'germination' immediately, if the ambient conditions are suitable.

Whilst within their resistant coats, the resting stages can withstand extreme conditions such as freezing and desiccation. Statoblasts have been found to be still viable after being kept dry for over 4 years, and various gemmules are known to survive dormancy in nature for several years and to 'germinate' when temporary water returns to their arid habitat. Some can survive passage through the gut of aquatic consumers, such as amphibians, reptiles and birds, and presumably thereby gain long-distance dispersal. A special dispersal function may also be served by several types of statoblast that possess flotation mechanisms — colonies of *Plumatella* can release about 1 000 000 statoblasts m^{-2}, and extensive drifts of these are commonly observed around the margins of ponds and lakes. Release from dormancy or 'germination' seems most usually to be under temperature control (although light regimes are also implicated in the case of phylactolaemes), and this culminates in the escape of the totipotent cells, or of their descendents, through a pore in the enclosing coating (in gemmules), or between the valves (in the bivalve statoblasts). A new module is assembled from these cells, and this then initiates colony formation, as described above (Section 8.3.1).

Although this is the most specialized form of overwintering, in less harsh conditions many modular colonies achieve the same end result simply by regressing to a few perennating modules (stolons, etc.) and then re-initiating the asexual multiplication phase, when the environment ameliorates. Comparably, unitary organisms may survive adverse periods and achieve a degree of passive dispersal as a quiescent egg or pupa, or in an encysted state, or the periods in question may be spent in a completely different habitat as a result of migration, or in the form of a larval stage.

8.4 THE LARVAL FORM

For free-living species, two major functional significances of the presence of larval stages in the life history have been proposed and debated: (i) food acquisition (together with reduction of parent–offspring competition); and (ii) dispersal (see Crisp, 1976). (Species of aquatic parasite are a separate case, in which larvae serve as the agents of host finding, and often of asexual multiplication.) This is a subject area in which there have been very few critical measurements and experiments. We are woefully ignorant, for example, of how far individual planktonic larvae do disperse (see Todd *et al.*, 1988), and of the food requirements and mortality rates of such stages. In a number of cases we do not even know which larva belongs with which adult organism. Most of the arguments presented below are, therefore, mainly theoretical, tempered by the scanty hard data available.

8.4.1 Dispersal

Most permanently aquatic organisms are benthic when adult, and most benthic organisms are sessile, anchored in or attached to the substratum, or sedentary in that they inhabit tubes or burrows, live on or in other organisms, or forage within a limited home range. Dispersal away from parents and areas of high population density, with concomitant potential avoidance of intraspecific competition for food and/or space, can therefore only be achieved by the possession of a specific mobile stage in the life history (*mobile* — able to be moved). These stages need not be *motile* (capable of moving under their own power), because, like fungal and plant spores or seeds in the air and the resistant modules in the water that we have just considered, they could be distributed passively by movements of the overlying water mass. In fact, the vast majority of aquatic dispersal stages — and all larvae — do have locomotory abilities, but these may be mainly of significance in small-scale prospecting for a suitable area of substratum on which to settle and metamorphose or grow.

It seems self-evident that the aquatic larval stages of benthic adults do serve a dispersal function, particularly in the case of non-feeding ('lecithotrophic') larvae, provided with a food store in the form of yolk. Although movement away from the immediate point of release is likely to be advantageous, there would be no general necessity to move far away. Indeed, under most circumstances the probability of finding an appropriate place to settle will decrease with distance travelled. A suitable rule of thumb might therefore be: if your parents have survived to breeding age in a given area, then that region can be regarded as of guaranteed suitability for you too. Short distance dispersal only requires a larval life of a few hours at most: hence, there is not necessarily any advantage to be gained by the possession, during the larval stage, of feeding organs.

The dispersal advantages of non-feeding larvae are applicable to all types of aquatic habitat; lecithotrophic larval stages are represented in benthic faunas from the Arctic to the tropics and from the deep sea to shallow ponds. In some habitats, however, especially in shallow waters, microhabitats may be ephemeral. Patches of bare rock, for example, may appear (as water and sediment movements abrade rock surfaces) and disappear again (as the available space is preempted by organisms). Suitable habitat patches may always be present somewhere, on scales measurable, say, in square kilometres, but they may be unpredictably located on smaller areal scales. The problems of colonizing such ephemeral and unpredictable patches are akin to those associated with forest clearings or areas of disturbed ground on land, and aquatic solutions are similar: the dispersal propagule must be capable of being transported over relatively large distances and must be potentially long lived. In the sea, however, where the problem is more widespread, the solution is more difficult to achieve than on land, because of tidal water fluxes.

In shallow regions, tidal flows oscillate in direction, and hence water masses pulse to and fro, roughly on a 6 (or in some regions, 12) hour cycle, and these pulses override any general directional flow (Table 8.3). Transport over distances greater than those that can be achieved in 6 or 12 hours, therefore, requires a disproportionately long period of time; although unlike most terrestrial propagules, marine ones can potentially sustain themselves by feeding *en route*. The longer these tiny organisms spend in the water dispersing, however, the greater will be their chance of being consumed by pelagic predators, and hence the fewer that are likely to survive and the more that must be produced initially to counterbalance such losses. Many will also fall on the marine equivalent of the proverbial stony ground (Thorson, 1950; Underwood & Fair-

weather, 1989), exacerbating the need to produce very large numbers of larvae that serve a long-distance dispersal function. A large number of young correlates with small individual size; small individual size equates with the provision of little or no foodstore to the egg; and no foodstore in the egg requires that the larva hatches at a small size and fuels its own subsequent development by feeding. The argument can clearly become circular (Fig. 8.6), and it is difficult to determine what is evolutionary advantage and what is automatic consequence. Does long-distance dispersal necessitate that the larvae feed in the plankton, or is the ability to obtain external supplies of food the real selective advantage (see below) and the long-distance dispersal suffered an incidental, even disadvantageous, consequence?

It has been argued (e.g. Strathmann, 1985) that dispersal cannot be the only function of the larval stage, because planktonic animals also have larvae and in their case, there is no reason to suppose that the larvae are dispersed to any greater extent than are the adults. It is, however, questionable how many planktonic organisms really do have larvae, although their juvenile stages are often so termed. A larva is defined as a stage in the life history that is morphologically so different from the adult (Fig. 8.7) as to require a traumatic intervening metamorphosis to convert from one to the other. This anatomical difference between larva and adult reflects their contrasting lifestyles and habitats. Many 'larvae' of planktonic animals, for example those of the arrow worms (Chaetognatha), are merely juvenile versions of the adults, into which they develop without any metamorphosis. Other planktonic animals can themselves be regarded as permanent (paedomorphic) larvae in evolutionary origin — for example, ctenophores, appendicularian tunicates, pteropod molluscs, copepod and cladoceran crustaceans (see de Beer, 1958, for a classic review) — and again no metamorphosis occurs and the so-called larvae differ relatively little from the adult stage. In direct contrast to the first statement of this paragraph, it could be argued that the permanent members of the plankton prove the general dispersal rule, in that organisms that are planktonic when adult do *not* normally have larval stages as defined above.

Table 8.3 The relationship between length of marine planktonic larval life and the distance likely to be transported away from the point of release. From Crisp (1978)

Duration of planktonic phase	Approximate distance transported (km)
3–6 hours	0.1
1–2 days	1
1–2 weeks	10
0.5–3 months	100
1 year	1000

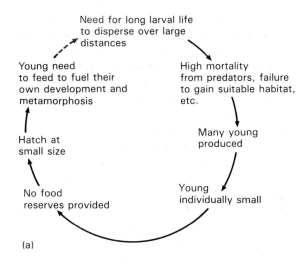

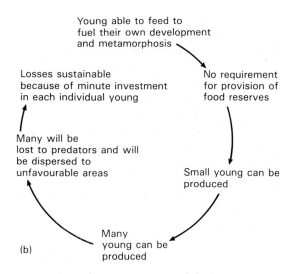

Fig. 8.6 Alternative explanations of the existence of long-lived planktonic larvae involving the same pieces of evidence linked in different chains of cause and effect: (a) in which the selective advantage is assumed to be dispersal; (b) in which it is assumed to be greater feeding opportunity.

8.4.2 Food acquisition

In the larger natural bodies of water — seas and deep lakes — phytoplankton productivity of the surface water is by far the most important source of fixed food energy for the entire ecosystem. Much, usually most, of this pelagic production is con-

sumed *in situ*, leaving the benthos to be dependent on such little as remains (faecal pellets, cast exoskeletons and the occasional more nutritious item of fall-out) (see Chapter 4). It follows automatically from this that, per unit consumer, utilizable food is more abundant in the photic production zone than anywhere else, and it is least abundant in aphotic benthic regions, primary chemosynthesis apart. This is evident, for example, from the greater productivity per unit biomass of pelagic organisms than of benthic ones: comparing like with like, pelagic species produce on average some 2–5 times more per unit of their standing biomass than do even relatively shallow water benthic forms, whether in fresh waters or in the sea.

We noted earlier, when considering the advantages of increase in size (Section 8.3.3), that small individuals are usually at greater risk from consumers than are large ones (see Table 8.4); hence there will be a selective advantage in juvenile animals growing as fast as possible to achieve some minimum size threshold. To do this they will require a source of easily obtainable food — as found in areas or at times of resource superabundance per unit of interested consumer. Usage of such 'nursery grounds', in which rapid growth of young is possible, is widespread in aquatic systems; for example, shallow and often brackish areas by flatfish, coastal lagoons by prawns and, arguably, planktonic zones by benthic animals and fresh and brackish waters by otherwise terrestrial insects (see Margalef, 1963).

On this argument, planktotrophic (plankton-feeding) larvae gain not only long-distance dispersal — if it is an advantage — but a more abundant and less contested source of food, than would be available on or in the substratum inhabited by adults. Such advantage, however, will be critically dependent on the presence of a persistent and predictable supply of planktonic food; observations that lend support to a feeding hypothesis of larval role are therefore: (i) that planktotrophic larvae are the norm in shallow, tropical waters where productivity is predictably year-round and fairly constant in magnitude; (ii) that they decrease in frequency as latitude increases and hence the pulse of net phytoplanktonic productivity becomes shorter, seasonal and, in some areas at least, somewhat unpredictable in timing and duration; and (iii) that they become

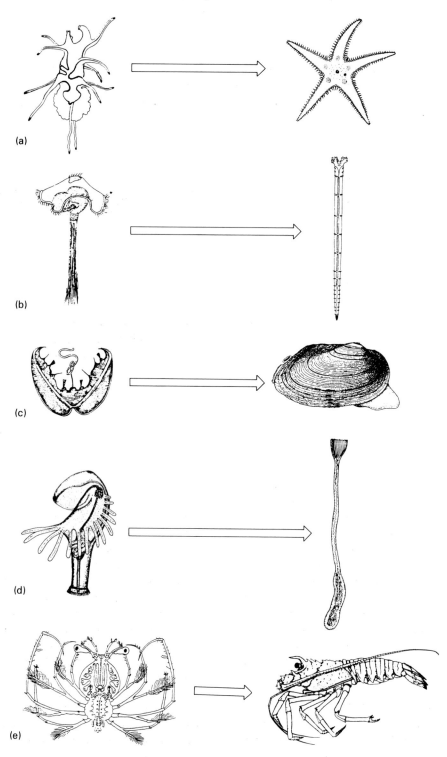

(a)

(b)

(c)

(d)

(e)

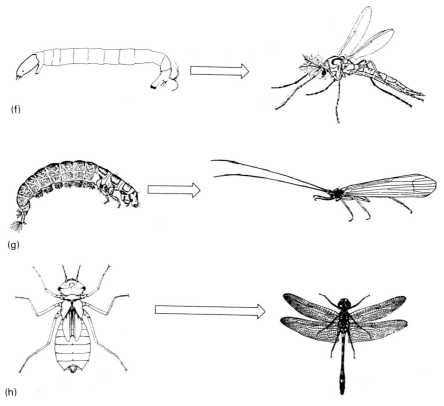

Fig. 8.7 Larval and adult morphology compared: (a) starfish (Echinodermata); (b) oweniid worm (Polychaeta); (c) swan mussel (Mollusca); (d) phoronan (Phorona); (e) crawfish (Crustacea); (f) chironomid midge (Diptera); (g) caddisfly (Trichoptera); and (h) dragonfly (Odonata). From several sources.

rare or absent at depth, when the distance between adult habitat and potential nursery ground becomes large and the intervening area is unfavourable, with little food per unit consumer and large numbers of predators. The evolutionary consequences of plank-totrophy and long-distance dispersal are many, including broad geographical range, low rates of speciation and low rates of extinction (see Levinton, 1988, and Chapter 9).

Terrestrial insects provide a direct antithesis to the large majority of aquatic organisms: the adults are more mobile and are therefore able to disperse, whilst the juveniles are sessile or sedentary and many do not move far (Table 8.5). This role reversal extends even further, particularly in those insects with aquatic juvenile stages. For example, the naiads and larvae of dragonflies (Odonata), mayflies (Ephemeroptera), caddisflies (Trichoptera) and true flies (Diptera) are the main feeding stage in the life history, leaving the adults to be mostly or entirely concerned with reproduction and dispersal; the aquatic juvenile stage occupies by far the greater proportion of the lifespan.

Thus, most of the lives of these insects is spent as a larval feeding machine and the adult has been relegated to a small dispersal appendage to the life cycle. If the function of the larval stage is to exploit a food-rich habitat unused by the adult, we may ask why the juvenile ever returns to the food-poor adult environment and/or why the adult has not evolved so as also to take advantage of the food bonanza? Other insects, most notably the water bugs (Hemiptera) and beetles (Coleoptera), have moved into fresh and brackish waters as adults too, and emerge only to disperse.

Many bodies of continental water (both ponds

Table 8.4 Variation of mortality rates with age and size, as exemplified by the plaice (*Pleuronectes platessa*), each female of which can produce up to 600 000 eggs per season, and the mud snail *Hydrobia ulvae*, which produces only 350–3500 per female. In each case, the number of larvae metamorphosing into benthic juveniles has been set at 100 000; in both species metamorphosis takes place after about 4–6 weeks. From several sources

| Age (months) | *Pleuronectes* | | *Hydrobia* | |
	Size (cm)	No. dying per monthly interval	Size (mm)	No. dying per monthly interval
Hatchling	0.7		0.15	
Metamorphosling	1.3	1 550 000	0.30	875 000
2	2.7	80 000	0.40	67 000
3	3.9	14 000	0.50	14 000
4	4.6	3500	1.75	6500
5	5.4	1300	1.80	4500
6	6.4	550	1.85	3500
7	6.8	220	1.90	2000
8	7.3	100	1.95	1100
9	7.4	50	2.00	500
10	7.5	20	2.05	300
11	7.6	17	2.10	200
12	7.7	16	2.15	150
	No. still alive after 1 year	130	No. still alive after 1 year	400

and streams) are ephemeral, and hence the retention of a non-aquatic dispersal phase may be advantageous. The long adult flight periods of many desert stream insects, for example, can be seen as adaptations to survive unfavourable periods during flash floods and stream drying (Townsend, 1989).

The seas of the world, on the other hand, form one single, interconnecting and permanent water mass. Like the beetles and water bugs above, several adult members of otherwise exclusively benthic marine groups have also invaded the larval habitat, moving up into the water column and becoming planktonic or nektonic — for example, the pelagic hoplonemertines, many polyclad turbellarians, phyllodocidan polychaetes, elasipod sea cucumbers, portunid and natantian decapod crustaceans and most cephalopod molluscs. In these cases, paedomorphosis has short-circuited the benthic stage out of the life history and rendered the whole of the lifespan pelagic. These examples not only conform to predictions of the feeding explanation of larval role, but are not in accord with a dispersal hypothesis.

The terrestrial–freshwater life cycle of the insects is curiously reminiscent of the much debated alternation of sexual and asexual stages shown by cnidarians. The planula larvae of cnidarians (like the larvae of sponges) are not really larvae at all, but are free-living embryos — hollow (in some sponges) or solid blastulas. The planulae settle on the substratum and grow into the polyp stage, which is functionally equivalent to a larva, although, unusually amongst entirely aquatic animals, it is benthic and sessile, or effectively so. The sexually reproducing adult stage is the medusa, which is mobile and hence capable of dispersing. This simple life history is complicated only by the ability of the benthic stage to produce clones of separate or modularly colonial polyps by repeated asexual budding (as do the larvae of many endoparasites), and by both widespread paedomorphosis and much rarer loss of the polyp stage during evolution. The often-posed question of 'which is the ancestral form, the polyp or the medusa?' may be as meaningful as 'which came first, the juvenile or the adult?'!

Nevertheless, the cnidarian life history is aberrant in at least one respect. The functional adult is

Table 8.5 Contrasting lifestyles of aquatic larvae and their adult stages

Larva	Adult	Example
Free-living, mobile; non-feeding, finds and infects new hosts; short-lived	Sedentary, endoparasitic; feeds; long-lived	Miracidium and cercaria of digenean flukes
Sedentary, endoparasitic; feeds, multiplies asexually; individually short-lived	Sedentary, endoparasitic; feeds, multiplies sexually; long-lived	Redia of digenean flukes
Sedentary, endoparasitic; feeds	Mobile, free-living; non-feeding	Nematomorph worms
Free-living, travels long distances; feeds; medium–long-lived	Free-living, travels short distances or sessile; feeds; long-lived	Most marine invertebrates
Free-living, mobile; non-feeding; short-lived	Free-living, travels short distances or sessile; feeds; long-lived	Many aquatic invertebrates
Sedentary, endoparasitic; dispersed by fish hosts	Free-living, sedentary filter feeders	Unionid bivalves
Free-living, sedentary or sessile; feeds; long-lived; inhabits 'new' environment	Free-living, mobile; feeds or non-feeding; short-lived; inhabits ancestral medium	Insects with aquatic larvae
Free-living, travels short distances; inhabits ancestral anamniote medium	Free-living, travels long distances; inhabits 'new' terrestrial environment	Amphibians
Free-living, travels long distances; inhabits ancestral medium	Free-living, travels short distances; inhabits 'new' terrestrial environment	Land crabs

budded from the larva rather than being formed in a metamorphosis. This is probably a direct reflection of the prevalent clonal/colonial state. Distinctive larvae are particularly associated with unitary organisms, and colonial groups derived from primitively unitary ones tend to lose any ancestral larval stage. Colonial organisms can simply bud off feeding or dispersive modules, as required (see Section 8.3), so that there is no advantage in the evolution or retention of a larva. There is no evidence that cnidarians ever possessed a true larval stage and their polyp–medusa alternation can be regarded as an independent parallel acquisition of what, in the bilateral phylogenetic line, is the larva–adult cycle.

8.4.3 Loss of the larval stage

We noted in Section 8.2 that the presence of a larval stage in the life history is the primary condition in most — and in all primitive — animal groups. Several later phylogenetic lines lost this primary larva, however; although some, including arguably the more successful ones (such as the endopterygote insects and many endoparasites) subsequently evolved replacement 'secondary' larvae. We also saw in the previous sections that larvae are particularly associated with feeding and/or dispersal; hence we might expect that forms which have evolved direct development instead: (i) would be mobile, so that at least short-distance dispersal can take place at any or all periods of the lifetime; and/or (ii) do not have any potential access to an adjacent under-exploited food source, requiring divergent adaptations for its utilization — the juveniles must then of necessity feed in the same general habitat as the adults. By and large, these two expectations are borne out by the distribution of larvae. Where juveniles and adults exploit the same resources (except, maybe, for differences related to the size of materials captured and/or ingested), intraspecific competition between young and old might be expected to occur, and therefore there would be all the usual selective advantages associated with the provision of resources to offspring, favouring large offspring at birth, and hence direct development. The same evolutionary end results could also be responses to intense predation pressure on small animals (where again large initial size might be advantageous) and to environments of absolute food scarcity, like the deep sea, in which juveniles may have to subsist on their own, parent-contributed resources for considerable periods of time, until a supply of food becomes available or can be located.

Somewhat paradoxically, however, the primary larva appears both to be a highly-conserved evolutionary feature and an easily lost one; many species in most aquatic animal phyla still possess the ancestral larval type, after presumably hundreds of millions of years of uninterrupted ancestry, but numerous species in those same phyla have lost the free-living larval phase and undergo direct development. In many cases, however, suppression of the larva may be more apparent than real — often the larval stages are simply passed through within the egg capsule.

Excision of the free-living larva from the life history is not always easily explicable in terms of habitat differences, not least because it often does not conform to the classic density-dependent r/K dichotomy of MacArthur and Wilson (1967) and later authors (for which, see Boyce, 1984). Thus pairs of species, sometimes even sibling species, often occur in the same habitat and differ only in the size of egg that they produce, and consequently in the number of eggs laid, and in the form of the hatchling. Amongst gastropod molluscs for example, the winkle *Littorina saxatilis* produces few,

large eggs that hatch to liberate young benthic snails, whereas *L. littorea* lays up to 300 times as many eggs, which are much smaller and develop into planktonic larvae; yet both species have otherwise similar life histories, live on the same rocky shores, feed on the same sort of foods, have apparently the same needs for dispersal and invest the same proportion of their available annual energy in egg production. Similarly, the mud snail *Hydrobia ulvae* has retained small eggs and planktonic larvae, whereas *H. ventrosa*, which occurs with *H. ulvae* in many European brackish lagoons, has abandoned the larval stage and lays few, large eggs; again, both appear to invest similar proportions of their available energy in reproduction. The same phenomenon is shown *within* a number of gastropod species (e.g. *Rissostomia membranacea* and *Brachystomia rissoides*) and even within members of the same population (e.g. in *Tenellia adspersa* and *Elysia cauze*) (see Eyster, 1979, for a detailed example, and Hoagland & Robertson, 1988, for a critical review).

Similar series of examples could be cited from within the echinoderms, polychaetes and several other aquatic groups. In all these cases, individuals of the same or of different species, with different juvenile strategies, live together in the same habitat, and species with the same strategy may also hail from widely contrasting environments. The major problem besetting analysis here, is that the quantity of energy devoted to reproduction (the fundamental feature underlying semelparity versus iteroparity, and r- versus K-selection) and the manner in which unit energy can be deployed are, for many organisms, entirely separable variables. As yet, the second of these two variables has received less critical attention.

One must also always remember that theories (and textbooks) often stress idealized and unnaturally discrete sets of circumstances; the conditions provided by different habitats form a continuum, however, and it is only to be expected that in some instances the balance of advantages between direct development and the production of larvae is roughly equal. It is, perhaps, under such circumstances that the gastropod examples cited above, and others like them, evolved. There is even evidence that in some gastropod species (e.g. *Spurilla neapolitana*)

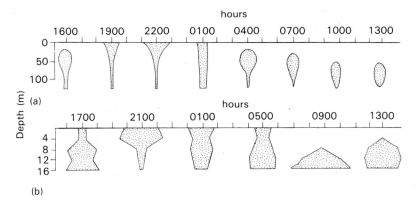

Fig. 8.8 Characteristic vertical movements undertaken by zooplanktonic copepods: (a) adult females of the marine *Calanus*; and (b) adults of the freshwater *Diaptomus*. From Barnes & Hughes (1988); Zaret & Suffern (1976).

the form of the juvenile is dependent entirely on local and temporary food availability, indicating the ease with which the balance can change, and the flexibility of some responses.

8.5 VERTICAL MIGRATION

We saw above that the mobile, fluid part of all environments may be used for dispersal — water in the case of marine organisms and air for continental ones. Hence, all environments are three-dimensional, yet in aquatic habitats the third, vertical, dimension encompasses the permanent habitat of numerous suspended species, to an extent unimaginable in terrestrial systems. Most such pelagic species also exploit the vertical dimension by moving between or amongst several depth horizons in the water column, sometimes in a rhythmic manner. The latter forms the distinctive behaviour pattern known as vertical migration, in which animals move actively over maximum distances of 10–50 000 times their own body length up and down in the water, each day.

Although vertical migration is far from being a stereotyped phenomenon, or of universal occurrence, when it does occur it generally approximates to the following pattern: during the daylight hours, organisms are relatively deep in the water column (100–1000 m depth in the sea, although less deep — often necessarily! — in ponds and lakes), but at dusk they ascend to near the surface. They may

then disperse, somewhat, through the surface water mass during the night, before re-aggregating near the surface at dawn and descending again to the daytime depth as light intensity increases (Fig. 8.8). Movement, both up and down, is by active swimming, with sustained upward speeds of 12–200 m hour^{-1} and downward ones some three times faster (most evidence indicates that counterintuitively, the energetic cost of this is negligible). There are many exceptions to this generalized picture — some organisms or life-history stages do not migrate at all; the extent of the movement varies in time, in space and with the nature of the mover; the timing may be reversed such that the organism is near the surface during the day, and at depth in the night; and so on — many of these exceptions are helpful in indicating likely selective advantages of the movements.

Given that the aquatic environment has a large and inhabitable vertical dimension, it is hardly surprising that organisms should exploit it. Further, it is not surprising that they should use rhythmic environmental variables to cue and time their movements; terrestrial species also make use of such signals. For species inhabiting the photic zone (and vertical migration is particularly marked near the surface), the daily change in light intensity forms just such a cue, and it appears to be used almost universally as the external timing setter (Forward, 1988). Whether reaction to light intensity *per se* provides any direct selective advantage (e.g. as a predictor of productivity), as has been claimed, is far less obvious.

Many of the numerous 'explanations' of vertical migration that have been proposed are characterized chiefly by a failure to distinguish between cause and effect, and, as in the case of reaction to light, between proximate (i.e. mechanistic) and ultimate (i.e. evolutionary) causation. Several hypotheses can be seen to be covertly group-selectionist (benefitting the group or species but not the individual), and hence although they may describe possible effects, they cannot provide causal selective advantages. These ideas include: (i) that vertical migration promotes gene flow, prevents the establishment of isolated subpopulations, and reduces speciation; (ii) that it reduces population fluctuations by permitting intrinsic self-regulation of population density in relation to food availability; and (iii) that the pulsed grazing pressure that results (a) conserves the food supply, (b) induces a greater phytoplanktonic productivity than would result under a continual grazing regime, and (c) is timed so as to avoid consumption of actively dividing cells.

A further three general types of explanation for vertical migration, two of them associated with feeding strategy and the other with predator avoidance, are potentially causal and espoused by significant numbers of aquatic biologists, although the first one to be considered is probably the least likely to have wide applicability. This, the 'energy bonus hypothesis' of McLaren (1963) (and see Enright, 1977), draws on the observations that metabolic rate is temperature dependent and that number of young produced is often a function of body size (all other things being equal). Where, therefore, the water column is stratified thermally and food is

abundant near the surface, consumers can obtain large quantities of food, but 'waste' a considerable proportion of their energy intake in fuelling a high basal metabolism. If they descend to cooler depths having fed, however, their metabolic rate will fall, less of their assimilated energy will have to be respired, more can be devoted to increase in somatic and reproductive tissues and the resultant increased numbers of potential young will thereby provide a selective advantage for migrants versus non-migrants. All experimental tests of this model (which have mainly used the freshwater cladoceran *Daphnia*) have found that, on the contrary, the fitness of consumers that remain permanently in the zone of photosynthesis is greater than that of migrants, even under conditions most favourable to the gaining of an energy bonus. By remaining continually to feed, their total food intake is larger, leading to a greater net intake notwithstanding their higher basal metabolic rate, their growth rate is faster at the surface and time to first reproduction is reduced so that intergeneration time is decreased — reduction of intergeneration time is a more potent way of increasing potential numbers of descendants than is increasing clutch size. If food is available (and again all other things being equal), the race would appear to go to the non-migrant.

The second hypothesis concentrates on the problems of foraging faced by planktonic consumers. Especially in the sea, phytoplankton concentrations are typically patchy and planktonic consumers cannot find their way to the next underexploited patch by movement in the horizontal plane (Fig. 8.9 and see Chapter 1). Hence zooplanktonic grazers (and so on up the food chain) face a problem when they have reduced the food concentration in the surrounding water mass to some critical level: how are

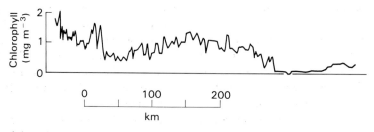

Fig. 8.9 An example of phytoplankton patchiness in the sea: a transect across the northeastern Atlantic on 11–12th May 1968. From Steele (1978).

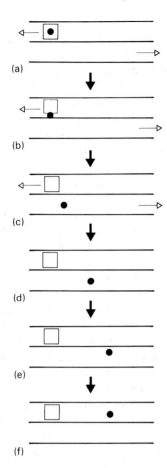

Fig. 8.10 Diagrammatic representation of how vertical migration between two water masses, flowing in different directions and/or at different speeds, can effect a change in the surface water mass inhabited by the migrating organism. From Barnes & Hughes (1988).

they to find more, when their powers of locomotion are (by definition) insufficient to permit them to make headway against current flow? Unlike terrestrial species, they cannot simply move parallel to the earth's surface.

As pointed out more than fifty years ago by Hardy and Gunther (1935) and later developed by Hardy (e.g. 1953), marine water masses flow at different speeds and in different directions at different depths below the surface, if for no other reason than because of the Ekman spiral (see, for example, Duxbury & Duxbury, 1984). Hence, by moving vertically from one water mass to another, and then

back again, a planktonic organism will of necessity find itself in a different body of surface water (Fig. 8.10), maybe kilometres away from its starting point, for very little travel time and expenditure of energy on locomotion. If a vertically-migrating consumer was to leave a given water mass when the food concentration around it had declined to the average level for the local surface water as a whole, then it would simply be behaving in accord with the familiar marginal value theorem (see Krebs & McCleery, 1984): it would be foraging optimally in a manner appropriate to a three-dimensional habitat. Of course, by aggregating in food-rich patches and then remaining to graze them down, the marine consumers are themselves important generators of horizontal patchiness and local food shortage (Fig. 8.11).

This foraging hypothesis predicts that non-feeding stages of organisms should not migrate, that migration should cease when the surface waters are devoid of food (as during the winter in high latitudes) and that it should also be suppressed in the presence of abundant food. By and large, these sorts of prediction have been borne out by studies in the sea, although not by those in fresh water. Accordingly, the hypothesis is favoured by marine biologists, but dismissed by freshwater ones. This is probably not just a reflection of divergent tradition in the two subdisciplines; the habitat difference may well be a real one, resulting from the disparate size and self-containment of the two. Variation of current speed and direction with depth is not nearly so well-developed in ponds and lakes (removing the transport system) and neither is large-scale patchiness developed to the same extent in fresh waters.

As a special case of the usage of different currents for horizontal transport, it is possible to use daily and/or seasonal vertical migrations as a means of remaining on station and avoiding the otherwise inevitable, current-induced displacement of near-surface water masses, which results wherever currents flow at 180° to each other at and below the surface, as in estuaries and around Antarctica, for example (see Bosch & Taylor, 1973, for an estuarine example).

The third hypothesis is that vertical migration occurs to minimize predation; that is, it keeps potential prey populations in the zones of low light

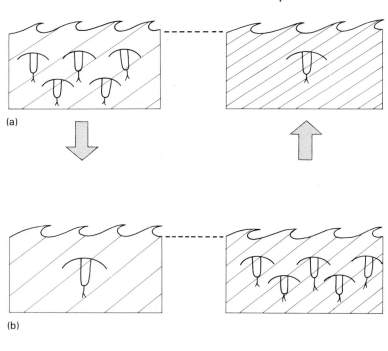

(a)

(b)

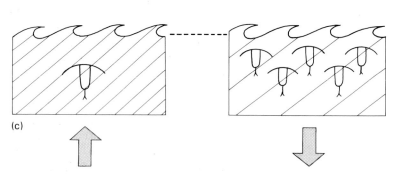

(c)

Fig. 8.11 Diagrammatic representation of the cyclic generation of patchiness as a result of the behaviour of the consumers. (a), (b) and (c) represent the same water masses at different intervals of time; arrows show the net change in numbers (i.e. the net direction of movement) of zooplankton consumers; and the intensity of the shading indicates the concentration of those species being consumed. From Barnes & Hughes (1988).

intensity, in which the capture efficiency of visually-hunting predators is lowest. If marine planktonic consumers must migrate when food supplies in their immediate vicinity have declined, their movement away would seem to be most advantageously timed to be during daylight hours rather than 'wasting' safe night-time in movement (i.e. timed so as to minimize predation risk whilst moving). *Pseudocalanus* in Dabob Bay, Washington, for example, is known to move out of synchrony with movements of its potential predators. Although predation pressure may be responsible for selecting at what time of day marine migration occurs, it is unlikely

that reduction in the profitability of predation provides the selective advantage of the movements themselves. Thus, for example, in the sea, many migrants bioluminesce (which would appear to defeat the object of moving into the dark to hide from predators!); many move hundreds of metres down below the level at which visually-hunting predators can operate (were it not for bioluminescence); and others are transparent and yet move to the same extent as visually-obvious prey (see Longhurst, 1976).

In lakes and ponds, however, there is abundant experimental and correlative evidence to suggest that the selective advantage *is* reduction in fish predation pressure (Zaret & Suffern, 1976; Gliwicz, 1986). It is known that the introduction of fish into lakes can dramatically alter zooplankton composition (Chapter 7) and can result in the persistence of only a few prey species, specifically those that can prevent their populations from being eaten out, by maintaining themselves in zones of minimum predator efficiency. In Polish lakes that lack predatory fish, plankton show no vertical migration; where fish have been recently introduced, significant migration is undertaken; and at sites where such fish have occurred for centuries, full-scale and marked regular migrations are observed. In Gatun Lake, Panama, to cite a tropical example, the fish *Melaniris* will avidly consume the copepod *Diaptomus* on the rare occasions when it can be obtained (probably only on fully moonlit nights), but otherwise this migrating species forms less than 5% of the food intake of the fish, whilst comprising 33% of the potential prey individuals. Even in strictly predator-structured systems, however, there may well be necessary compromises to be made between avoiding predators on the one hand and obtaining food on the other. Hungry prey are thus more likely to remain feeding at the surface during daylight hours than are recently fed individuals, for example.

Demonstrations that vertical migration in fresh waters does not occur in the absence of predation, or else is very variable and inconsistent in its expression under such circumstances, indicate the possibility that this behaviour pattern has evolved entirely independently in the two major aquatic environments, convergently using the same light intensity cues for triggering and timing the various phases. The predominantly continental-water copepod family Pseudodiaptomidae is interesting in this respect. Species of *Pseudodiaptomus* typically live a demersal existence, resting on the bottom during the day, and swimming up into the water to feed at night and during heavy cloud cover. Some coastal marine species are also known; they live in very shallow water and behave just like their freshwater/hypersaline/brackish relatives. One such species has been shown: (i) to produce as many eggs under an intermittent feeding regime as under one of continuous feeding opportunity; and (ii) to be

particularly susceptible to predation by the visually-hunting fish *Aldrichetta* when in the water column. In both these respects it contrasts with the marine copepod *Acartia*, with which it occurs (Fancett & Kimmerer, 1985). The experimental evidence suggests a predator-avoidance explanation of the vertical movements of the marine copepod with the freshwater ancestry, but not for those of the more thoroughly marine species.

8.6 ENDPIECE

This chapter has investigated such apparently diverse phenomena as clonal modular colonies, vertical migration and the occurrence of larval stages in the life history, and yet much of the discussion of each has revolved around the same biological requirements: obtaining food, avoiding predators and dispersal. These are clearly all-pervasive necessities. Indeed, it can be argued that in most habitats (and certainly all the larger aquatic ones) the pressures of finding sufficient food, of avoiding becoming the food of others and, for many, of finding, preempting and retaining adequate space in which to achieve the other two necessities are the *only* ones influencing survival. In several habitats, the inanimate environment, in the form of seasonal adversity or periodic physical disturbance, may also be a powerful agent of mortality, but dispersal — for example, via migration or recolonization from a pool of planktonic larvae — is the solution here too, even if in some isolated systems recolonization is a rare and chancy event (Barnes, 1988; 1991). The topics considered in this chapter are therefore the specialist aquatic solutions to what are the major fundamental problems of short-term individual survival.

FURTHER READING

Barnes, R.S.K. (1988) The faunas of land-locked lagoons: chance differences and the problems of dispersal. *Est. Coast. Shelf Sci.* **26**: 309–18.

Barnes, R.S.K. (1991) Macrofaunal community structure and life histories in coastal lagoons. In: Kjerfve, B. (ed) *Coastal Lagoon Processes*. Elsevier, Amsterdam. (In press.)

Barnes, R.S.K. & Hughes, R.N. (1988) *An Introduction to Marine Ecology*. 2nd edn. Blackwell Scientific Publications, Oxford.

Begon, M. (1985) A general theory of life-history variation. In: Sibly, R.M. & Smith, R.H. (eds) *Behavioural Ecology*, pp. 91–7. Blackwell Scientific Publications, Oxford.

Bosch, H.F. & Taylor, W.R. (1973) Diurnal vertical migration of an estuarine cladoceran, *Podon polyphemoides*, in the Chesapeake Bay. *Mar. Biol.* **19**: 172–81.

Boyce, M.S. (1984) Restitution of *r*- and *K*-selection as a model of density-dependent natural selection. *Ann. Rev. Ecol. Syst.* **15**: 427–47.

Buss, L.W. (1987) *The Evolution of Individuality*. Princeton University, New Jersey.

Crisp, D.J. (1976) The role of the pelagic larva. In: Spencer Davies, P. (ed) *Perspectives in Experimental Biology: 1, Zoology*, pp. 145–55. Pergamon Press, Oxford.

Crisp, D.J. (1978) Genetic consequences of different reproductive strategies in marine invertebrates. In: Battaglia, B. & Beardmore, J.A. (eds) *Marine Organisms: Genetics, Ecology and Evolution*, pp. 257–73. Plenum Press, New York.

de Beer, G. (1958) *Embryos and Ancestors*, 3rd edn. Clarendon Press, Oxford.

Duxbury, A.C. & Duxbury, A. (1984) *An Introduction to the World's Oceans*. Addison-Wesley, Massachusetts.

Enright, J.T. (1977) Diurnal vertical migration: adaptive significance and timing. Part 1. Selective advantage: a metabolic model. *Limnol. Oceanogr.* **22**: 856–72.

Eyster, L.S. (1979) Reproductive and developmental variability in the opisthobranch *Tenellia pallida*. *Mar. Biol.* **51**: 133–40.

Fancett, M.S. & Kimmerer, W.J. (1985) Vertical migration of the demersal copepod *Pseudodiaptomus* as a means of predator avoidance. *J. Exp. Mar. Biol. Ecol.* **88**: 31–43.

Forward, R.B. Jr (1988) Diel vertical migration: zooplankton photobiology and behaviour. *Oceanogr. Mar. Biol. Ann. Rev.* **26**: 361–93.

Ghiselin, M.T. (1987) Evolutionary aspects of marine invertebrate reproduction. In: Giese, A.C., Pearse, J.S. & Pearse, V.B. (eds) *Reproduction of Marine Invertebrates* Vol. 9, pp. 609–65. Blackwell Scientific Publications, Boston.

Gliwicz, M.Z. (1986) Predation and the evolution of vertical migration in zooplankton. *Nature* **320**: 746–8.

Hardy, A.C. (1953) Some problems of pelagic life. In: *Essays in Marine Biology*, pp. 101–21. Oliver & Boyd, Edinburgh.

Hardy, A.C. & Gunther, E.R. (1935) The plankton of the South Georgia whaling grounds and adjacent water, 1926–1927. *Discovery Rep.* **11**: 1–456.

Harper, J.L. (1977) *Population Biology of Plants*. Academic Press, London.

Hoagland, K.E. & Robertson, R. (1988) An assessment of poecilogony in marine invertebrates: phenomenon or fantasy? *Biol. Bull.* **174**: 109–25.

Hughes, R.N. (1989) *A Functional Biology of Clonal Animals*. Chapman & Hall, London.

Jackson, J.B.C. (1979) Morphological strategies of sessile animals. In: Larwood, G. & Rosen, B.R. (eds) *Biology and Systematics of Colonial Organisms*, pp. 499–555. Academic Press, London.

Jackson, J.B.C., Buss, L.W. & Cook, R.E. (eds) (1985) *Population Biology and Evolution of Clonal Organisms*. Yale University, New Haven.

Krebs, J.R. & McCleery, R.H. (1984) Optimization in behavioural ecology. In: Krebs, J.R. & Davies, N.B. (eds) *Behavioural Ecology. An Evolutionary Approach*. 2nd edn, pp. 91–121. Blackwell Scientific Publications, Oxford.

Levinton, J. (1988) *Genetics, Paleontology, and Macroevolution*. Cambridge University Press, Cambridge.

Little, C. (1983) *The Colonisation of Land*. Cambridge University Press, Cambridge.

Little, C. (1990) *The Terrestrial Invasion*. Cambridge University Press, Cambridge.

Longhurst, A.R. (1976) Vertical migration. In: Cushing, D.H. & Walsh, J.J. (eds) *The Ecology of the Seas*, pp. 116–37. Blackwell Scientific Publications, Oxford.

Lopez, G.R. (1988) Comparative ecology of the macrofauna of freshwater and marine muds. *Limnol. Oceanogr.* **33**: 946–62.

MacArthur, R.H. & Wilson, E.O. (1967) *The Theory of Island Biogeography*. Princeton University, New Jersey.

Mackie, G.O., Pugh, P.R. & Purcell, J.E. (1987) Siphonophore biology. *Adv. Mar. Biol.* **24**: 97–262.

McLaren, I.A. (1963) Effects of temperature on growth of zooplankton and the adaptive value of vertical migration. *J. Fish. Res. Bd Can.* **20**: 685–727.

Margalef, R. (1963) On certain unifying principles in ecology. *Am. Nat.* **97**: 357–74.

Margulis, L. (1970) *Origin of Eukaryotic Cells*. Yale University, New Haven.

Margulis, L. & Sagan, D. (1986) *Origins of Sex*. Yale University, New Haven.

Nieuwkoop, P.D. & Sutasurya, L.A. (1981) *Primordial Germ Cells in the Invertebrates*. Cambridge University Press, Cambridge.

Olive, P.J.W. (1985) Covariability of reproductive traits in marine invertebrates: implications for the phylogeny of the lower invertebrates. In: Conway Morris, S., George, J.D., Gibson, R. & Platt, H.M. (eds) *The Origins and Relationships of Lower Invertebrates*, pp. 42–59. Clarendon Press, Oxford.

Ryland, J.S. & Warner, G.F. (1986) Growth and form in modular animals: ideas on the size and arrangement of zooids. *Phil. Trans. R. Soc. Lond. B* **313**: 53–76.

Sebens, K.P. (1987) The ecology of indeterminate growth in animals. *Ann. Rev. Ecol. Syst.* **18**: 371–407.

Sibly, R. & Calow, P. (1985) Classification of habitats by selection pressures: a synthesis of life-cycle and *r/K* theory. In: Sibly, R.M. & Smith, R.H. (eds) *Behavioural Ecology*, pp. 75–90. Blackwell Scientific Publications, Oxford.

Steele, J.H. (1978) Some comments on plankton patches. In: Steele, J.H. (ed) *Spatial Pattern in Plankton Communities*, pp. 1–20. Plenum Press, New York.

Strathmann, R.R. (1985) Feeding and non-feeding larval

development and life-history evolution in marine invertebrates. *Ann. Rev. Ecol. Syst.* **16**: 339–61.

Thorson, G. (1950) Reproductive and larval ecology of marine bottom invertebrates. *Biol. Rev.* **25**: 1–45.

Todd, C.D., Havenhand, J.N. & Thorpe, J.P. (1988) Genetic differentiation, pelagic larval transport and gene flow between local populations of the intertidal marine mollusc *Adalaria proxima* (Alder & Hancock). *Funct. Ecol.* **2**: 441–51.

Townsend, C.R. (1989) The patch dynamics concept of stream community ecology. *J. Am. Benth. Soc.* **8**: 36–50.

Underwood, A.J. & Fairweather, P.G. (1989) Supply-side ecology and benthic marine assemblages. *Trends Ecol. Evol.* **4**: 16–20.

Zaret, T.M. & Suffern, J.S. (1976) Vertical migration in zooplankton as a predator avoidance mechanism. *Limnol. Oceanogr.* **21**: 804–13.

9
Speciation and Biogeography

J.H.R. GEE

9.1 INTRODUCTION

If ecology can be said to have a principal aim, it is to arrive at an understanding of the factors that determine the species composition of natural communities. Consistent spatial trends in diversity, and the far from random distributions of the individual species that form these assemblages, present us with evidence of the end product of some form of organization. There are few species or communities for which the answers to the questions begged by the patterns are more than tentative. To be sure, some part of the distribution of species is due to underlying patterns in the distribution of resources or physical conditions; species do not succeed unless the environment offers them the material necessities of life and a tolerable climate. It is also likely that the distribution of some species, and the composition of some assemblages, is determined by interactions between the potential members: competition may limit the similarity of species able to coexist in a community; predation may restrict the distribution of vulnerable species, or permit the coexistence of competitors that otherwise would be mutually exclusive; mutualistic interactions may demand the coexistence of particular species pairs.

All of these factors operate to select a subset of species from the set of potential community members; however, the nature of the subset is also determined by the content of the set from which it is drawn. The size and composition of the set of potential species in any locality depends on the balance between the rate of loss of species through emigration or, more probably, extinction, and the rate of appearance of new species through speciation or immigration. Although rates of immigration are determined partly by the inherent

vagility of the species, movements are also constrained by natural barriers, such as land masses (in the case of aquatic species) or zones of inimical conditions, such as extremes of temperature or salinity.

On a geological time-scale, speciation, and the subsequent spread of new species away from their site of origin, are events that have taken place on a moving stage. The biogeography of aquatic species is a result of the interplay between speciation and dispersal on the one hand, and on the other the changes in the distribution of water bodies and in their physical conditions, wrought by continental drift and other geomorphic events. The fundamental processes leading to speciation and biogeographic patterns are common to both terrestrial and aquatic biota. However, the end results of these processes differ, in part because of the differences in the physical properties of air and water, and in part because the areas that represent barriers to dispersion and colonization by terrestrial organisms are often immigration corridors or suitable habitats for aquatic species.

9.2 SPECIATION

Speciation is the splitting of one set of interbreeding populations, constituting a species, into two or more sets, each reproductively isolated. It requires the development of a genetic difference between a group of individuals and the remainder of the parent species. Furthermore, it is necessary for there to be some restriction in gene flow between the group sharing this genetic feature and the rest of the ancestral stock. Without this restriction, the genetic distinctiveness of the divergent group would soon be lost.

9.2.1 Allopatric speciation

Evolutionary biologists have described several ways in which speciation might take place. They differ mainly in the nature of the restriction in gene flow between the incipient species. In allopatric speciation, the scenario is as follows: first, the populations constituting an existing species are separated by a physical barrier, so that they occupy two non-overlapping areas (a condition known as allopatry). This might happen through a change in oceanic circulation patterns, altered sea level during glaciation, a lowering of lake level that isolates a bay, or by earth movements that disrupt a drainage basin. The barrier prevents migration between the populations and so restricts or eliminates gene flow. Each population then evolves in response to local selection pressures, which are likely to differ between the two areas, or diverges as a result of chance. In time, the populations will become genetically distinct.

When migration is subsequently re-established between the areas by removal of the barrier, several outcomes are possible. If the populations have diverged genetically to only a small degree, it is likely that the differences will disappear gradually as a result of gene flow between them. If the populations have been separated for a long time, or have been subjected to strong selection or rapid genetic drift, so that the genetic differences are extensive, it is probable that hybrids between them will not be viable. Should hybridization occur, the hybrids or their offspring are likely to be infertile. In this case the populations have become reproductively isolated and allopatric speciation has taken place. If the genetic differences are not sufficient to prevent the production of viable, fertile hybrids, but the hybrids are less fit than the non-hybrids, then it is possible that reproductive isolating mechanisms would evolve as a result of selection against individuals that mate outside their group. A difference in timing of reproduction, in habitat selection, in courtship behaviour or in the shape of genitalia might evolve in this way and result in reproductive isolation.

On North Atlantic coasts there are five related species of small intertidal isopods of the genus *Jaera*. The females are indistinguishable morpho-logically, but the males differ in detailed structure of the pereiopods. Forced matings between species in the laboratory (Solignac, 1977) have shown that many of the crosses can produce viable offspring, although F1 and F2 generation progeny often show reduced viability. Despite the frequent co-occurrence of species under the same stone, natural hybrids are rare, apparently because females do not respond to the mating stimuli of males of another species. This is an example of a behavioural mechanism of reproductive isolation.

9.2.2 Sympatric speciation

An alternative mechanism, sympatric speciation, is believed to occur without the need for geographic separation of populations. In this case, a genetic or behavioural change affecting a subgroup results in the partial restriction or complete cessation of gene flow, despite the fact that it continues to occupy part of the range of the ancestral stock. A reproductive isolating mechanism must therefore be in place at an early stage, and genetic divergence follows. Sympatric speciation is still treated with scepticism by some evolutionary biologists; they suggest that some cases of sympatric speciation in fact involve microgeographic separation of individuals that do not normally stray far from home territory, and that some represent allopatrically-derived species that subsequently became sympatric due to a shift in their distribution.

9.2.3 Genetic variability and larval dispersal

Given the importance of a restriction in gene flow to the genetic divergence of populations, it is not

Table 9.1 Dispersal ability, measured as length of larval life, and genetic variability in five unnamed species of *Capitella*. From Grassle & Grassle (1977)

Species	Length of larval life	% electrophoretic loci polymorphic
IIIa	None	50
I	Several hours	50
II	6–24 hours	25
III	≥2 weeks	20
Ia	2 months	12.5

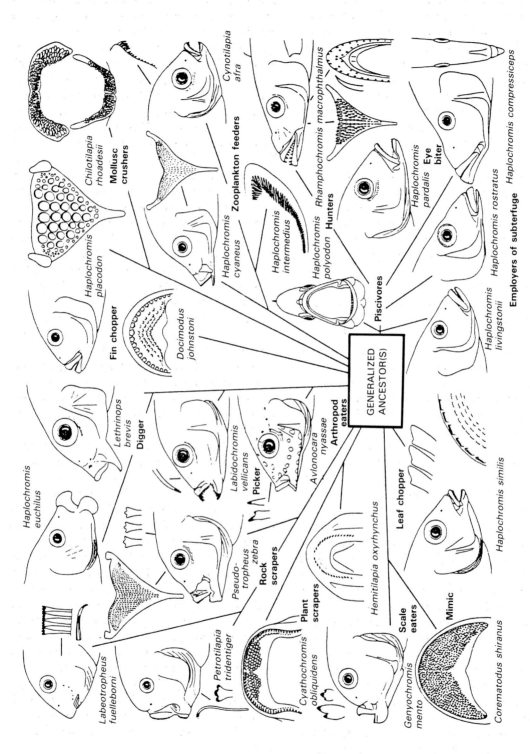

Fig. 9.1 Dietary specializations in the haplochromine cichlids of Lake Malawi, represented by outline drawings of heads, jaws (U-shaped structures), pharyngeal bones (triangular), individual teeth and a gill raker. Dietary habits are mostly self-explanatory: 'mimics' resemble herbivorous species, but in fact scrape scales from other fish; 'employers of subterfuge' are fish predators that lure prey by lying on the bottom and resembling a carcass. From Fryer & Iles (1972).

surprising that genetic variability within species is often related to their dispersal ability. For marine species with planktonic larvae, the length of the larval stage gives an indication of dispersal ability. Five sibling species of the infaunal polychaete *Capitella*, previously thought to be the same species, have larval lifespans which vary greatly in duration (Table 9.1). Genetic variability, measured by the proportion of loci that are polymorphic, is greater the more mobile the larvae.

9.2.4 Cichlid fish of the African Great Lakes

In comparison with the sea, fresh water is undoubtedly deficient in major taxa (see Chapter 1). At the species level, the story is quite different. For instance, 41% of the known species of teleost fish are found in fresh water, although fresh water represents only 0.0093% of the volume of the aquatic realm (Moyle & Cech, 1982). Many of the world's largest and oldest freshwater lakes have been the sites of fish speciation on a impressive scale. Since individual river systems are effectively isolated from each other by the stretches of salt water between their estuaries, there is often little dispersal of species between lakes, unless they are located on the same system. Lakes, therefore, have developed very high levels of endemicity in their fish faunas. Lake Baikal in southern USSR, the world's deepest and seventh largest, has 18 species of endemic cottids. There is the same number of endemic species of cyprinids in Lake Lanao in the Philippines, and 14 endemic cyprinodontid species of the genus *Orestias* in Lake Titicaca in Peru.

By far the most spectacular examples of speciation and adaptive radiation are to be found amongst the cichlid species of the African Great Lakes. Each of these lakes (Victoria, Malawi and Tanganyika) contains more coexisting fish species than any other lake outside the African Rift Valley. The exact numbers are a matter of debate, but at least 200 species of cichlid fish are known from the Mwanza Gulf of Victoria, and the species list for Malawi runs to more than 400. For comparison, the whole of the European freshwater fish fauna consists of just 192 species (Lowe-McConnell, 1987).

Most of the speciation seen in these African lakes has occurred within the group known as the haplo-chromine cichlids. It has resulted in an extraordinary diversity of morphology, particularly in the mouthparts (Fig. 9.1). Many of the species are specialized in diet, and restricted in their distribution with respect to substrate type and depth range. Ecologically equivalent types have evolved in each of the three Great Lakes; for instance, mollusc crushing, with associated massive pharyngeal bones, has appeared in *Haplochromis placodon* in Lake Malawi, *Lamprologus tretocephalus* in Lake Tanganyika and *Labrochromis* species in Lake Victoria. Colour polymorphisms are common within species; in some cases electrophoretic studies and close observation of behaviour and ecological characteristics have indicated that these morphs are separate species. Some well-established sibling species, which differ little in morphology, are most easily distinguished by differences in male breeding colours (e.g. *Oreochromis squamipinnis* and *O. saka* in Lake Malawi).

The rate of speciation is generally a matter for conjecture, but in one instance there is evidence that it is rapid. Lake Nagugabo was isolated from Victoria by a sand bar about 4000 years ago. In that time it has acquired seven cichlid species, of which five are endemic and distinguished in colour and morphology from their closest relatives in Lake Victoria.

Haplochromine cichlids are not the only taxon to occur commonly in the Great Lakes, but none of the others has undergone such explosive speciation. The closely related tilapias have been much more conservative. There is no single characteristic of haplochromine cichlids that explains their rapid speciation, but several may be contributory. Possession of pharyngeal jaws has made it possible for the mouth to be adapted for specialized methods of food collection. An unspecialized body shape has facilitated occupation of a wide range of habitats and adoption of many different patterns of trophic behaviour. Short generation times and the production of several broods in a single year may have permitted more rapid reshuffling of genetic material than in other groups. The habit of maternal mouth-brooding of the young, or, in some cases, the guarding of eggs and young at a nest site, tends to restrict gene flow between groups occupying different areas.

There is still no consensus on the mechanism of speciation amongst the African cichlids. Proponents of allopatric speciation originally suggested that the large number of sympatric cichlid species was due to successive colonizations of the Great Lakes by species that evolved allopatrically in neighbouring rivers and lakes. Others (Fryer, 1977; Ribbink *et al.*, 1983) now believe that the species originated within the lakes, but that this apparent sympatric speciation is in fact a case of microgeographic allopatric speciation. Changes in lake level probably resulted in the temporary isolation of groups of individuals, which rapidly adapted to local conditions. When lake levels rose again, the groups remained genetically isolated. This mechanism would have been enhanced by the fact that many species show natural tendencies to remain within very localized areas (i.e. are highly philopatric), and are associated with particular types of habitat. There may have been as many as 20 cycles of glaciation in the last 2 million years, each of which could have affected climate and lake levels in Africa.

In Lake Malawi, *Pseudotropheus zebra* is present in four colour morphs, blue-black bars (BB), orange-black (OB), blue (B) and white (W). Electrophoretic studies have shown that in places where all four morphs live side-by-side, there is assortative mating, forming two groups BB/OB and B/W that are at least partially isolated genetically (McKaye *et al.*, 1984). Assortative mating and sexual or social selection based on colour might provide sufficient genetic isolation to lead to sympatric speciation. Equally, the rapid changes in colour that are possible under sexual or social selection reduce the length of time for which populations need to be isolated spatially, in order for allopatric speciation to take place.

9.3 BIOGEOGRAPHIC PATTERNS

9.3.1 Plate tectonics and biogeographic provinces

Prior to the Permian (which commenced about 280 million years BP) the terrestrial world consisted of a single large land mass, Pangaea. Surrounding this ancient continent was an immense continuous expanse of ocean. Movements of the crustal plates first resulted in the bisection of Pangaea by the Tethys Sea, and later caused the fragmentation of the northern and southern parts (Laurasia and Gondwanaland). This fragmentation, which produced many of the land masses that are familiar today, did not occur simultaneously. Sections of Laurasia and Gondwanaland broke off successively and eventually were carried by plate tectonics to their latter day positions.

Throughout the known history of the seas, the marine environment has been divisible into a number of biotic areas, each distinguished by the presence of a characteristic set of endemic species. These areas are known by biogeographers as floral or faunal provinces. The total number of species in the world's marine biota must depend partly on the number of distinct provinces and partly on the diversity within each province. Provinces seem to be delimited by the major oceanic circulation systems, which restrict the free movement of organisms between them. Restriction according to patterns of water movement is most evident in species which are planktonic or have planktonic propagules. In turn, the circulation patterns are dependent on the disposition of land masses which act as barriers to water movements. Some large expanses of water, which are not partitioned by currents or continental barriers, can accommodate more than one province, apparently because population movements over immense distances are so slow as to permit some differentiation.

With a knowledge of the factors that define provinces, it is possible to predict the number of marine provinces at various stages in the fragmentation of Pangaea. In the Permian, the predicted number of continental shelf provinces is eight: two polar, two tropical and four temperate. Late Permian brachiopod faunas appear to have had a pattern of provincialization close to this prediction (Schopf, 1974). Modern shelf regions divide into 18 predicted provinces, which are close in number and distribution to the actual biogeographic provinces shown, for example, by present day bryozoans.

The distribution patterns of freshwater taxa over land masses are much more akin to those of the terrestrial than the marine biota. Although many insects that are fully aquatic in their larval stages have adults that are winged, the dispersal of many species has depended on the presence of land connections. Most mayflies (Ephemeroptera) have fully functional wings, but their long-distance dispersal is poor (Edmunds, 1972), being limited to the occasional windblown female or, perhaps, an egg mass. Females in flight readily succumb to desiccation. Hence, oceanic islands are generally depauperate in mayflies; if present at all, they are often represented only by the robust genera *Baetis* and *Cloeon*. *Cloeon* is represented by two phyletic lines in Australia, but is absent from New Zealand and New Caledonia. The Baetidae of New Guinea apparently evolved from a single invading species. Amongst other freshwater organisms, there are notable exceptions to this requirement for land connections. Species that have small resting eggs, resistant to desiccation, may be carried long distances by the wind. Consequently, many genera, and even some species of copepods, cladocerans and rotifers, are cosmopolitan.

Evidence from the biogeography of ephemerids indicates the likely sequence of fragmentation of Gondwanaland. The last connection between the southern continents appears to have been between Australia and South America via Antarctica. Comparisons of the ephemerid faunas of these land masses and those of New Zealand and New Caledonia, suggest that the latter two broke away from Australia and the rest at an earlier date. Many mayfly species are dependent on cool-water habitats for survival. Their absence from some of the southern hemisphere land masses that drifted north, may have resulted from a gradual warming of the climate. Thus Siphlonuridae are represented by several species in New Zealand, but are absent from New Caledonia. The greater similarities between the ephemerid faunas of Australia, New Zealand and South America, than between these three and Africa suggests either that Africa parted company sooner, or that it too suffered a loss of cool-water species as it drifted northward.

In the northern hemisphere, mayflies exhibit the classic biogeographic patterns shown by many terrestrial species. There is a distinct Holarctic fauna, with close similarities between the Palaearctic and Nearctic groups (Europe and North America were united at least as late as the Palaeocene — 65 million years BP). There is also evidence of exchange of species along the land connections between the Oriental and Palaearctic, and between the Oriental and Ethiopian provinces.

9.3.2 Freshwater biogeography and terrestrial vegetation

All freshwater bodies are influenced to some degree by the terrestrial vegetation that lines their banks or clothes their catchments. The smaller the water body, the greater the proportional influence of bankside, or riparian, vegetation. In small streams running through deciduous forest, the vegetation imposes particular environmental conditions. These streams are subject to heavy shading in summer, which keeps water temperatures relatively low, and only light shade in winter. They also receive a substantial input of organic matter from the vegetation, principally in the form of falling leaves in autumn. A large fraction of the trophic energy budget of such streams is provided by this externally-derived (allochthonous) material. The nature of riparian and catchment vegetation determines the extent of soil erosion, which in turn influences the dynamics of dissolved nutrients. It is also well-established that catchment vegetation affects the proportion of precipitation that appears as surface runoff, through variations in evapotranspiration with plant type.

It is not surprising, therefore, that there are close correlations between the geographic distributions of some freshwater taxa and some terrestrial biomes. The 14 species of the cased caddis genus *Pycnopsyche*, insects that are usually found only in streams of less than about 3 m breadth, are restricted mainly to the area occupied by the temperate deciduous forest of eastern North America (Fig. 9.2). Ross (1963) suggests that this set of species, and each of a number of other caddis groups that share similar distributions, evolved from a single ancestor that became isolated in the eastern deciduous forest in the middle Tertiary. Of course, correlations such as these do not prove that the distribution of the caddis is caused by the conditions imposed by the forest. It could be that both caddis and forest are responding to other environmental cues; but the fact that species of caddis typical of larger streams, in which the forest influence is less pronounced, have geographic

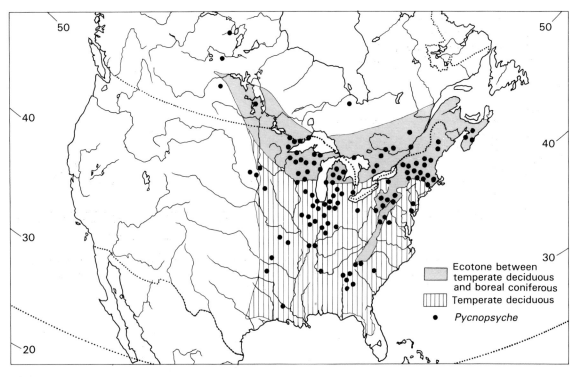

Fig. 9.2 Distribution of the caddisfly genus *Pycnopsyche* in North America in relation to the temperate deciduous forest and the ecotone between temperate deciduous forest and boreal coniferous forest, or taiga. From Ross (1963).

distributions that cross the boundaries of several terrestrial biomes, infers a direct causal link.

9.3.3 Latitudinal trends in species diversity

One of the most consistent biogeographic patterns is the tendency for the number of species in either particular taxa or specific communities to increase with decreasing latitude. With only a few exceptions, diversities peak near the Equator. Many of the reported cases involve terrestrial organisms, but it is clear that similar latitudinal trends in diversity can be found in the aquatic environment. Figure 9.3 illustrates latitudinal trends shown by: (i) benthic invertebrates from freshwater streams; and (ii) by marine predatory gastropods.

Such consistent patterns demand ecological explanations. Pianka (1988) lists 10 hypotheses, many of which are discussed in other general textbooks of ecology. It is fair to say that none of these explanations is entirely satisfactory. The hypoth-

eses based on differences in environmental stability (i.e. the degree of temporal constancy in conditions) between the tropics and temperate regions, find some support in studies on aquatic organisms. For instance, Taylor and Taylor (1977) found a marked step in the diversity of predatory prosobranch molluscs occurring at around latitude 40°N. This latitude marks the transition between northern zones, in which primary productivity is markedly seasonal, and southern zones, in which production is more or less continuous. Coincident with this transition was a change in the composition of gastropod diets. In the northern zone, the dominant families are the Buccinidae and the Turridae. The former are generalist predators and scavengers of carrion; they prey on bivalves, polychaetes and crustaceans. The latter are specialist predators on infaunal polychaetes which consume detritus in the sediments. By the time organic matter produced in the surface layers has accumulated in the sediments, the seasonal fluctuations in productivity are greatly at-

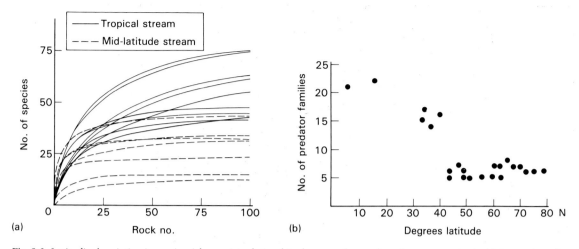

Fig. 9.3 Latitudinal variation in species richness: (a) relationships between the number of stones examined and the number of species of insects found in tropical and mid-latitude streams; (b) latitudinal change in the number of predatory gastropod families on the eastern Atlantic shelf. From Stout & Vandermeer (1975); Taylor & Taylor (1977).

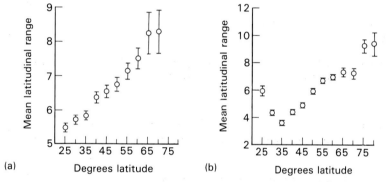

Fig. 9.4 Latitudinal gradients in the mean distributional ranges of North American marine shelled molluscs (a) and coastal and freshwater fish (b). Bars indicate standard errors. From Stevens (1989).

tenuated, so it is a near-constant food supply for benthic detritivores. The northern predators are therefore either generalists, or specialists on unusually stable prey populations. To the south of 40°, the predatory gastropod fauna consists of a larger number of families whose members are dietary specialists. For instance, the Cassidae prey on echinoids, the Tonnidae on holothurians and the Harpidae specialize on decapod crustaceans. These restricted diets may be made possible by the more constant availability of prey which results from the reduced seasonal variation in primary productivity. So, southern gastropod assemblages seem to consist of a larger number of species, with narrower trophic niches, than their northern counterparts.

In parallel with the relationship between species diversity and latitude, runs a relationship between the geographical range of species and the latitude which marks the centre of their distribution. Species found at higher latitudes tend to have wider latitudinal ranges (Fig. 9.4). Once again, environmental variability is suggested as the root cause of this relationship. Seasonal variation in temperature, and in other climatic variables, is greater at higher latitudes. Species of high latitudes are selected for tolerance of this variability, and thus are physiologically adapted to living conditions over a wide geographic range. Conversely, there is no selective advantage in an organism resident near the Equator being tolerant of fluctuating conditions.

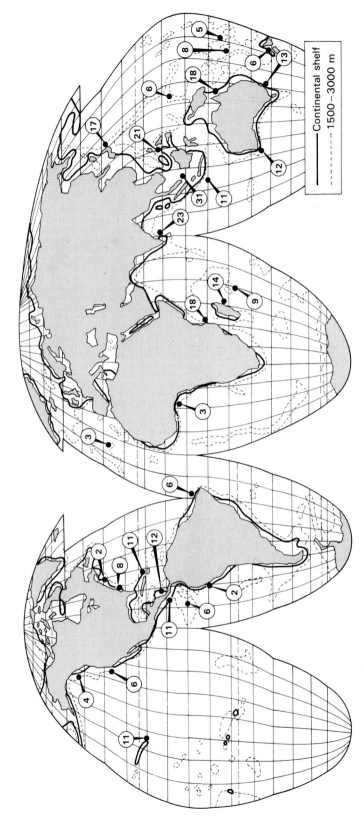

Fig. 9.5 The numbers of species of clypeasteroid echinoderms found in various geographic regions. From Ghiold & Hoffman (1986).

This effect may provide yet another explanation of the latitude–diversity relationship. If reduced seasonal variation in climate leads tropical organisms to be less tolerant of climatic change, then a tropical organism will perceive any environment to be more variable spatially than will a climatically-tolerant temperate species. A small shift in location is likely to take a tropical organism out of its optimal climatic range, but will have less impact on a temperate species. Stevens (1989) has suggested that the number of species in tropical communities is inflated by the presence of species from neighbouring areas. These intruders will not be well-adapted to local conditions, but their populations may nevertheless persist due to continuous immigration.

9.3.4 Longitudinal trends in diversity

In addition to variations with latitude, many marine invertebrate taxa show consistent differences in diversity between ocean basins. The diversity distribution of the clypeasteroid echinoderms is typical (Fig. 9.5). Species richness peaks in the Indo–West Pacific area, and falls to a minimum on the east coast of the Atlantic. The reasons for this distribution may be twofold. Firstly, in the Indo–West Pacific there is a large area of shelf that is subdivided by a complex series of islands and deep troughs. This morphology leads, no doubt, to opportunities for genetic isolation and speciation. In contrast, the East Atlantic shelf area is more continuous. The north–south orientation of the major land masses now prevents species generated in the tropical Indo–West Pacific from gaining access to the Atlantic. Secondly, temperature variation in the Indo–West Pacific shelf area is less extreme than in the Atlantic, because the latter is more heavily influenced by the the climate of the adjacent continents. As we have already seen (Section 9.3.3), there is often an inverse correlation between environmental variability and species richness.

9.3.5 Patterns in the third dimension — depth

Surveys of the variation in diversity of benthic invertebrates with depth, in oceans and lakes, reveal patterns that are consistent but different (Fig. 9.6). In the deep oceans, diversity first increases with depth, reaching a maximum at around 2000 m, and then decreases. This pattern has been explained as a result of two opposing influences. A decrease in food availability leads to an exponential decline in benthic biomass and population density with depth. This leads to a gradual loss of species, particularly those of larger body size. Against this trend, the increase in the temporal stability of the benthic environment with depth may facilitate dietary specialization and enhance diversity.

Lakes show no comparable initial increase in benthic species diversity with depth. Instead, there is a marked decline, which is more rapid the more eutrophic the lake. This pattern may be the result of several factors. As in the marine situation, food availability decreases with depth, causing a similar change in benthic population density. Many freshwater invertebrates require occasional access to the water surface, either for respiration, or to enable them to pupate or emerge as adults: this may limit their depth range. Thermal stratification can lead to stressful conditions in the hypolimnion, such as low oxygen tensions, which seem to be tolerated by a relatively small number of species. Finally, habitat diversity decreases rapidly from the littoral zone to the profundal, so that the deeper parts offer fewer ecological opportunities.

9.3.6 The area effect

On oceanic islands, the larger the land area, the greater the number of species it supports. Many studies have shown a direct linear relationship between species diversity and island area, when these are expressed logarithmically. A similar relationship, but usually with a lower slope, applies to successively larger areas within a particular habitat patch. Species–area relationships have been found not only for oceanic islands, but also for many other types of isolated habitat patches. In this context, water bodies count as wet islands surrounded by dry land (Fig. 9.7).

Although the existence of a relationship between area and species number is well-established, its ecological basis is not clear (Williamson, 1981). It is probable that when 'islands' of different size are

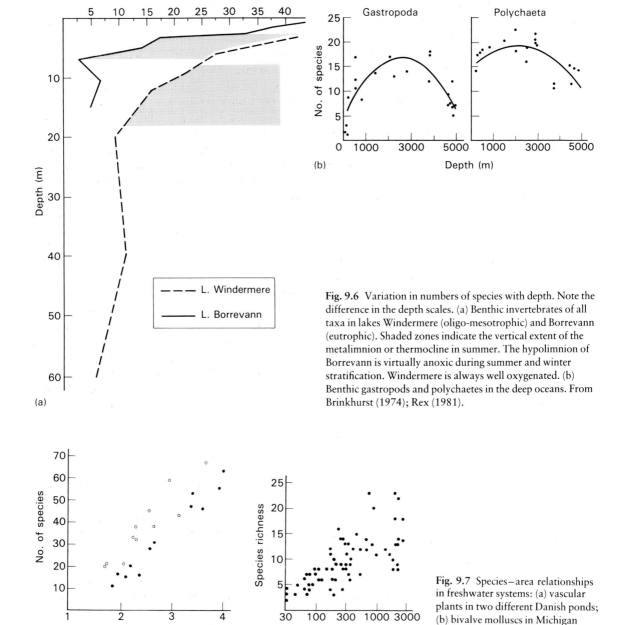

(a)

(b)

Fig. 9.6 Variation in numbers of species with depth. Note the difference in the depth scales. (a) Benthic invertebrates of all taxa in lakes Windermere (oligo-mesotrophic) and Borrevann (eutrophic). Shaded zones indicate the vertical extent of the metalimnion or thermocline in summer. The hypolimnion of Borrevann is virtually anoxic during summer and winter stratification. Windermere is always well oxygenated. (b) Benthic gastropods and polychaetes in the deep oceans. From Brinkhurst (1974); Rex (1981).

Fig. 9.7 Species–area relationships in freshwater systems: (a) vascular plants in two different Danish ponds; (b) bivalve molluscs in Michigan streams. From Moller & Rordam (1985); Strayer (1983).

compared, they also differ in the number of distinct microhabitats that they offer. Larger areas may thus contain more species, simply because there is greater environmental heterogeneity and are more exploitable niches. Some species may be lacking from small islands, because the area is not large enough to support a viable population. This explanation is most likely to apply to large predators. The Macarthur and Wilson (1967) theory of island biogeography explains the lower diversity found on small islands in terms of the higher extinction rates suffered by small populations. Providing that the rate of arrival of immigrant species remains constant, higher extinction rates should result in a lower equilibrium number of species. Finally, the larger an island the greater the probability that there will be reproductive isolation of groups of individuals within species, and the greater the rate of speciation.

In the case of lakes as islands, the area effect is confounded by differences in age. The fate of many lakes is to become shallower, and eventually disappear, as their basins fill with sediment. On the whole, lake basins of larger area tend also to be deeper. Therefore, it is often true that the largest lakes are also the oldest. The length of time that has been available for existing species to colonize and for speciation to take place, may be as important a factor in explaining the high diversities of Lake Baikal and the Great Lakes of the African Rift Valley, as is their immense size.

The area effect may help explain part of the relationship between species diversity and latitude. Schopf *et al.* (1977) show that the area of continental shelf decreases from the Equator to the Poles. The greater area available near the Equator may favour more rapid speciation, leading to greater diversity, providing there is no concomitant increase in the rate of species extinction.

9.3.7 Victims of climatic change

Palaeoclimatic changes in particular locations appear to have been due to movements of land masses as a result of plate tectonics, and to the oscillations in world climate that resulted in glaciations in temperate zones. These two processes operate on very different time-scales. Changes in the African climate, as the continent drifted northwards, have already been noted (Section 9.3.1); they resulted in the restriction of the cool-water fauna to high altitude streams.

Glaciation has had more recent and more dramatic effects on the distribution of freshwater organisms. It seems highly probable that the British Isles were rendered fishless by the last glaciation, in which only areas in the southeast remained ice-free (Wheeler, 1977). The only truly indigenous, stenohaline fish species now in Britain are the few that colonized the rivers of the southeast during the short postglacial period, when there was a land bridge to Europe. The rest of the indigenous fish fauna consists of species that were able to recolonize the rivers via the sea.

As the ice retreated northwards, and the burden was lifted from previously glaciated areas, these areas rose in relation to sea level. This had the effect of isolating low-lying lakes and lagoons that had been connected with the sea. The organisms they contained gradually became adapted to the progressive reduction in salinity, and now constitute a glacial relict fauna. Populations which had been interbreeding previously, became genetically isolated. In many cases this has led to evolutionary divergences in body form. A good example is the copepod *Limnocalanus*, which is to be found in the Baltic Sea and in a number of lakes that became isolated from the Baltic with the eustatic rise of Scandinavia. These isolated populations have diverged in appearance from the ancestral form. The degree of divergence is related to the length of time since isolation, itself a consequence of the height of each lake above sea level.

Changing temperatures have also served to isolate relict faunas in other ways. Danish springs contain at least three different classes of thermal relicts (Nielsen, 1950). Arctic species, such as the caddis *Apatania muliebris*, survive in some springs due to their consistently low temperatures. Others contain oceanic species that colonized Denmark during an interglacial, and persist because the springs offer warm winter conditions. The third group, continental relicts, entered Denmark in a period of more continental climate and are still found in springs, where shallow waters exposed to the sun provide high summer temperatures.

9.3.8 The hand of man

The present distribution patterns of many species bear the mark, not of natural processes such as speciation, dispersal and the splitting of species

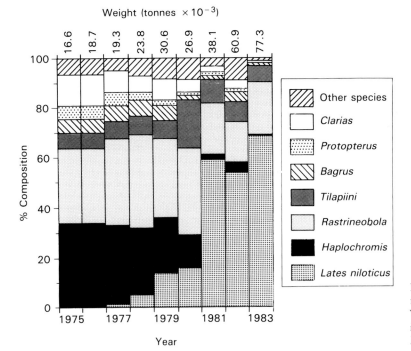

Fig. 9.8 Changes in the landings of fish from the Kenyan waters of Lake Victoria. The Nile perch *Lates* was first introduced in 1960. From Lowe-McConnell (1987).

distributions by the intrusion of natural barriers (vicariance), but of human intervention. The fact that stenohaline fish species are to be found in the northern fresh waters of Britain and Ireland, is almost entirely due to deliberate introductions or to escapees from ornamental waters. In general, these modifications of the British ichthyofauna have served to increase fish diversity. Elsewhere, similar introductions, intended to improve fisheries, have had less desirable results. The recent release of the predatory Nile perch *Lates niloticus* into Lake Victoria has already had disastrous consequences for the indigenous haplochromine cichlids (Fig. 9.8). This is an irreversible ecological experiment on a vast scale; the final consequences for the unique Victorian fish fauna are not yet known.

Although less isolated, and probably less vulnerable, marine communities have also been changed by human hands. Deliberate introduction of the American cord-grass *Spartina alterniflora* to Britain, resulted first in hybridization with a native European species, then the development of a vigorous polyploid species *S. anglica*. This new species is so successful that it has colonized large areas of pre-

viously open estuarine mud-flats, causing dramatic changes in the resident fauna. In what is frequently described as a more environmentally sympathetic era, but in which man's potential for ecological mischief is unsurpassed, we may hope that a less cavalier approach to nature will prevail.

FURTHER READING

Brinkhurst, R.O. (1974) *The Benthos of Lakes.* Macmillan, London.

Edmunds, G.F. (1972) Biogeography and evolution of Ephemeroptera. *Ann. Rev. Entomol.* 17: 21–42.

Fryer, G. (1977) Evolution of species flocks of cichlid fishes in African lakes. *Z. Zool. Sytem. Evol.* 15: 141–65.

Fryer, G. & Iles, T.D. (1972) *The Cichlid Fishes of the Great Lakes of Africa.* Oliver & Boyd, Edinburgh.

Ghiold, J. & Hoffman, A. (1986) Biogeography and biogeographic history of clypeasteroid echinoids. *J. Biogeogr.* 13: 183–206.

Grassle, J.F. & Grassle, J.P. (1977) Life histories and genetic variation in marine invertebrates. In: Battaglia, B. & Beardmore, J.A. (eds) *Marine Organisms: Genetics, Ecology, and Evolution*, pp. 347–64. NATO Conference Series IV, Marine Sciences; Vol. 2. Plenum Press, New York.

Lowe-McConnell, R.H. (1987) *Ecological Studies in Trop-*

ical Fish Communities. Cambridge University Press, Cambridge.

Macarthur, R.H. & Wilson, E.O. (1967) *The Theory of Island Biogeography*. Princeton University Press, New Jersey.

McKaye, K.R., Kocher, T., Reinthal, P., Harrison, P. & Kornfield, I. (1984) Genetic evidence for allopatric and sympatric differentiation among colour morphs of a L. Malawi cichlid fish. *Evolution* 38: 215–19.

Moller, T.R. & Rordam, C.P. (1985) Species numbers of vascular plants in relation to area, isolation and age of ponds in Denmark. *Oikos* 45: 8–16.

Moyle, P.B. & Cech, J.J. (1982) *Fishes: An Introduction to Ichthyology*. Prentice Hall, New Jersey.

Nielsen, A. (1950) On the zoogeography of springs. *Hydrobiologia* 2: 313–21.

Pianka, E. (1988) *Evolutionary Ecology*. Harper & Row, New York.

Rex, M.A. (1981) Community structure in the deep-sea benthos. *Ann. Rev. Ecol. Syst.* 12: 331–53.

Ribbink, A.J., Marsh, B.A., Marsh, A.C. & Sharp, B.J. (1983) A preliminary survey of the cichlid fishes of rocky habitats in L. Malawi. *S. Af. J. Zool.* 18: 160 pp.

Ross, H.H. (1963) Stream communities and terrestrial biomes. *Arch. Hydrobiol.* 59: 235–42.

Schopf, T.M. (1974) The role of biogeographic provinces in regulating marine faunal diversity through geologic time. In: Gray, J. & Boucot, A.J. (eds) *Historical Biogeography, Plate Tectonics, and the Changing Environment*, pp. 449–57. Oregon State University Press, Corvallis.

Schopf, T.M., Fisher, J.B. & Smith, C.A.F. (1977) Is the marine latitudinal diversity gradient merely another example of the species area curve? In: Battaglia, B. & Beardmore, J.A. (eds) *Marine Organisms: Genetics, Ecology, and Evolution*, pp. 365–86. NATO Conference Series IV, Marine Sciences; Vol. 2. Plenum Press, New York.

Solignac, M. (1977) Genetics of ethological isolating mechanisms in the species complex *Jaera albifrons* (Crustacea, Isopoda). In: Battaglia, B. & Beardmore, J.A. (eds) *Marine Organisms: Genetics, Ecology, and Evolution*, pp. 637–64. NATO Conference Series IV, Marine Sciences; Vol. 2. Plenum Press, New York.

Stevens, G.C. (1989) The latitudinal gradient in geographical range: how so many species coexist in the tropics. *Am. Nat.* 133: 240–56.

Stout, J. & Vandermeer, J. (1975) Comparison of species richness for stream-inhabiting insects in tropical and mid-latitude streams. *Am. Nat.* 109: 263–80.

Strayer, D. (1983) The effects of surface geology and stream size on freshwater mussel (Bivalvia, Unionidae) distribution in south-eastern Michigan, USA, *Freshwater Biol.* 13: 253–64.

Taylor, J.D. & Taylor, C.N. (1977) Latitudinal distribution of predatory gastropods on the eastern Atlantic shelf. *J. Biogeogr.* 4: 73–81.

Wheeler, A. (1977) The origin and distribution of the freshwater fishes of the British Isles. *J. Biogeogr.* 4: 1–24.

Williamson, M. (1981) *Island Populations*. Oxford University Press, Oxford.

10
Specialist Aquatic Feeding Mechanisms

J.H.R. GEE

10.1 INTRODUCTION

When it comes to acquiring food, many aquatic organisms have terrestrial counterparts. In both aquatic and terrestrial ecosystems there are species that lie in wait to pounce on passing prey, that scavenge in sediments or soils for the dead and dying, or that grind up decaying plant material and its microbial entourage. Evidently there are methods of feeding that are effective in either situation, despite the differences in the physical properties of terrestrial and aquatic environments.

Nonetheless, some feeding mechanisms are unique, or nearly so, to aquatic heterotrophs. Two of these are particularly important, in the sense that they account for a large proportion of the food energy gained by many species in a wide range of habitats and geographical locations. They form the subject of this chapter. The first is suspension feeding, in which the organism gathers small organic particles suspended in the water. In the second method, at least a portion of the organism's daily energy gain is provided by smaller, autotrophic bodies resident within its tissues. Although these two mechanisms are very different, they both depend on particular physical properties of aquatic environments. The differences in these properties between aquatic and terrestrial environments help explain why the mechanisms are virtually restricted to aquatic organisms.

Suspension feeding requires an exploitable food supply suspended in the surrounding medium. For a consumer to make a living, the food supply has to be sufficiently abundant to provide a net gain in energy, after allowing for the costs of its acquisition. It is also desirable that the food supply should be available all year round, or at least be predictable in space and time.

Water is a relatively dense liquid, which provides buoyant support for planktonic phototrophs. Since it is a polar solvent, it can also meet their need for a supply of dissolved inorganic nutrients. Water often moves sufficiently fast, in rivers, strong tidal currents and waves approaching gently-shelving shores, to sweep small benthic organisms and fragments of organic detritus into suspension. Together, these sources of suspended organic matter provide a predictable energy base for a vast assemblage of heterotrophs, many of which are suspension feeders. In contrast, the atmosphere provides little in the way of buoyancy or nutrients. The aerial 'plankton' is sparse, dependent on ground-based primary productivity, and is distributed in space and time at the whim of the weather.

It is also likely that there is a limited range of environmental conditions in which it is feasible for an organism to act as a host to an internal autotroph. Photosynthetic autotrophs require adequate light, which in turn demands that the epidermis of the host is translucent and not shaded from the sun. In a terrestrial environment, this requirement would render a thin-skinned host vulnerable to desiccation. The marine animals that contain symbiotic phototrophs (e.g. flatworms, sea anemones and sponges) are generally intolerant of desiccation. When they, or related non-symbiotic forms, occur intertidally, it is almost always in pools or in moist, dark crevices. Many species of sublittoral mollusc contain phototrophic symbionts, but intertidal molluscs that tolerate emersion have to conceal their soft tissues within a shell to avoid drying out.

A similar argument can be made for the hosts of chemosynthetic autotrophs, although here there is a need for a source of a reduced chemical substrate, rather than light. A large area of permeable epithelium is probably necessary, to allow adequate uptake of the substrate. It is difficult to imagine how an animal might meet these requirements in a dry or oxygen-rich terrestrial environment.

10.2 SUSPENSION FEEDING

Since the ultimate source of food in deep water is suspended algae, there is a trivial sense in which most aquatic heterotrophs could be called suspension feeders. However, some benthic species must await the deposition of suspended particles onto the sediment. Here they feed by selecting organic material from amongst the inorganic particles, or by ingesting the sediment *en masse*. These species are deposit feeders.

By raising a food-collecting appendage above the sediment surface, it is possible for an organism to intercept suspended particles. This has the advantage that the food is not heavily admixed with sediment, although some sorting of particles according to their food value may still be required. Moreover, the organism can then feed in situations where the currents are so strong that little deposition takes place. Such an organism would be regarded as a suspension feeder, although its feeding apparatus might differ very little from that of a deposit-feeding relative. Amongst the holothurian echinoderms there are species that use similar adhesive tentacles; in one case to sweep up food from the sediment surface, and in another to collect suspended particles from the water above.

By definition, suspension feeders differ in size from their food particles by a factor of about 100, or two orders of magnitude. The importance of this is that it is generally uneconomic for suspension feeders to pursue, capture and handle food particles individually. This sort of feeding behaviour would be described as raptorial. Instead they must have some means of processing large volumes of water and extracting many particles simultaneously.

10.2.1 Suspended food

Particles suspended in water are collectively known as seston. The organic fraction of seston (bioseston) contains the food of suspension feeders. Since many organisms discriminate between particles only on the basis of size, they may also ingest considerable quantities of suspended inorganic particles, or abioseston. In addition, some of the bioseston is likely to be resistant to digestion and of little food value.

Seston is a ragbag of particles of diverse sizes and origins. In headwater streams, the coarsest organic particles (much larger than 1 mm) are fragments of terrestrial plant litter that fall or are blown into the water. Further downstream, and in the littoral zones of lakes and coastal waters, decomposing aquatic macrophytes are an important source of coarse seston. In oceans, delicate organic aggregates, up to tens of millimetres in size, are common. These flakes of 'marine snow' probably start life as strands of mucus shed by zooplankton. In time they acquire a rich community of bacteria, diatoms, dinoflagellates and protozoa, together with non-living material that settles on their surfaces (Silver *et al.*, 1978). The absence of similar large particles in lakes may be due to the scarcity of mucus-secreting organisms in the freshwater plankton. Some of the largest seston particles consumed by suspension feeders are the Antarctic krill *Euphausia superba*, marine crustaceans that are about 5 cm in length and themselves suspension feeders. They are the chief food of the blue whale.

At the other end of the seston size spectrum are organic particles of around 1 μm in length. Many heterotrophic bacteria are of this order of size, although some may be as small as 0.1 μm. Suspended organisms with cell diameters from 0.2 to 2 μm are termed picoplankton, and include phototrophic cyanobacteria and eukaryotic algae, together with heterotrophic bacteria and protists. Recent research has shown these to be abundant and ubiquitous in both fresh water and the sea, and to constitute an important microbial loop in aquatic food webs.

Almost invariably, much of the organic matter in natural waters is in dissolved form. We use the term dissolved organic matter (DOM) generally to mean anything that will pass a 0.5 μm (or sometimes 1 μm) pore-size filter, and may include material that is actually particulate or colloidal. Dissolved organic matter often aggregates to form small particles of non-living bioseston. This can happen when there is a change in the physicochemical properties

Table 10.1 Concentrations of particulate and dissolved organic matter (POM and DOM, respectively) in a variety of water bodies. Data from Jorgensen (1966) and Wetzel (1983)

Locality and type of water body	POM concentration (mg dry wt l^{-1})	DOM concentration (mg dry wt l^{-1})
Oceanic		
NE Pacific		
Surface	0.32	—
Depth > 300 m	0.07–0.11	3.3
Coastal		
Wadden Sea		
Offshore	2.5	—
Inshore	4.0	—
Lake		
Lunzer Untersee (oligotrophic)	0.003–0.05	—
Lake Erken (mesotrophic)	0.01–1.0	—
Lawrence Lake (mesotrophic)	0.30–1.0	10–15
Lake Krzywe (eutrophic)	3.0–7.0	—

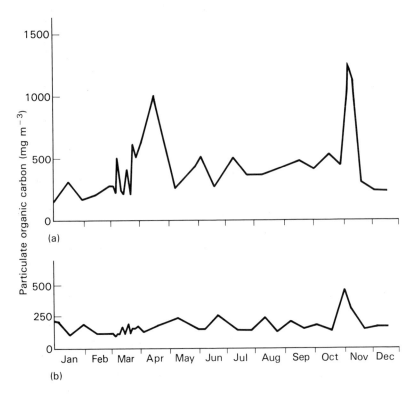

Fig. 10.1 Variation in the concentration of particulate organic carbon with season and with depth: (a) 2.5 m; (b) 30 m, in the Bedford Basin, Nova Scotia, 1974. From Mayzaud *et al.* (1984).

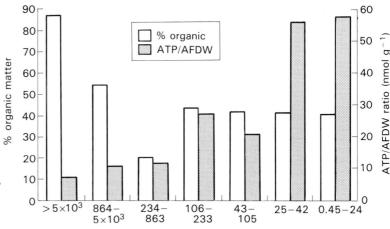

Fig. 10.2 The organic matter and ATP content of seston in a southern Appalachian stream. AFDW, ash-free dry weight. From Wallace *et al.* (1982).

of the water (e.g. pH or salinity), an event that is more likely to occur in rivers, lakes or estuaries than in the sea. In addition, DOM often collects at an air−water interface. Agitation of the interface, or the collapse of an air bubble, can cause an organic particle to form. Increased concentrations of organic particles in streams have been found below waterfalls, apparently due to the agitation of the water surface. Bacteria often become associated with these particles, but are not necessary for their formation.

Measurement of the availability of particulate organic matter is plagued by methodological problems. Apparent variation in availability within and between aquatic habitats may be due as much to differences in techniques as to anything else. Nevertheless, it is clear that the concentration of particulate organic matter is much higher in coastal waters than in the deep oceans (Table 10.1). This is undoubtedly due to variations in photosynthetic productivity. Similarly, concentrations of suspended organic particles vary with the trophic status of lakes, and span as wide a range as those recorded in the sea.

In general, the particle concentration is higher in the euphotic zone than in the abyssal depths, although it may rise again close to the sediment surface whenever water movements resuspend deposited material. The decrease in particle concentration with depth (Fig. 10.1) is due to the remineralization of organic material by bacteria and suspension feeders in the upper part of the water column. Typically, less than 10% of production in the euphotic zone reaches the bottom of the deep sea. Processing of organic particles also affects their distribution along the course of a river. In general, large particles dominate the seston in the headwaters, where the influence of material shed by bankside vegetation is greatest. Further downstream, particle size declines as the material is broken up by detritivores and by physical abrasion. Small seston particles often have a lower organic content than large particles, but their greater surface area for microbial colonization may lead to higher concentrations of adenosine triphosphate (ATP) (Fig. 10.2).

Bioseston concentrations in lakes and the sea tend to vary seasonally, in a way that reflects the changing rate of primary production in the euphotic zone (Fig. 10.1). Within the larger oceans, bioseston concentration is generally greater at high latitudes than near the Equator (Fig. 10.3). To a first approximation, the concentration of bioseston — expressed as volume or weight of particles per unit volume of sea water — appears to vary little with particle size. Large particles are encountered less frequently than small ones, but the mass per unit volume of water is about the same. In rivers, the concentration and mean particle size of seston increases dramatically during spates. Despite this

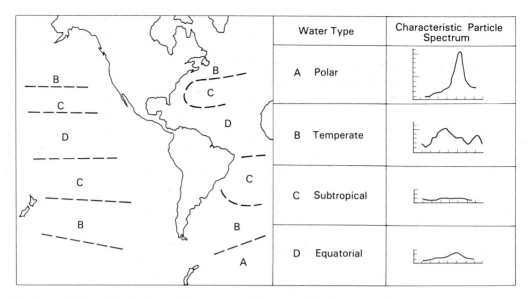

Fig. 10.3 Geographical variation in the particle size spectra of the seston of surface waters (about 4 m depth) in the sea. The vertical axis indicates particle concentration in ppm (by volume), whilst the horizontal axis shows particle size on a logarithmic scale from 1–128 μm. From Sheldon *et al.* (1972).

stochastic variation, there is often an underlying pattern which reflects the seasonality of the input of coarse particles from terrestrial and aquatic macrophytes.

For a population of suspension feeders, the rate at which the store of organic particles is replenished when depleted is as important as the standing crop. The little that is known about the natural turnover rates of living and non-living bioseston suggests that the picture of relative availability, based on standing crop information, would not be changed much if turnover was taken into account.

The value of suspended particles as food varies with particle type. In one study of suspended organic particles in the sea (excluding phytoplankton) it was shown that whilst 16–52% of particles collected at 1 m depth were biodegradable, all of the material collected at 200–1000 m was refractory. Amongst planktonic algae, there are species that are readily digested in the guts of suspension feeders and others that appear to pass through unharmed, or even to benefit from the experience (Porter, 1977). Algae protected by a mucilaginous outer layer are particularly resistant to digestion, and of least value to suspension feeders.

10.2.2 Methods of particle capture

All suspension feeders require that a current of water carrying potential food items passes their feeding apparatus. A simple functional distinction between suspension feeders is based on the means by which that current is generated. Active suspension feeders expend energy in driving water through their feeding structures, either by moving the water or moving themselves. Passive suspension feeders rely on a current driven by some other external force, such as wave action or the gravitational movement of water in a river. A few groups, including some sponges, barnacles and tunicates, have the ability to switch between active and passive modes as external conditions change.

Early views of suspension-feeding techniques were strongly influenced by experiences with kitchen colanders and sieves. In these devices, water passes through the apertures with little resistance and any suspended particles are captured with complete efficiency if their diameter exceeds the aperture size. The picture changes if the particles are sufficiently abundant to cause serious clogging of the sieve, but suspension feeders probably experience such conditions rarely. However, for a variety of reasons, this is too simplistic a view of the method of particle capture used by many suspension feeders.

In the late 1960s, aquatic biologists became aware that intuition was an unreliable guide to the behaviour of water currents near the feeding appendages of small organisms. When water (or any other fluid) moves slowly around small objects, its flow is influenced more by its viscosity than by its inertia. Friction between the water and the surface of the object becomes more important than the resistance felt as the object deflects the water, and in so doing changes its velocity. The relative importance of inertial and viscous forces is measured by the dimensionless Reynolds number, *Re*. Its magnitude depends upon the dynamic viscosity of the water, the critical dimension (e.g. pore size) of the feeding appendage and the velocity of the water through it. A domestic sieve might operate at a *Re* of more than 1×10^3, at which value viscous forces are nearly negligible. In contrast, the feeding system of a bivalve mollusc operates at a *Re* of about 1×10^{-4}, at which value viscosity predominates. In these circumstances, each part of the feeding apparatus is surrounded by a shell of near static water, and a current can be made to pass through the apertures only with difficulty. Thus the task of collecting food particles from water at such low *Re* has been likened to fishing peas from treacle with a table fork, an activity that takes place at a similar *Re*.

In addition to the straightforward sieving of particles too large to pass between the fibres of a filter, an individual fibre may collect particles, provided it is sticky. A filter composed of sticky fibres can collect particles which are much smaller than the gaps between the fibres. This process, first identified for air filters, is called aerosol filtration; in an aquatic context 'hydrosol' would be more appropriate. The stickiness may be due to an adhesive material on the fibres, or to an electrostatic or molecular attraction. A particle that is carried by the water current sufficiently close to touch a fibre may be caught in this way (direct interception), but particles at a greater distance may also be captured (Fig. 10.4). This may occur if the momentum of a particle causes it to hit the fibre, even though the streamline along which it is being carried is deflected

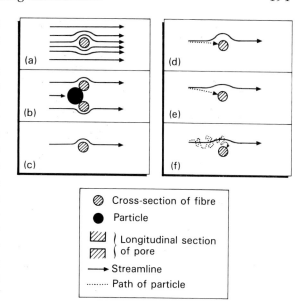

Fig. 10.4 Possible modes of operation of a suspension-feeding device: (a) flow pattern around a single fibre; (b) simple sieving by a pair of fibres; (c) to (f) particle acquisition by a single sticky fibre: (c) direct interception; (d) inertial impaction; (e) gravitational attraction; (f) diffusional deposition. From Rubenstein & Koehl (1977).

around the fibre (inertial impaction). It may also occur if gravity takes a hand and the particle falls across streamlines onto the fibre (gravitational deposition), or if the particle is propelled towards the fibre by Brownian motion (diffusional deposition).

It is often difficult to determine precisely which method an organism is using to collect suspended particles. Perhaps the best method is to track the paths of individual particles as they approach and contact the feeding apparatus. This is not easy when the organism is small and motile, and virtually impossible when the feeding apparatus is concealed within a shell or body chamber. In these cases it is the relationship between the aperture size of the apparatus and the size range of particles effectively removed from suspension that gives a clue to the method being used. Simple sieving mechanisms generally do not collect many particles smaller than the minimum aperture of the sieve; there is usually a sharp cut-off in the efficiency of particle capture at this size. Other methods may permit particles much smaller than the minimum aperture size to be

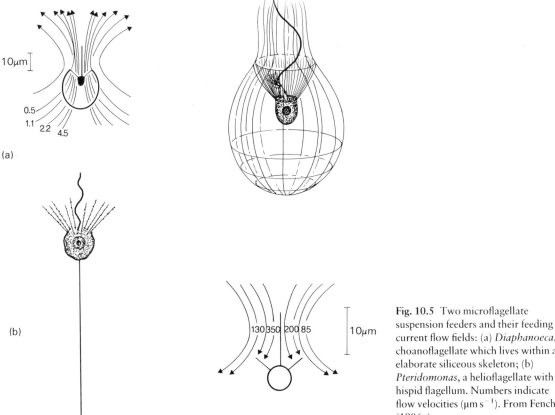

Fig. 10.5 Two microflagellate suspension feeders and their feeding current flow fields: (a) *Diaphanoeca*, a choanoflagellate which lives within an elaborate siliceous skeleton; (b) *Pteridomonas*, a helioflagellate with a hispid flagellum. Numbers indicate flow velocities ($\mu m\,s^{-1}$). From Fenchel (1986a).

collected, and efficiency tends to decrease gradually with particle size at the lower end of the size range.

10.2.3 Simple sieves

Many of the organisms capable of feeding effectively on particles of bacterial size (<1 μm), appear to do so by simple sieving. Since the apertures of their filters are so small, they must operate at very low Reynolds numbers. In these conditions, simple sieving is only feasible if the current speed through the filter is very low (<50 $\mu m\,s^{-1}$). Higher current speeds could only be achieved by maintaining a pressure gradient across the filter that would be energetically costly for the organism. Unless the filter area is large, relative to the mass of the organism, or the suspended food present at high concentration, the rate of food capture at such low current

speeds may not be sufficient to meet metabolic demands. Thus simple scaling factors impose limits on the activity of such suspension feeders.

The principal consumers of bacterial and algal picoplankton appear to be the microflagellate protozoa, although some ciliates can also feed on particles in this size range. Choanoflagellates are often so common in the plankton that they may be the most numerous phagotrophs on earth. Their filtering surface is composed of a collar of microvilli surrounding an apical flagellum (Fig. 10.5). Typically, the microvilli are only about 0.1 μm thick and the spaces between them, the pores of the filter, vary in width from 0.1 to around 0.5 μm. The beating of the flagellum draws a current of water through the collar at about 10 $\mu m\,s^{-1}$ or less, and then drives it away from the organism. Suspended particles collect on the outside of the collar and are incorporated into food vacuoles by phagocytosis.

A similar set of structures is possessed by helio-flagellids such as *Pteridomonas*, although here the hispid (bristly) flagellum drives water in the opposite direction. *Pteridomonas* has larger spaces between its microvilli than have choanoflagellates, and for this reason they can maintain higher water velocities at the cost of losing some of the smaller food particles. Sticky 'extrusosomes' on the microvilli may enable *Pteridomonas* to collect some of these small particles by aerosol filtration.

The bacterivorous ciliates strain their food through a screen of cilia. In *Cyclidium*, a water current is driven by a group of membranelles (cilia placed so closely that they move together as if joined) towards a row of cilia about $0.25\,\mu m$ apart. This spacing corresponds with the lower limit of particle size that *Cyclidium* is able to clear from suspension. Although water leaves the membranelles at relatively high velocities ($>100\,\mu m\,s^{-1}$), the large surface area of the ciliary filter ensures that water passes it at only about $20\,\mu m\,s^{-1}$. This type of collection mechanism is sometimes called downstream retention, in order to distinguish it from situations in which the membranelles that generate the current also serve to collect the food. In upstream retention mechanisms, food particles accumulate on the upstream side of the membranelles, as in *Uronema* (Fenchel, 1986b).

Sponges comprise one of the few groups of metazoans that can collect suspended particles in the bacterial size range. Larger particles ($5–50\,\mu m$) are engulfed by the pinacocytes surrounding the incurrent pores, or ostia. Particles of $1\,\mu m$ size or smaller are filtered out by the choanocytes that line the feeding chambers of the sponge. The choanocyte filter is composed of a ring of microvillae which surrounds the flagellum of each cell. These choanocytes are identical in many respects to individual choanoflagellates, and function in the same way. Although the velocity of the water passing through the ostia, and particularly that passing through the excurrent pore, or osculum, is high, the cross-sectional area of the feeding chambers is much greater than that of either the ostia or osculum. In consequence, the water velocity through the collars of the individual choanocytes is low (about

$5\,\mu m\,s^{-1}$) and requires only a small pressure differential between the sides of the filter.

Apart from sponges, the urochordates (tunicates) and cladoceran crustaceans seem to be the only animals capable of clearing natural bacteria from water at rates which indicate that bacterioplankton might be a significant part of their diet. The method used by cladocerans will be discussed later, but urochordates appear to feed by sieving. All urochordates feed on suspended particles using a net composed of a mesh of mucus threads. This net is supported on a basket-like pharynx, into which water is drawn by ciliary activity. The net is produced continuously by a pharyngeal gland, the endostyle. At the opposite side of the pharynx, the used net is rolled up and ingested, together with the trapped food it contains. The sizes of the rectangular apertures of the net range from 0.2 to $0.5\,\mu m$ in width and from 0.5 to $2.2\,\mu m$ in length. The operation of such a fine-meshed filter appears to be facilitated by the fact that the threads from which it is constructed are remarkably thin ($10–40\,nm$), thus minimizing frictional resistance. It is possible that some of the material is captured by adhesion to the mucus threads rather than by simple sieving.

Pelagic tunicates of the class Larvacea have an unusual additional feeding structure. In *Oikopleura*, this is part of an elaborate mucus 'house' within which the tadpole-like larvacean resides (Fig. 10.6). When the larvacean beats its tail, water is drawn into the house through two relatively coarse, in-current filters ($40–170\,\mu m$ apertures). This filtered water then traverses the tail chamber and enters two winglike 'feeding filters', the walls of which have rectangular pores about 0.2 by $1\,\mu m$ in size. Recent evidence strongly suggests that these function not to capture particles, but rather to exclude water, so that the water current leaving the feeding filters contains a much more concentrated suspension of food. This current then enters the pharyngeal mucus filter for particle capture and ingestion. The particle-free water that passes through the feeding filters is ejected from the house through a posterior exit port, and serves to provide a propulsive force. When the feeding filters in the house become clogged, the larvacean exits, abandons the old house and sets about secretion of a new one. This process may occur several times in one day.

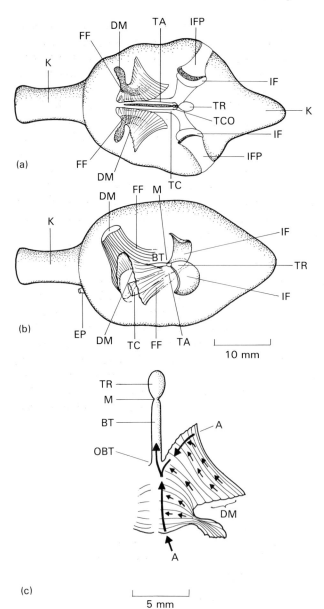

Fig. 10.6 *Oikopleura vanhoeffeni* inside its house, showing dorsal (a) and lateral (b) views of an individual, together with a dorsal view (c) of the flow pathway through one wing of the feeding filter. A, streamlines originating from the lateral openings at the base of the wing; BT, buccal tube; DM, distal margin; EP, exit port; FF, feeding filter; IF, incurrent filter; IFP, incurrent filter passageway; K, keel; M, mouth; OBT, opening of buccal tube; TA, tail; TC, tail chamber; TCO, tail chamber opening; TR, trunk. From Deibel (1986).

There are a host of other passive and active suspension feeders which, in the absence of evidence to the contrary, appear to use simple sieves. Most have been much less thoroughly investigated than the examples above. Generally, the minimum size of particle they collect is much larger than a free-living bacterial cell. They include the net-spinning caddis larvae of freshwater streams, which use silken nets to strain food items from the passing current.

Planktivorous fish, like the menhaden *Brevoortia tyrannus*, filter water through fine bony combs or gill rakers as it passes through their mouths, crosses the gills and exits through the gill opercula (Friedland, 1985). In a similar way, the keratinous baleen plates of the blue whale are used to strain out the krill collected in gargantuan mouthfuls ($60\,m^3$) of water. Although little is known in detail of the feeding mechanism of the larger whales, it seems likely that they are the only suspension feeders that operate at high Reynolds numbers.

10.2.4 Sticky screens

In theory, sticky filter screens may capture particles that are smaller than the minimum aperture size, by any one of four alternative methods (see Section 10.2.2). In practice, direct interception seems to be the method most likely to be of importance in capturing organic particles, in a medium as dense and as viscous as water. A screen that catches a substantial proportion of particles by a non-sieving method will have a minimum capture size that does not correspond with the dimensions of its apertures. Furthermore, since its operation depends on the adherence of particles to the elements of the screen, efficiency of capture is likely to depend on the surface properties of the particles as well as their diameter.

The brittlestar *Ophiopholis aculeata* captures small particles of suspended food from water currents passing over marine sediments. When feeding, its arms project above the sediment surface and are held approximately perpendicular to the current.

The food particles adhere to the mucus-covered tube feet, arranged in rows along the arms. At intervals, the most distal tube feet bend over, and are wiped clean of adhered food by adjacent, more proximal, tube feet. This process is repeated sequentially along the arm, and results in both the formation of a bolus of mucus-bound food particles and its movement towards the mouth. There is no active selection of individual food particles, and the bolus is either consumed or rejected as a whole. By stealing a bolus from a brittlestar, it is possible for a researcher to compare the particle size distribution of captured food with that of the water current passing the arms.

The minimum distance between tube feet in *Ophiopholis* is 440 μm. Figure 10.7 shows the results of an experiment in which all the offered and captured particles (synthetic beads) were smaller than 360 μm in diameter. There is no sharply-defined minimum size of captured particles, and the size distributions of offered and captured particles were similar, although not identical. Simple sieving could not account for these results; instead, the rows of tube feet are acting as a sticky screen. The stickiness is due to the acid mucopolysaccharide

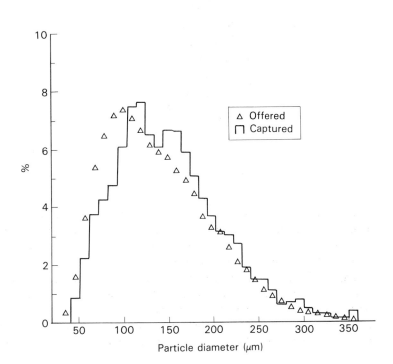

Fig. 10.7 The size distributions of sephadex beads offered to and captured by the suspension feeding brittlestar *Ophiopholis aculeata*. From LaBarbera (1984).

which covers the feet and has negatively charged groups attached to its molecular surface. Further experiments with beads carrying different charges showed that the size distribution of captured beads varied, as expected, with their surface charge. Again, this is a result that is consistent with the sticky screen mechanism of particle capture, but not with simple sieving.

Amongst freshwater organisms, there are few that appear as elegantly adapted for suspension feeding as the larvae of the blackfly *Simulium*. These feed on particles carried past by the current, but to call them passive feeders is misleading. *Simulium* larvae live attached by a posterio–ventral circle of hooks to a silken pad, which they spin on a solid surface, often a large stone. Tens or hundreds of larvae may be found on a single stone, usually spaced in a regular pattern within a favoured area. The drag of the current causes the tapered bodies of the larvae to trail downstream from their attachment pads. Whilst feeding, they twist their bodies through nearly 180°, so that the ventral side of the head faces upwards. The food catching organs are a pair of cephalic fans which are extended into the current and then retracted alternately to be cleaned of food by the mouthparts.

In natural situations, *Simulium* larvae capture food particles in the size range 0.1–350 μm, with the bulk of material frequently less than 30 μm in diameter. This ability to capture particles of bacterial, or even colloidal, dimensions was difficult to understand, when the minimum distance between fan rays is about 35 μm. Each fan ray bears a row of fine spines, or microtrichia, which are spaced at 0.2 μm intervals, but the discovery of mucosubstance on the rays led to the suggestion that the fans function as sticky screens, rather than as sieves. In this interpretation, the role of the microtrichia is to increase the surface area to which passing particles might adhere.

It is possible to record the speed and direction of water movements around a *Simulium* larva by photographing it, under strobe illumination, in a water current containing reflective aluminium flakes. The slightly bulbous posterior of the larva generates two swirling currents, or vortices, on the downstream side (Fig. 10.8). One of these vortices remains in the slow-moving boundary layer of

water close to the stone surface, whilst the other is drawn upwards along the larval body. As it ascends, it draws with it food particles from close to the stone surface. Faeces, shed from the upstream, posterior of the animal, are also entrained into the vortex, but are denser than food particles and soon escape to be carried away downstream. The vortex then passes through the lowermost of the cephalic fans, which captures a proportion of the particles. The other cephalic fan samples a different stream of water, passing at a greater height above the stone surface. It seems probable that the close proximity of *Simulium* on the stone surface enhances feeding by concentrating food particles in narrow streams between adjacent larvae.

Strobe photography cannot resolve the movements of water around individual cephalic fan rays. A convenient feature of hydrodynamics is that flow patterns around objects of similar shape but differing sizes are identical, providing the Reynolds numbers are the same. Instead of attempting to visualize the flow around a real *Simulium* head fan, it is much easier to work with a large-scale model. The *Re* of the model can be made the same as the real head fan by using a medium with a higher viscosity. In this way it has been shown that there is virtually no movement of water between the microtrichia, even at high flow rates. At low flow rates, there is little water movement between fan rays, and at moderate to high flow rates, particles passing between the rays settle out into the near-stationary boundary layer around the microtrichia (Braimah, 1987).

Hence, a structure that at first sight looks well-adapted as a simple sieve appears to operate as a sticky screen. It is possible that it operates in both ways, because the efficiency of capture of particles is greatest for those with diameters that exceed the fan ray spacing. Given the ability of the larvae to produce and manipulate vortices by virtue of morphology and behaviour, it seems uncomplimentary, at the very least, to describe them as passive suspension feeders, despite their dependence on existing water currents.

The production and manipulation of vortices is by no means restricted either to simuliids or to passive suspension feeders in general. Tubicolous phoronid worms are active suspension feeders, but

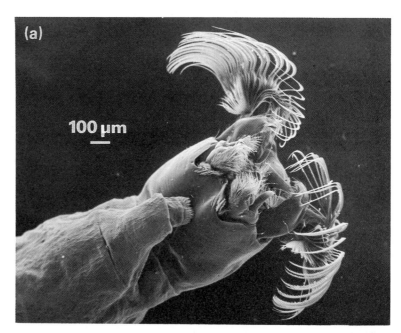

Fig. 10.8 Suspension feeding by blackfly (*Simulium*) larvae: (a), scanning electron micrograph of the ventral aspect of the head, showing the labral fans in the extended position; (b), light macrograph of a larva feeding in a current containing pigmented particles, showing entrainment of particles from close to the substratum into the upper vortex. Flow is from right to left at $150\,\mathrm{mm\,s^{-1}}$; overall length of the larva is 8 mm. lv, lower vortex; uv, upper vortex; uf, upper labral fan (lower labral fan is obscured by particles in the lower vortex); arrow indicates the pigment flow up and over the posterior abdomen. From Chance & Craig (1986).

the upward movement of water and suspended materials, from the substratum surface towards the ciliated tentacular crown, is not dependent on ciliary action. These movements are generated, instead, by the interaction between the morphology of the worm and its tube and the ambient currents across the substratum (Johnson, 1988).

10.2.5 Scan and trap

The abundance of calanoid copepods, in both marine and freshwater pelagic communities, makes them one of the more important groups of suspension-feeding organisms. Until recently, it was believed that calanoids fed by passing a stream of water continuously through the setae of the stationary second maxillae. When a method was devised for tethering copepods, so that they could feed normally but remained immobile, it was possible to use high speed cinematography to observe and record the fate of dye streams and individual food particles as they were drawn towards the animal. These studies have shown that calanoids have several different feeding behaviours to capture particles of differing sizes, but that none of the methods appears to involve simple sieving through the maxillary setae.

A calanoid, feeding on algal cells about 10–50 μm in diameter, draws water towards it by flapping movements of its second antennae, mandibular palps, first maxillae and maxillipeds. During this process, in which the copepod scans the passing water for food, the second maxillae remain stationary but water does not pass through them. When a food particle is drawn near, the feeding appendages flap asymmetrically, which in an untethered animal would result in a turn towards the approaching particle. The food particle is trapped by a rapid 'fling and clap' movement of the second maxillae. As the maxillae are flung apart, the parcel of water containing the particle is drawn between them. The maxillae then close around the parcel and the excess water is squeezed out between the maxillary setae (Fig. 10.9). With the aid of the endites of the first maxillae, the captured particle is removed from the setae and transferred to the mouth. In this way, a calanoid can scan a large volume of water at relatively little energetic cost,

and restrict the energetically expensive trapping movements to occasions when food comes within range. The mechanism by which the copepod senses the proximity of food has not been identified with certainty, but it seems probable that chemosensors and mechanoreceptors on the first antennae are important. In the low Re world of the feeding copepod, a shell of water adheres to small food particles. In principle at least, it should be possible to detect such particles at a distance, by mechanoreception of the hydraulic disturbance caused by the shell, or by sensing metabolites contained within it (Strickler, 1985).

When feeding on larger algae ($>50\,\mu m$), calanoids use a similar mechanism, although more complex motions of the second maxillae are required to orientate the food for ingestion. The mechanism used to capture cells smaller than about 10 μm is not well-understood, but it differs in that it involves continuous, low amplitude movements of the second maxillae. These movements seem to cause food particles to be funnelled across the maxillary surface towards the mouth. Many of these particles pass between the maxilllary setae. Calculated values of the thickness of the boundary layers around setae suggest that there should be little or no flow between them (Price & Paffenhofer, 1986). Hydrodynamic theory may be only an approximate guide to the behaviour of fluids around oscillating structures of complex shape.

10.2.6 *Daphnia* and *Mytilus*: controversial cases

The planktonic *Daphnia* and the benthic *Mytilus* are representatives of two important groups of suspension-feeding organisms, the Cladocera and the bivalve Mollusca. Perhaps because their feeding appendages are concealed from view within the carapace or shell, their feeding mechanisms are still a matter of debate. Whilst the nature of their feeding structures seems well enough understood, the manner of their deployment and the hydrodynamics of the system they comprise is much less certain.

There is no doubt that *Daphnia* can collect suspended particles as small as 1 μm in diameter. They achieve this feat by drawing a current of water across setose appendages held within a partially open bivalve carapace. The size of particle collected

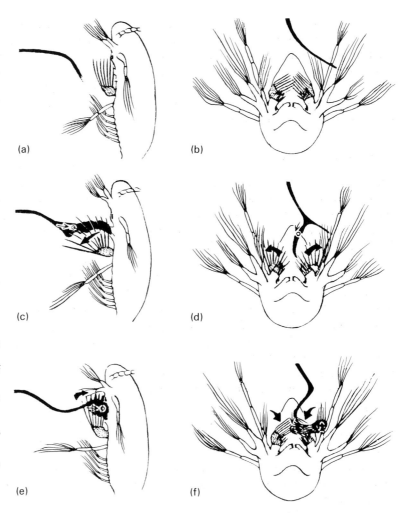

Fig. 10.9 Capture of a food particle by *Eucalanus pileatus*, viewed from the left side (a), (c) & (e) and from the anterior end (b), (d) & (f). In (a) and (b), the black streak marks the path of dye released from a micropipette and carried past the feeding appendages in a normal scanning current. In (c) and (d) an alga is drawn between the second maxillae by the outward 'fling' movement, and in (e) and (f) the inward 'clap' of the maxillae results in capture. From Koehl & Strickler (1981).

most rapidly by a daphnid is related to the size of the apertures between adjacent setae (Fig. 10.10). These observations suggest that daphnid feeding appendages operate as simple sieves. On the other hand, the performance of the feeding apparatus can be changed by altering the surface charge of the particles, or by adding a surfactant chemical to the water (Gerritsen & Porter, 1982). This suggests that adhesion between the particles and individual setae may be important. Attempts to model the hydrodynamics of daphnid feeding often indicate that the boundary layers around the setae would prevent water movement between them. Proponents of the sieving mechanism, however, point out that the setose appendages operate within the restricted

space of the carapace, so that the usual hydrodynamic models do not apply (Brendelberger *et al.*, 1986). Recent evidence from cinematography of tethered *Daphnia* seems to show that most of the water does indeed flow across the surface of the feeding appendages, rather than between the setae, so sieving may be less important than once was thought (Gerritsen *et al.*, 1988).

In *Mytilus* and other bivalve molluscs, particulate food is extracted from the water that passes between the filaments of the gills (Fig. 10.11). The water current is driven by bands of lateral cilia on the opposing faces of the filaments. Two other sets of ciliary structures are involved in food collection. Each filament bears two rows of laterofrontal cirri

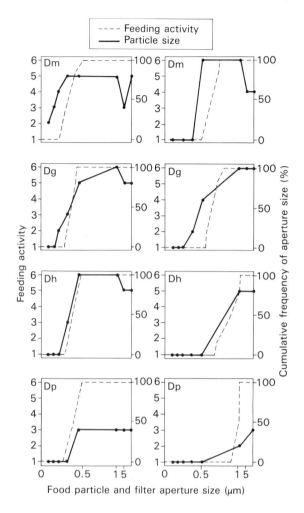

Fig. 10.10 The relationship between the cumulative frequency distributions of the meshes of the filter combs of eight daphnids and the variation in the rate of collection of different sizes of particles. The offered particles were inert artificial spheres in the size range 0.1–5 μm and *Scenedesmus* cells (25–35 μm). Feeding activity was assessed by eye on an arbitrary scale (from 0, no activity, to 6, highest activity). Two individuals (one with large meshes, one with small) of four species were used (Dm, *Daphnia magna*; Dg, *D. galeata*; Dh, *D. hyalina*; Dp, *D. pulex*). From Gophen & Geller (1984).

on its upstream edge. The cirri are compound structures, composed of many cilia stuck together in an arrangement reminiscent of a set of pan pipes. At the distal end, each cilium is free from its neighbours and the tip is bent at a right angle to the plane of the 'pipes'. Alternate tips are bent in opposite directions. A row of cirri, together with their outturned cilial tips, thus forms a grid with apertures about 1 μm in size. The positioning of the rows on adjacent filaments means that there is the potential for the cirri to form a fine-mesh filter through which all of the water moving between the filaments must pass. Frontal cilia on the upstream edge of each filament comprise the second set of food collecting structures. These transport food particles from the

vicinity of the cirral rows, along the edges of the filaments, towards the palps and mouth.

Disagreement on the mechanism of food capture centres on the pressure difference necessary to maintain water movement through the supposed cirral sieve. An alternative suggestion is that the complex oscillating currents, generated by movements of the lateral cilia and the laterofrontal cirri, cause particles to migrate out of the water current towards the frontal cilia, under the influence of shear forces (Jorgensen, 1983). Shear forces arise at the interface between laminar currents, and can cause particles to move perpendicularly to the streamlines. It is also possible that shear forces explain the coherence of the streams of particles

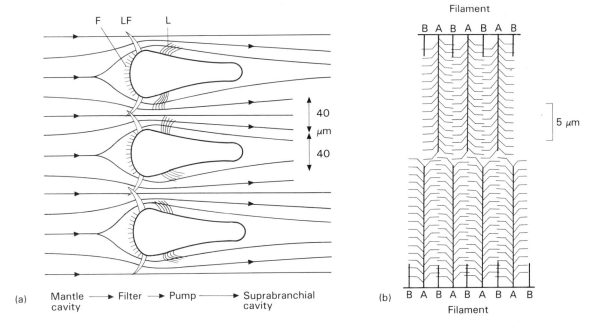

Fig. 10.11 The structure of *Mytilus* gills: (a) cross-section of three filaments showing frontal cilia (F), lateral cilia (L) and laterofrontal cirri (LF); (b) frontal view of the space between two filaments showing the filter grid formed by laterofrontal cirri. For most of the beat cycle, alternate cirri (A) are extended across the space, whilst intervening cirri (B) are bent towards the frontal surface.

carried by the frontal cilia. Often it is assumed that the particles stay within the streams because they are bound in mucus. However this explanation seems at odds with the ability of the palps to sort particles according to size, as they pass across a complex series of ciliated ridges and troughs (Silvester & Sleigh, 1984). There also seems to be a conflict between empirical estimates of water velocity between the filaments and the conceivable maximum rate of pumping by the lateral cilia. The water moves much faster than seems possible, so perhaps another pumping mechanism is at work (Silvester, 1988).

10.2.7 Spatial distribution of suspension feeders

In general, the spatial distribution of motile, planktonic suspension feeders reflects the distribution of their food. Numbers tend to be highest in the euphotic zone, where food availability is greatest. Small-scale distribution patterns of individual species are often modified by predator avoidance tactics. Many planktonic species aggregate at night, near the water surface, to feed on the high densities of algal cells, but descend to deeper water during the day, probably partly to avoid detection and consumption by sight-hunting predators (see Chapter 8). In the littoral zones of lakes, daphnids such as *Bosmina* gather in dense swarms when threatened by predators. Although swarming reduces the probability that any individual will be eaten, this advantage is partially offset by the food depletion that occurs within the swarms. This trade-off leads to the dispersal of swarms when food is in short supply (Jakobsen & Johnsen, 1988).

The spatial distribution of some suspension feeders can be explained in terms of the performance characteristics of their food collecting devices. At the very least, the rate of food capture must provide sufficient energy income to support expenditure in respiration. The maximum clearance rates achieved by some ciliates necessitate relatively high densities of suspended bacterioplankton, for energy income to match or exceed expenditure. This suggests a simple economic reason for the scarcity of bacterivorous ciliates in the plankton, where particle concentrations are frequently below the threshold level, and their abundance in and around sediments, where bacterial populations are more dense (Fenchel, 1980).

Most suspension-feeding benthic macroinvertebrates in the sea are either sessile or, if motile, do not move far in search of food. Their limited ability to move as adults places great importance on the selection of suitable sites by their motile, planktonic larval forms. In comparison with deposit feeders, benthic suspension feeders are usually found in shallower water, where water movements are more pronounced. Passive suspension feeders are dependent on currents to bring them food and are typical of high energy habitats where the substratum is either bare rock or coarse sediment. Being less dependent on water movements, active suspension feeders penetrate to greater depths and colonize finer sediments (Pearson & Rosenberg, 1987). Species which can operate in either mode, such as acorn barnacles, grow more rapidly the faster the passing current. This is likely to be due, in part, to the greater energy expenditure required for any specified rate of food capture when feeding actively, than passively. Differences in the distribution of active and passive suspension feeders can even be detected on very small spatial scales. The diverse epizoite fauna of the erect, branching hydroid *Nemertesia antennina* includes both passive and active suspension feeders. Passive forms predominate on the distal parts of the hydroid, where water movements are greatest, whereas the parts of the hydroid closest to the sediment surface support mainly active forms (Hughes, 1975).

Many of the suspension-feeding invertebrates in rivers are passive feeders, such as the larvae of the blackfly *Simulium* and the net-spinning caddis *Hydropsyche*. The greatest densities of these species often occur below lake outlets, where there is an abundant supply of high quality, particulate organic matter. Each species appears to feed most effectively within a narrow range of current speeds. In net-spinning caddis, there is a close relationship between preferred speed for a species and the characteristic size, shape and mesh width of its net

(Georgian & Wallace, 1980). At any one spot on the substratum of a river, flow conditions are unpredictable and liable to rapid and extensive change. In the headwaters, particles up to the size of boulders may move during spates. The meandering lower parts of a river's course are also prone to extensive movements of finer sediments and even complete changes of channel. Faced with abrupt changes in local conditions, simuliids and hydropsychids can relinquish their feeding stances and move in search of more suitable sites.

10.2.8 Suspension feeders as optimal foragers

Optimal foraging is a concept that we have found useful in explaining the behavioural adaptations of organisms to particular patterns of food availability. An optimal foraging strategy is a suite of behavioural tactics that maximizes the net rate of energy gain by an organism. The concept is usually framed in terms of energy, although it is recognized that fitness may also be influenced by the rate of acquisition of specific nutrients. For most animals, the benefits of optimal foraging have to be weighed against risks of predation or other hazards, such as desiccation, that a foraging strategy may entail.

A motile, raptorial organism has many options available to it whilst feeding. It may choose to pursue only items of a particular size (energy content) and ease of capture or digestibility, and to reject others. When the range of potential food items changes, it can change its preferred type, in an attempt to maintain its rate of energy gain. If its rate of food capture is not satisfactory in a particular location, it may elect to spend time and energy moving to a more profitable patch. In contrast, a suspension feeder has fewer choices. If it feeds by simple sieving, particles can be discriminated only by size. A mechanism for post-capture rejection may prevent ingestion of unprofitable particles, but energy will already have been spent on their capture, and the rejection process itself may be costly. The aperture of the sieve determines the size range that is captured effectively, and changes in aperture size may only be possible through evolution. If the organism is sessile, it cannot respond to decreasing food availability by moving. Altering the rate at which it passes water through its filter (clearance rate) may be the only adjustment that it can make to its performance.

Mathematical models of suspension feeding have been used to determine the relationship between food particle concentration and clearance rate for an optimally-foraging suspension feeder (Lam & Frost, 1976; Lehman, 1976). Below a threshold, food concentration clearance should decline rapidly or cease as feeding becomes unprofitable. This response could help stabilize predator–prey population dynamics by providing a respite or 'refuge' for the prey when it becomes scarce. As food concentrations increase above the threshold, clearance rates should also increase, up to the incipient limiting food concentration. This is the point at which ingestion rates are limited by the rate at which food can be processed by the gut. Above the incipient limiting concentration, clearance rates should decrease, so that the ingestion rate just keeps the gut filled.

Very few experimental studies of suspension feeders have shown a threshold food concentration. It may be that the operating costs are so low that it is profitable to continue filtering, even at extremely low food concentrations. In some cases respiratory requirements may necessitate continuous filtering, even in the absence of food. Perhaps a suspension feeder, once it shuts down its filter in the face of a declining food supply, has no other way of detecting an improvement in conditions. Several studies confirm that clearance rates generally increase with food concentration (e.g. Porter *et al.*, 1982; Hart & Latta, 1986). However, there is only limited evidence of a reduction in clearance rates at very high food concentrations. At high densities of algal cells, daphnids maintain their clearance rate, but increase their post-capture rejection rate. Daphnids, being motile, have the potential to respond to gradients of food availability by moving, but there is little evidence of orientation towards patches of high food density, or adjustment of swimming speed (Porter *et al.*, 1982). The predictions of optimal foraging models have thus had limited support from studies of daphnids and simuliids.

Rejection of unsuitable particles by daphnids necessitates the loss of an entire bolus of food. In contrast, copepods can handle particles individually and discriminate between them, on the basis of nutritional value. Optimal foraging models predict that in a mixture of particles of different value, an animal should be more selective at high than at low particle densities. When calanoids are presented with mixtures of equal numbers of two species of algae, they discriminate more strongly against the poorer, mucilagenous species, the higher the absolute abundance of cells (Fig. 10.12).

It is now clear, as we discussed in previous sec-

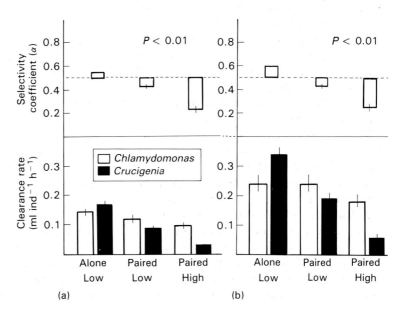

Fig. 10.12 Selection between *Chlamydomonas reinhardi* (high quality) and *Crucigenia tetrapedia* (low quality) by C3 and C4 copepodites (a), and C5 copepodites and adults (b), of *Eudiaptomus* species. The algae were offered either alone at low density, or in species pairs at low or high density. Selection coefficients <0.5 indicate a preference for *Chlamydomonas*. Bars are standard errors. From DeMott (1989).

tions, that a relatively small number of organisms feeds simply by sieving. Copepods can feed selectively, choosing to trap only a proportion of what they scan, although the cues that they use to discriminate between particles can only be surmised. Organisms that feed by flapping setose appendages may, in general, have more control over what they catch than is generally appreciated. Theoretical studies show that a simple setose limb can have the potential to act either as a sieve or a paddle, depending on the speed with which it is moved (Cheer & Koehl, 1987).

10.3 SYMBIOSES WITH INTERNAL AUTOTROPHS

All heterotrophs are dependent, ultimately, on the fixation of inorganic carbon by autotrophs. Within any locality, aquatic primary productivity is often limited by the availability of dissolved inorganic nutrients, principally phosphorus and nitrogen. Just as heterotrophs need autotrophs for carbon fixation, autotrophs are often dependent on heterotrophs for the recycling of inorganic nutrients bound in organic compounds. The most common form of these reciprocal transactions is perhaps the least efficient; heterotrophic animals broadcast nutrients into the water in the form of excretory products, and then expend energy in capturing the carbon-rich autotrophs in which the nutrients eventually are bound. Clearly there is potential for much tighter coupling of heterotrophic and autotrophic processes. One way in which this coupling has been achieved has been the evolution of symbioses, in which a unicellular autotroph is located within the cells or tissues of a heterotrophic host.

A restrictive definition of symbiosis specifies that there should be mutual advantage in the association of the species. Heterotrophic hosts can derive part, or even all, of their energy requirements from organic compounds released by their autotrophic partners. To do this, they must not only acquire a viable population of suitable symbionts, but also suspend or limit the processes that would normally lead to the digestion of foreign bodies within their tissues. This would be the equivalent of killing the goose that laid the golden eggs. For their part, internal autotrophs benefit from the release of

nitrogen and phosphorus in inorganic form by the catabolic processes of the host. They may find additional advantage in the protection from predation that residence within a larger organism can provide. A sessile host can also offer an opportunity for a symbiont to remain stationary in a place that offers appropriate conditions (e.g. adequate light) for autotrophic processes. In contrast, planktonic free-living autotrophs are at the mercy of water circulation patterns, and attached forms risk being blanketed by sediment.

10.3.1 Photoautotrophy

In the clear, nutrient-poor waters around tropical coasts there are many marine invertebrates that harbour photosynthetic symbionts. Endodermal cells of almost all of the reef-forming, or hermatypic, corals contain large numbers of symbiotic unicellular algae. Other anthozoan coelenterates, such as the solitary sea anemones of the genus *Anthopleura*, together with some nudibranch gastropods, also contain intracellular algal symbionts. Although the appearance of the algae can differ between hosts, it is believed that only one species is involved in these relationships, the dinoflagellate *Symbiodinium microadriaticum*. The same species also occurs intercellularly in the mantle tissues of many giant clams of the family Tridacnidae. Symbiotic dinoflagellates are often referred to as zooxanthellae, a term of convenience that is used for all brown or yellow-brown symbiotic algae and has little taxonomic significance. Besides dinoflagellates, sponges may contain intra- or intercellular cyanobacteria, whilst some tunicates contain the somewhat equivalent photosynthetic prochlorophytes.

Symbioses between algae and invertebrates are much less common in temperate waters. On some sandy beaches at the western end of the English Channel, large numbers of the flatworm *Convoluta roscoffensis* often form green patches at low tide. These worms contain the green alga *Platymonas*, principally within their subepidermal cells. Symbiotic green algae are often termed zoochlorellae.

In fresh water, the overwhelming majority of algal–invertebrate symbioses involve zoochlorellae, usually of the genus *Chlorella*. The best known hosts of these algae are the green hydras, but zoo-chlorellae are also found in some sponges, flat-worms, bivalve molluscs and protozoa.

The photosynthetic elements of all these associa-tions are essentially entire algae. In many cases they can be isolated from their hosts ad cultured under appropriate laboratory conditions. However, the dark green opisthobranch mollusc *Elysia viridis* feeds by extracting the cell contents of macroalgae, such as *Codium*, and retains only the intact and functional chloroplasts within the cells of its diges-tive gland tubules.

Most of the hosts of algal symbionts retain the ability to feed in the same way as their non-symbiotic relatives. The coelenterates remain capable of capturing zooplankton by use of their nematocysts, and the sponges and bivalves continue to feed on suspended organic matter. This facility may be important in maintaining an adequate supply of nitrogen, phosphorus and other essential nutrients.

The persistence of these symbioses in populations requires a means of transmission of algae between successive generations of the host. Hosts that re-produce by asexual budding pass symbionts directly to their progeny at cell division. The buds of *Hydra* and the growing polyps of hermatypic corals are invariably 'infected' in this way. When *Hydra* pro-duce sexual eggs, these too are infected with maternal zoochlorellae. Such infection at birth is not universal. Newly-hatched *Convoluta* contain no zoochlorellae and seem to acquire them by in-gestion of free-living forms (Holligan & Gooday, 1975). Similarly, the eggs and trochophore larvae of tridacnid clams lack zooxanthellae. These are ingested later by the planktonic, suspension-feeding veliger larvae (Fitt & Trench, 1981). The chloro-plasts sequestered by *Elysia* are unable to divide or grow and must be acquired by each individual during feeding. Since they also appear to have a limited lifespan in the host cells, it is likely that a continuous supply is necessary to maintain numbers (Trench, 1975).

Hydra can be induced to shed their zoochlorellae by treatment with glycerol. These aposymbionts provide an opportunity for study of the mechanism of reinfection and the control of the numbers of zoochlorellae. If algae are injected into the gastro-vascular cavity of an aposymbiotic *Hydra*, the gas-trodermal cells take up the algal cells into vacuoles by phagocytosis. After a period, phagocytosis ceases and a process of sorting ensues. Only certain strains of symbiotic *Chlorella* become established; algal cells isolated from symbiosis with *Paramecium* and sponges do not, neither do free-living forms. Un-wanted or excess algae are either expelled from the gastrodermal cells or digested. The remaining zoochlorellae migrate to the bases of the cells and multiply by division. When the reinfection process is complete (10–20 days), a constant number of zoochlorellae is maintained in each gastrodermal cell. The method by which division of the zoo-chlorellae is controlled is not understood, but it may be due to nutrient limitation. *Hydra* kept in nutrient-rich conditions often die following un-regulated growth of their algal symbionts. Keeping well-fed *Hydra* for prolonged periods in the dark causes a reduction in the number of zoochlorellae per cell, but they are not entirely eliminated and the animal remains pale green indefinitely (Muscatine *et al.*, 1975; Neckelman & Muscatine, 1983).

Evidence that symbionts pass fixed carbon to their hosts, and that this translocation is beneficial to the host, takes a variety of forms. Experiments in which algal–invertebrate associations are cultured in water containing radioactively-labelled carbon, in the form of $NaH^{14}CO_3$, often show significant amounts of labelled organic compounds in the host tissue within a matter of hours. In *Hydra*, most of the carbon fixed by the zoochlorellae and trans-located to the host is in the form of the sugar maltose. Strains of chlorellae that do not release extracellular maltose are incapable of establishing infections in aposymbiotic *Hydra*. Maltose may be important in preventing the fusion of *Hydra* lyso-somes with the vacuoles in which the zoochlorellae reside. Within 48 hours of the exposure of symbiotic *Hydra* to a pulse of $NaH^{14}CO_3$, much of the labelled maltose is incorporated by the host tissues in pro-teins, glycogen and lipids. Fixation of inorganic carbon by non-symbiotic *Hydra* is negligible by comparison. Maltose is not a universal extracellular product of symbiotic algae. In the *Convoluta–Platymonas* association, the principal compounds translocated from alga to host are the amino acids alanine and glutamate. Zoochlorellae in sponges and *Codium* chloroplasts in *Elysia* release glucose. Zooxanthellae in corals and tridacnid clams pass organic carbon to host tissues mainly in the form of glycerol (Cook, 1983).

Observations on growth and survival of hosts also suggest that the possession of symbiotic algae is beneficial. Although the growth rates of normal symbiotic *Hydra*, aposymbiotic *Hydra* and non-symbiotic species are identical when the animals are kept in the light and well-fed on brine shrimp, symbiotic *Hydra* grow faster than the others on reduced rations, and survive longer when starved. Symbiotic *Convoluta* flatworms can be sustained through many generations in sea water containing only inorganic nutrients and vitamins, provided they have access to light and carbon dioxide. Without their symbionts, immature *Convoluta* fail to develop and eventually die, despite voracious feeding. The growth and survival of tridacnid clams is greater when they possess zooxanthellae than when they do not. Juvenile clams containing zoo-xanthellae will grow successfully under illumina-tion in membrane-filtered sea water for more than 10 months, but juveniles without zooxanthellae die in less than 3 weeks. Zooxanthellae contri-bute to the energy requirements of hermatypic corals, but in addition they are important in the deposition of calcium carbonate to form the coral skeleton. Corals without symbionts, or symbiotic corals kept in the dark, show rates of carbonate deposition that are lower than under normal conditions.

The proportion of photosynthetically-fixed car-bon that is translocated from the alga to the host under normal circumstances is difficult to measure. Estimates range from less than 5% in the case of zoochlorellae in the sea anemone *Anthopleura xanthogrammica*, to more than 95% by zooxan-thellae in the hermatypic coral *Stylophora pistillata*. The contribution of the translocated photosynthetic products to the metabolic energy requirements of the host may also be substantial. Zooxanthellae in clams appear to be capable of supplying more than 50% of the carbon required by their hosts. In *Stylophora*, all of the energy requirements of the coral may be met by photosynthesis in well-lit situa-tions, but the exudate from the zooxanthellae is deficient in nitrogen. Much of this 'junk food' is

converted to mucus, and secreted by the coral. It is likely that the uptake of dissolved organic nitrogen from the water, and the capture of nitrogen in zooplankton, is important for growth of coral tissues (Falkowski *et al.*, 1984). Mucus secretion attracts fish to the reef to feed; it is possible that the nitrogen excreted by the fish benefits the coral–alga association.

10.3.2 Chemoautotrophy

Before 1980, invertebrates dependent for food on internal chemoautotrophic symbionts were un-known. In the late 1970s, exploration by deep submersible vehicles of hydrothermal vents near sites of active sea floor spreading yielded one of the most exciting of recent biological discoveries. Many of the vents were surrounded by whole communi-ties of invertebrates, often of species new to science, at densities that could not be supported by the meagre rain of organic matter descending from the sunlit surface to depths often exceeding 3000 m. The water emerging from these vents is rich in reduced inorganic compounds, including sulphides. Initially, it was proposed that the vent communities were subsisting on free-living sulphur bacteria, which fix inorganic carbon by utilizing the energy released in the oxidation of reduced sulphur. For some invertebrates this hypothesis was challenged when it became clear that their structure was not consistent with either suspension feeding or the absorption of dissolved organic matter from the water.

One of the most astonishing of the vent organisms is the pogonophoran worm *Riftia pachyptila*. This large tube-dweller (up to 1.5 m in length and 40 mm in diameter), like other adult pogonophorans, has no mouth, digestive tract or means of capturing suspended particles. It was inconceivable that an organism of this size and surface area could survive and grow, simply by absorption of the low concen-trations of organic matter around the vents. The importance of internal chemoautotrophs was sus-pected when it was discovered that the trophosome tissue, which accounts for about half the worm's wet weight, contained many bacteria-like intra-cellular inclusions. Analysis of the base composi-tion of their RNA and DNA suggested that these inclusions are related to the free-living, purple, sulphur-oxidizing bacteria.

Evidence that the symbionts are capable of chemoautotrophy is provided by enzyme assays. Cell-free extracts of the trophosome tissue show the presence of enzymes of the Calvin–Benson cycle of CO_2 fixation, notably ribulose-1,5-biphosphate carboxylase (RuBPCase). That reduced sulphur is likely to be the energy source for carbon fixation is shown by the presence of adenosine-5′-phosphosulphate (APS), which is diagnostic of sulphur oxidation. Finally, particles of elemental sulphur are common in the trophosome tissue. Elemental sulphur is an intermediate product of the oxidation of sulphides or thiosulphates, and is known to occur in the cells of free-living, sulphur-oxidizing, chemoautotrophic bacteria. The activity of RuBPCase and APS, whilst strongly inferring chemoautotrophy by sulphur-oxidation, does not exclude the possibility that there is also direct uptake of dissolved organic matter from the surrounding water (Cavanaugh, 1985).

Chemoautotrophy by sulphur-oxidizing bacteria requires the simultaneous presence of reduced-sulphur compounds, CO_2 and O_2. For this reason, free-living sulphur bacteria are often found at the interface between anaerobic and aerobic zones in water or sediments. *Riftia*, and other symbiotic invertebrates found around the margins of hydro-thermal vents, are well-placed to provide exactly these conditions for their internal autotrophs. Since the discovery of symbiotic sulphur bacteria in the vent fauna, similar internal bacteria have been found in other invertebrates inhabiting chemical interfaces. The bivalve *Solemya velum* lives in burrows in the sandy sediment of coastal eelgrass beds. It has access to sulphide-rich pore water in the sediment and to well-oxygenated water above it. Like *Riftia*, it is host to symbiotic bacteria, this time located in the cells of the gills, and shows significant RuBPCase activity. In the laboratory, addition of reduced sulphur compounds stimulates fixation of $^{14}CO_2$ by excised gill tissue. Symbiotic bacteria capable of sulphur-based chemoautotrophy also occur in the tissues of *Phallodrilus leukodermatus*, a gutless oligochaete living in calcareous sands of the Bermudian reefs. The bulk of the *Phallodrilus* population is concentrated at the transition zone

between the oxygenated surface layer and the deeper sulphide-rich layer in the sediment (Giere, 1989). Symbioses with sulphur-oxidizing bacteria are now known or suspected in many other species within these three phyla, and from a wide range of marine habitats. The common factor in all these associations is that they occur in locations where both oxygen and reduced sulphur are readily available.

The use of deep submersible vehicles has led to the discovery of other strange communities, occupying special, restricted areas of the sea bed. Off the coast of Louisiana there are places where hydrocarbons seep from the sediments. Studies of mussels from beds adjacent to the seeps showed that they had intracellular bacteria in their gill tissues. These bacteria are of a type that can subsist on reduced carbon compounds, such as methane, a common constituent of seep water. The methane provides not only a carbon source, but also the energy for autotrophy. Like the sulphur-oxidizing bacteria, these methanotrophic symbionts are aerobic. The fact that the mussels will grow in the laboratory on a diet of oxygen and natural gas (Cary *et al.*, 1988), strongly suggests that they acquire organic compounds from their chemoautotrophic symbionts. Analyses of the relative abundances of the stable carbon isotopes carbon-12 and carbon-13 in mussel tissue, leads to the same conclusion. Mussel tissues are depleted with respect to the heavier isotope and in this resemble the methane from the seeps (Childress *et al.*, 1986). Particulate organic matter, derived from photosynthetic fixation of atmospheric carbon in the euphotic zone, has a much higher proportion of the heavier isotope. Thus it appears that the mussels gain at least part of their organic carbon, ultimately, from the seep. Methanotrophic bacteria have also been found in mussels collected from saline seeps off the Florida coast (Cary *et al.*, 1989). Once again, the stable carbon isotope ratios of mussel tissue suggest that methane is the carbon source. However, radio-carbon dating implies that the methane is not of great age and may have been generated by anaerobic bacteria metabolizing organic matter in the sediment.

Although the discovery of chemoautotrophy in marine invertebrates is recent, it now appears that many species derive at least part of their carbon and energy requirements in this way. No doubt many more remain to be detected. As yet, chemoautotrophy in freshwater invertebrates is unknown. Low sulphate concentrations prevent the formation of appreciable amounts of sulphides in most freshwater sediments; but methane (marsh gas) is ubiquitous and abundant, particularly in swamps and the sediments of productive lakes. Perhaps a freshwater invertebrate that possesses methanotrophic symbionts is waiting to be revealed.

FURTHER READING

Braimah, S.A. (1987) Pattern of flow around filter-feeding structures of immature *Simulium bivittatum* Malloch (Diptera: Simuliidae) and *Isonychia campestris* McDunnough (Ephemeroptera: Oligoneuridae). *Can. J. Zool.* **65**: 514–21.

Brendelberger, M., Herbeck, M., Lang, H. & Lampert, W. (1986) *Daphnia*'s filters are not solid walls. *Arch. Hydrobiol.* **107**: 197–202.

Cary, S.C., Fisher, C.R. & Felbeck, H. (1988) Mussel growth supported by methane as a sole carbon source. *Science* **240**: 78–80.

Cary, S.C., Fry, B., Felbeck, H. & Vetter, R.D. (1989) Multiple trophic resources for a chemoautotrophic community at a cold water brine seep at the base of the Florida Escarpment. *Mar. Biol.* **100**: 411–18.

Cavanaugh, C.M. (1985) Symbioses of chemoautotrophic bacteria and marine invertebrates from hydrothermal vents and reducing sediments. *Biol. Soc. Washington Bull.* **6**: 373–88.

Chance, M.M. & Craig, D.A. (1986) Hydrodynamics and behaviour of Simuliidae larvae (Diptera). *Can. J. Zool.* **64**: 1295–309.

Cheer, A.Y.L. & Koehl, M.A.R. (1987) Paddles and rakes: fluid flow through bristled appendages of small organisms. *J. Theor. Biol.* **129**: 17–39.

Childress, J.J., Fisher, C.R., Brooks, J.M., Kennicutt, M.C., Bidigare, R. & Anderson, A. (1986) A methanotrophic marine molluscan (Bivalvia, Mytilidae) symbiosis: mussels fueled by gas. *Science* **233**: 1306.

Cook, C.B. (1983) Metabolic interchange in algae–invertebrate symbiosis. *Int. Rev. Cytol. Supplement* **14**: 177–210.

Deibel, D. (1986) Feeding mechanism and house of the appendicularian *Oikopleura vanhoeffeni*. *Mar. Biol.* **93**: 429–36.

DeMott, W.R. (1989) Optimal foraging theory as a predictor of chemically mediated food selection by suspension-feeding copepods. *Limnol. Oceanogr.* **34**: 140–54.

Falkowski, P.G., Dubinsky, Z., Muscatine, L. & Porter, J.W. (1984) Light and the bioenergetics of a symbiotic coral. *BioScience* **34**: 705–9.

Fenchel, T. (1980) Suspension feeding in ciliated protozoa: feeding rates and their ecological significance. *Microb. Ecol.* **6**: 13–25.

Fenchel, T. (1986a) The ecology of heterotrophic microflagellates. *Adv. Microb. Ecol.* **9**: 57–97.

Fenchel, T. (1986b) Protozoan filter feeding. *Progr. Protistol.* **1**: 65–113.

Fitt, W.K. & Trench, R.K. (1981) Spawning, development, and acquisition of zooxanthellae by *Tridacna squamosa* (Mollusca, Bivalvia). *Biol. Bull.* **161**: 213–35.

Friedland, K.D. (1985) Functional morphology of the branchial basket structures associated with feeding in the Atlantic menhaden, *Brevoortia tyrannus* (Pisces: Clupeidae). *Copeia* **1985**: 1018–27.

Georgian, J. & Wallace, J.B. (1980) A model of seston capture by net-spinning caddisflies. *Oikos* **36**: 147–57.

Gerritsen, J. & Porter, K.G. (1982) The role of surface chemistry in filter feeding by zooplankton. *Science* **216**: 1225–7.

Gerritsen, J., Porter, K.G. & Strickler, J.R. (1988) Not by sieving alone: observations of suspension feeding in *Daphnia*. *Bull. Mar. Sci.* **43**: 366–76.

Giere, O. (1989) Meiofauna and microbes — the interactive relations of annelid hosts with their symbiotic bacteria. *Proc. Biol. Soc. Washington* **102**: 109–15.

Gophen, M. & Geller, W. (1984) Filter mesh size and food particle uptake by *Daphnia*. *Oecologia* **64**: 408–12.

Hart, D.D. & Latta, S.C. (1986) Determinants of ingestion rates in filter-feeding larval blackflies (Diptera: Simuliidae). *Freshwater Biol.* **16**: 1–14.

Holligan, P.M. & Gooday, G.W. (1975) Symbiosis in *Convoluta roscoffensis*. *Soc. Exp. Biol. Symp.* **29**: 205–27.

Hughes, R.G. (1975) The distribution of epizoites on the hydroid *Nemertesia antennina*. *J. Mar. Biol. Assoc. UK* **55**: 275–94.

Jakobsen, P.J. & Johnsen, G.H. (1988) The impact of food limitation on swarming behaviour in the waterflea *Bosmina longispina*. *Anim. Behav.* **36**: 991–6.

Johnson, A.S. (1988) Hydrodynamic study of the functional morphology of the benthic suspension feeder *Phoronopsis viridis*. *Mar. Biol.* **100**: 117–26.

Jorgensen, C.B. (1966) *Biology of Suspension Feeding*. Pergamon Press, Oxford.

Jorgensen, C.B. (1983) Fluid mechanical aspects of suspension feeding. *Mar. Ecol. Progr. Ser.* **11**: 89–103.

Koehl, M.A.R. & Strickler, J.R. (1981) Copepod feeding currents: food capture at low Reynolds number. *Limnol. Oceanogr.* **26**: 1062–73.

LaBarbera, M. (1984) Feeding currents and particle capture mechanisms in suspension feeding animals. *Am. Zool.* **24**: 71–84.

Lam, R.K. & Frost, B.W. (1976) Model of copepod feeding response to changes in size and concentration of food. *Limnol. Oceanogr.* **21**: 490–500.

Lehman, J.T. (1976) The filter-feeder as an optimal forager, and the predicted shapes of feeding curves. *Limnol. Oceanogr.* **21**: 501–16.

Mayzaud, P., Taguchi, S. & Laval, P.L. (1984) Seasonal patterns of seston characteristics in Bedford Basin, Nova Scotia, relative to zooplankton feeding: a multivariate approach. *Limnol. Oceanogr.* **29**: 745–62.

Muscatine, L., Cook, C.B., Pardy, R.L. & Pool, R.R. (1975) Uptake recognition and maintenance of symbiotic *Chlorella* by *Hydra viridis*. *Soc. Exp. Biol. Symp.* **29**: 175–203.

Neckelmann, N. & Muscatine, L. (1983) Regulatory mechanisms maintaining the *Hydra–Chlorella* symbiosis. *Proc. R. Soc. Lond.* B **219**: 193–210.

Pearson, T.H. & Rosenberg, R. (1987) Feast and famine: structuring factors in marine benthic communities. In: Gee, J.H.R. & Giller, P.S. (eds) *Organization of Communities: Past and Present*, pp. 373–95. Blackwell Scientific Publications, Oxford.

Porter, K.G. (1977) The plant–animal interface in freshwater ecosystems. *Am. Sci.* **65**: 159–70.

Porter, K.G., Gerritsen, J. & Orcutt, J.D. (1982) The effect of food concentration on swimming patterns, feeding behaviour, ingestion, assimilation and respiration by *Daphnia*. *Limnol. Oceanogr.* **27**: 935–49.

Price, H.J. & Paffenhofer, G.A. (1986) Capture of small cells by the copepod *Eucalanus elongatus*. *Limnol. Oceanogr.* **31**: 189–94.

Rubenstein, D.I. & Koehl, M.A.R. (1977) The mechanisms of filter feeding: some theoretical considerations. *Am. Nat.* **111**: 981–94.

Sheldon, R.W., Prakash, A. & Sutcliffe, W.H. (1972) The size distribution of particles in the ocean. *Limnol. Oceanogr.* **17**: 327–48.

Silvester, N.R. (1988) Hydrodynamics of flow in *Mytilus* gills. *J. Exp. Mar. Biol. Ecol.* **120**: 171–82.

Silvester, N.R. & Sleigh, M.A. (1984) Hydrodynamic aspects of particle capture by *Mytilus*. *J. Mar. Biol. Assoc. UK* **64**: 859–79.

Silver, M., Shanks, A.L. & Trent, J.D. (1978) Marine snow; microplankton habitat and source of small-scale patchiness in pelagic populations. *Science* **201**: 371–3.

Strickler, J.R. (1985) Feeding currents in calanoid copepods: two new hypotheses. *Symp. Soc. Exp. Biol.* **39**: 459–85.

Trench, R.K. (1975) Of 'leaves that crawl': functional chloroplasts in animal cells. *Symp. Soc. Exp. Biol.* **29**: 229–65.

Wallace, J.B., Ross, D.H. & Meyer, J.L. (1982) Seston and dissolved organic carbon dynamics in a southern Appalachian stream. *Ecology* **63**: 824–38.

Wetzel, R.G. (1983) *Limnology*. Saunders, Philadelphia.

PART 4
HABITAT TYPES PECULIAR TO AQUATIC SYSTEMS

11
Reefs

R.N. HUGHES

11.1 INTRODUCTION

A reef is a ridge, or block of substratum, that rises towards the water surface. Its framework may be derived abiotically, from bedrock, boulders, pebbles and sand, or in the case of man-made reefs, from concrete blocks, car bodies and other refuse; or it may be constructed biologically from the skeletal material of various organisms.

Reefs increase topographical complexity on a large scale, influencing hydrographic conditions and providing living places for many types of organism; consequently, they are usually associated with greater biological richness than adjacent habitats. Abiotically derived reefs abound in freshwater and marine environments, but biologically constructed ones are confined to the sea, and it is only the latter which we will consider in the present chapter.

Organisms which contribute substantially to the framework of a reef are often described as hermatypic. This term is applied most frequently to corals, where it corresponds to those tropical, shallow-water species that harbour symbiotic zooxanthellae (unicellular algae) in their tissues. The symbiosis enables these corals to secrete large amounts of calcium carbonate. Corals lacking zooxanthellae secrete smaller quantities of calcium carbonate and are described as ahermatypic, even though they are able to form small reefs in certain circumstances. Schuhmacher and Zibrowius (1985) recommend that symbiotic, reef-building and frame-building properties should be distinguished by the terms zooxanthellate, hermatypic and constructional, respectively. In the following account however, the terms zooxanthellate, reef-building and frame-building will be used. Reef-builders are organisms that contribute to reef formation; frame-builders are those which form major constructional units.

11.2 AZOOXANTHELLATE FRAME-BUILDERS

Corals lacking symbiotic algae, and living below the photic zone in cool, deep water, sporadically form thickets or reefs along the continental shelf margin, from northern Norway to northwestern Africa (Joubin, 1922). Relatively few species are involved, and they are of simple, branching form (Teichert, 1958).

Serpulids form small reefs on the open coast in certain cold-temperate localities, and in lagoonal and brackish-water habitats over a wider geographical range (Ten Hove, 1979). Aggregative larval settlement, augmented by adult fission in some species, results in an intertwining growth of the small calcareous tubes secreted by the worms. Aggregates may form a layer up to 1 m thick over several square metres, but never extensive reefs. These aggregates tend to occur in hydrographic conditions that impede larval dispersal. Under other circumstances the serpulids may be solitary, although some, such as the estuarine *Hydroides* species, habitually form small reefs.

Vermetid gastropods, particularly of the genus *Dendropoma*, form aggregations in wave-exposed, warm-temperate and subtropical localities throughout the world. After settlement of crawling young, the shells grow in a worm-like fashion, producing interwoven masses (Hughes, 1979). Often, vermetid aggregations merely form a veneer, several centimetres thick, over the rock surface, but in certain localities, such as the eastern Mediterranean, they build extensive ramparts (Safriel, 1966).

Oyster and mussel larvae are attracted during settlement by the presence of adults (Bayne, 1969; Seed, 1976). As a result, oyster and mussel beds may form extensive layers over the substratum. But on reaching a critical thickness, the aggregations usually become locally unstable and susceptible to dislodgement by waves or currents (Seed, 1969). Along the coast of Florida and Texas however, shells of *Crassostrea virginica* accumulate parallel to brackish-water tidal channels, building reefs some 8–10 km long, 150 m wide and 4 m thick (Stenzel, 1971). Also, large subtidal reefs of *Ostrea edulis* once flourished in the North Sea (Caspers, 1950).

The bryozoans, *Electra crustulenta* and *Alcyonella fungosa* form small reefs in brackish-water tidal channels and lagoons along the Netherlands coastline (Bockschoten & Bijma, 1982) and elsewhere. *E. crustulenta* grows very quickly, forming calcareous masses over 10 m wide and 1 m thick, whereas *A. fungosa* makes a lesser, locally-restricted contribution to reef formation. These bryozoans have only been described as reef-forming in brackish water.

Not all animals form reefs by secreting frame-building materials, some cement together pre-existing sediment particles to form what has been termed 'sand coral'. Throughout the world, the sabellariid polychaetes, for example, may construct massive sand-coral reefs covering several square kilometres. One colony near the island of Norderney in the southern North Sea incorporated 800–1000 m^3 of sand into its reef in a 6-month period (Linke, 1951).

All of the reefs described above have an enriching effect on the local biota. They increase topographical complexity, providing shelter and microhabitats for other organisms. Reefs overlying sediments form hard surfaces, suitable for colonization by sedentary organisms that would otherwise be absent from the area. For example, the Dutch bryozoan reefs increase the faunal diversity of muddy creeks by providing shelter and firm substrata for hydroids, nudibranchs, polychaetes, crustaceans and fish (Bockschoten & Bijma, 1982). The reef-building organisms themselves may be food for certain animals, thus oysters support muricid gastropods and bryozoans are grazed by nudibranchs. In other respects however, these reefs differ little from those of abiotic origin.

11.3 ZOOXANTHELLATE FRAME-BUILDERS

Tropical coral reefs present a different story. They can be of immense topographical and geological importance, supporting communities with a species richness and productivity far exceeding that of neighbouring habitats. The key to this profusion is a symbiotic relationship between limestone-secreting polyps and photosynthesizing zooxanthellae. By absorbing carbon dioxide for photosynthesis, zooxanthellae facilitate the deposition of calcium carbonate (Goreau, 1959). Moreover, photosynthates, translocated within the host's tissues, may enhance calcification by serving as specific substrates for the organic matrix of the skeletal material or as a general energy source (Pearse & Muscatine, 1971). Consequently, the polyps are able to secrete limestone in such quantities that reef frameworks of massive proportions are formed. Niches arise, which are exploited by diverse organisms, the huge surface area and biomass of living tissue resulting in rapid nutrient flux and high productivity (Section 11.4.5). As the zooxanthellae require plenty of light, and the symbiosis itself seems to require relatively high, stable temperatures, large coral reefs are confined to clear, tropical or subtropical waters (Fig. 11.1).

11.3.1 Geological succession of frame-builders

Massive reef formation has been an important phenomenon in warm seas since the Cambrian period, some 500–600 million years ago. During the course of time, particular frame-builders have been replaced by others, of different taxonomy. Most frame-builders appear to have had symbiotic associations with microphotosynthesizers, which, as in the case of modern corals, probably demanded clear, warm seas. Evidence strongly suggests that in each case, existing frame-builders were exterminated by a general cooling of the world climate; as occurred, for example, when Antarctica drifted over the South Pole, precipitating glaciation and altering oceanic circulation (Stanley, 1987). When

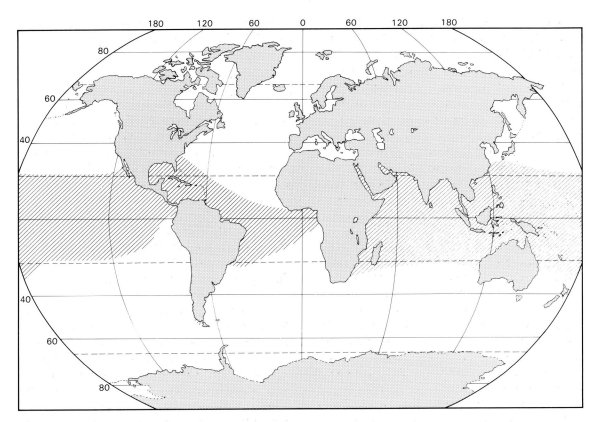

Fig. 11.1 Geographical distribution of recent coral reefs. Most reefs lie within the tropics, where the surface water temperature remains above 20°C throughout the year. From Fagerstrom (1987).

the seas warmed again, new reef-builders evolved (Table 11.1).

In Precambrian times, 600–2000 million years ago, cyanobacterial mats trapped mounds of sediment. Seen in cross-section, some of these mounds were composed of concentric layers and are known as stromatolites; others, the thrombolites, were not layered. These reefs were topographically relatively simple and, at that early stage in organic evolution, must have been associated with a modest diversity of other organisms.

Metazoans first began building reefs in the late Cambrian, when sponges joined the skeletal cyanobacteria as frame-builders. In the Ordovician, massive bryozoans became the frame-builders, later to be joined by tabulate and rugose corals, and by stromatoporoid sponges that secreted substantial limestone skeletons. Stromatoporoids still linger as a minor group in dimly lit caves in the Caribbean,

where they escape competition from the faster-growing modern corals. By the end of the Ordovician period, the frame-builders were decimated, but regained ascendancy in the Silurian–Devonian periods, when extensive reefs were built by tabulate corals and stromatoporoids, with calcareous algae making a secondary contribution. Some of these reefs became buried and fossilized, the porous limestone subsequently holding commercially important reservoirs of petroleum. During the late Devonian, the reef-builders were exterminated again, and it was about 100 million years before new reefs appeared. These were built in the Permian, by calcareous algae and calcareous sponges; then in the Triassic, by hexacorals, very similar to those living today. In the Cretaceous period, the hexacorals were replaced as the major frame-builders by strange bivalves in the family Hippuritacea, commonly known as rudists. Like modern giant clams, rudists

Table 11.1 Palaeological succession of frame-building organisms. From Fagerstrom (1987)

Period	Age (millions of years ago)	Frame-builders
Precambrian	2000–570	Stromatolite and thrombolite cyanobacteria
Early Cambrian	550–540	Calcareous cyanobacteria and Archaeocyatha
Middle Cambrian– Early Ordovician	540–500	Calcareous cyanobacteria and sponges
Middle Ordovician– Late Devonian	480–350	Calcareous algae, Stromatoporoidea, tabulate and rugose corals, Bryozoa
Late Devonian– Late Permian	360–260	Calcareous algae, sponges, Crinoidea, Brachiopoda, Bryozoa
Middle Triassic– Late Triassic	240–220	Calcareous algae, sponges
Late Triassic– Early Cretaceous	210–100	Calcareous algae, sponges, scleractinian corals (hexacorals)
Early Cretaceous– Late Cretaceous	95–65	Scleractinian corals, Hippuritacea
Paleocene–Eocene	60–40	Calcareous algae, scleractinian corals
Oligocene–Holocene	35–0	Calcareous algae, milleporine hydrozoans, scleractinian corals, Bryozoa

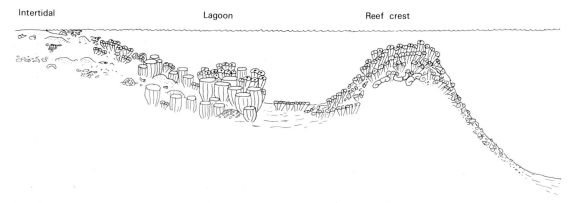

Intertidal   Lagoon   Reef crest

Fig. 11.2 Model of a rudist-dominated reef, such as existed in the Caribbean during the Cretaceous period. From Kauffmann & Sohl (1974).

bore zooxanthellae in their mantle tissue, the symbiosis promoting massive shell secretion. Evolution of an internally-folded hinge ligament enabled one valve to grow into a horn, cornet, or cup-shaped structure; the other valve fitting as a lid. In their variety of forms, often aggregated into great masses, the rudists built large reefs, of considerable topographical complexity (Fig. 11.2). At the end of the Cretaceous period, the rudists were exterminated, along with many other organisms. Hexacorals eventually replaced the rudists as the major frame-

builders, a role they have continued to play to the present day. During this geological succession of reef communities, species diversity has increased by some two orders of magnitude (Fig. 11.3).

11.4 MODERN CORAL REEFS

11.4.1 The coral organism

Hexacorals, whose tentacles and internal septa number in multiples of six, are the primary frame-

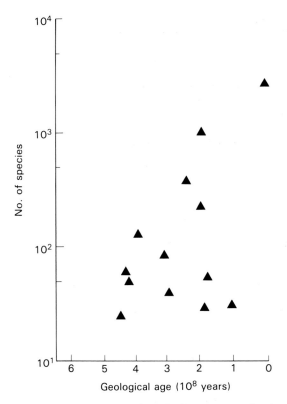

Fig. 11.3 Number of major frame-building species in fossil assemblages, plotted as a function of geological age. From Stanley (1987).

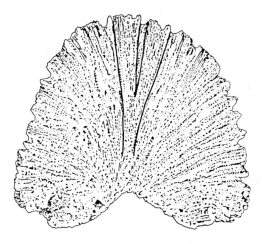

Fig. 11.4 Vertical section through a coral skeleton, showing arrangement of corallites. Two corallites, diverging from a common stem, where a polyp has undergone fission, have been outlined. From Hughes (1989).

builders of today's reefs. Corals are modular animals (Harper *et al.*, 1986; Hughes, 1989), which characteristically can assume diverse growth forms, from plates, mounds and pillars to bushes and trees (see Section 8.3). The basic module is the polyp. Polyps grow to a certain size, undergo incomplete binary fission, then resume growth. Continued modular iteration results in a colony of interconnected polyps, which form a layer of tissue, almost paper thin, overlying the exoskeleton.

Zooxanthellae, of the dinoflagellate species *Symbiodinium microadriaticum*, reside within the gastrodermal cells of the polyp, imparting various shades of brown or green. When placed in continual darkness, either in the laboratory or at depth in the sea, the zooxanthellae are unable to photosynthesize, begin to deteriorate, and are finally extruded by the polyps. Having jettisoned their

zooxanthellae, polyps lose the capacity to form large amounts of calcium carbonate (Goreau, 1959).

The exoskeleton, secreted by the basal layer of epidermis, forms a limestone substratum, intricately following the contours of the living tissue. As the colony grows, each polyp continues to secrete calcium carbonate, which forms a column, known as the corallite (Fig. 11.4). Adjacent corallites are usually cemented together, but may diverge when branches are formed.

Owing to their symbiosis with algae, corals have both autotrophic and heterotrophic elements of nutrition. Autotrophy depends on photosynthesis by the zooxanthellae. Incubation with $^{14}CO_2$ has shown that as much as 50% of the carbon-14 fixed daily by photosynthesis becomes incorporated into coral tissue, mostly as lipid and protein (Muscatine & Cernichiari, 1969; Kellogg & Patton, 1983). Heterotrophy depends on the capture of plankton. This is achieved by a mucociliary mechanism that also serves as a cleansing device (Fig. 11.5). The relative importance of autotrophy and heterotrophy varies among corals, some deriving over 95% of their energy from photosynthesis, others as little as 50%.

The more autotrophic corals have tree-like growth forms and relatively small polyps. Light is

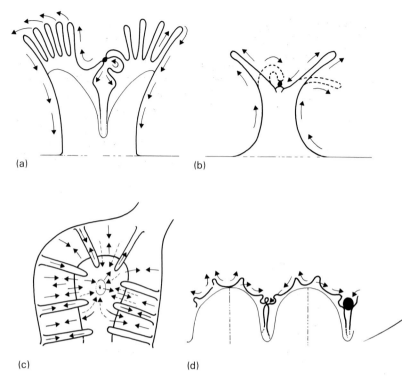

(a)   (b)

(c)   (d)

Fig. 11.5 Mucociliary cleansing and feeding mechanisms in corals. (a) Coral with large polyps; cilia beat away from the mouth. (b) Coral with small polyps; cilia beat up the column. In (a) and (b), particles are rejected from the tips of the tentacles unless the tentacles bend over to the mouth. (c) Brain coral with short tentacles; cilia can reverse beat towards or away from the mouth. (d) Massive corals with short tentacles; cilia carry all particles across the mouth; edible particles are ingested, inedible particles are rejected from the tips of the tentacles. From Yonge (1963).

extensively scattered by suspended particles and molecules in solution, so that in shallow water, there is a considerable horizontal component of diffuse light. A multilayered, branching growth form, projecting in three dimensions, yet minimizing self-shading, maximizes the interception of incident and scattered light (Porter, 1976). Thin tissue-layers and small polyps expose the maximum area of zooxanthellae to the light. Moreover, the amount of algal protein in the tissues approximately equals that of the polyps themselves.

The more heterotrophic corals have less algal material per unit mass of tissue, thicker tissue layers, larger polyps and their colonies tend to have a single-layered, spheroidal form. In the clear waters flowing over coral reefs, plankton densities are normally so low that after passing over a single layer of polyps, most food particles are removed. A branching colony would be inefficient, since inner layers of polyps would be deprived of food.

Growth form is also influenced by environmental conditions. Where light has attenuated at depth, or

in shaded places in shallow water, low photosynthetic rate limits calcium carbonate secretion. In these situations, corals tend to have foliose or plate-like growth. This minimizes the amount of skeletal material required per polyp, since a foliose colony essentially grows two-dimensionally, whereas a spheroidal colony would grow volumetrically. Corals such as *Montastrea annularis* (Fig. 11.6a) change from a volumetric growth form in well-lit shallow water, to a flattened form in poorly-lit, deeper water (Dustan, 1975). Some species can be induced, experimentally, to change their growth form (Fig. 11.6b). Other species have narrower environmental requirements and are architecturally less adaptable.

Modularity, in addition to promoting architectural diversity and flexibility, also compartmentalizes damage and enables colonies to regenerate from fragments (Jackson & Hughes, 1985). Severe storms and hurricanes hit most reefs about once a decade, breaking up many of the corals in shallow water. Most of the fragments die, but others survive,

(a)

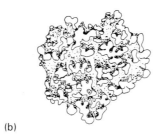

(b)

Fig. 11.6 Architectural changes in corals, associated with water depth. (a) *Montastrea annularis* with volumetric form at 5 m (left) and plate-like form at 35 m (right). (b) *Porites compressa* with stubby branches in shallows on the reef flat (left), and longer branches after transplanting to deeper water (right). From Maragos, cited in Hughes (1989).

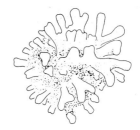

the proportion of survivors increasing with fragment size (Fig. 11.7). Both categories of fragment are important in reef development. Dead fragments contribute to the accretion of limestone, and in certain species the live fragments are an important source of propagation (Fig. 11.8).

Not all corals are able to propagate by fragmentation, however. Some are broken into small, unviable fragments, while others are seldom fragmented at all. Such corals rely more on the release of planula larvae, a few of which may survive to settle, metamorphose and develop into new colonies (Fig. 11.9).

Colonial life expectancies vary, depending on size and species. This relationship is further complicated by fragmentation and fission, which decouple size from age (T.P. Hughes & Jackson, 1980). In general, however, small fast-growing corals have life expectancies in the order of years and decades, whereas massive, slow-growing species may live for centuries, or even millennia.

11.4.2 Reef formation

Coral skeletal systems are major constructional units of most reefs. Other important contributors are calcareous algae, foraminiferans calcareous hydrozoans, molluscs and bryozoans. In deeper, cooler or climatically variable waters, where conditions are suboptimal for zooxanthellate corals, these other frame-builders may assume dominance. Sediments also play an extremely important role, since they may become bonded and transformed into rock. Indeed, the majority of reefs contain more sedimentary material than coral framework (Thomassin, 1978). The sediments are formed by physical and biological erosion of the limestone secreted by reef-builders.

Binding of sediments is initially biological, followed by cementation. On shallow Caribbean reefs, rubble is quickly colonized and stabilized by encrusting sponges, then by calcareous algae, which bind the particles solidly together. Within about 10

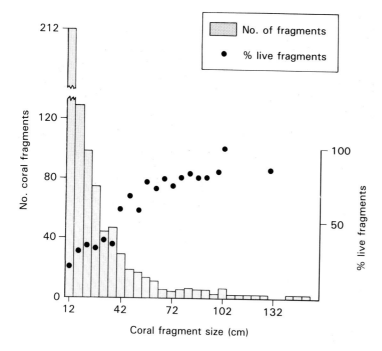

Fig. 11.7 Histogram to show size-frequency of fragments generated by a hurricane, also showing survivorship of fragments as a function of their size. From Highsmith *et al.* (1980).

Fig. 11.8 Living fragments of *Acropora cervicornis*, forming a thicket.

months, the bound rubble is colonized by corals, adding to the vertical growth of the reef (Wulff, 1984).

Cementation begins within the cavities, pores and interstices of coral fragments, rubble and finer sediments. We do not know exactly how the cement is formed. Precipitation of calcium carbonate may be initiated by the reduced partial pressure of dissolved carbon dioxide, as sea water is flushed through the cavities by wave surge. Flushing with large amounts of sea water would be necessary to produce volumetrically significant precipitation, and this may partly explain why cementation is most pronounced in shallow, turbulent water (Fagerstrom, 1987). On the other hand, the fact that cements are out of isotopic equilibrium with sea water, suggests a biological influence, perhaps involving bacteria (Liddell *et al.*, 1984).

Accretion, through limestone secretion, sediment binding and cementation, is accompanied by erosion. Physical destruction and chemical dissolution are greatest on windward reefs, whereas biological erosion is greater in sheltered areas. Biological erosion increases the porosity and permeability of the substratum, facilitating additional erosion by physical and chemical forces, where these are important. In other circumstances, the cavities and pores resulting from biological erosion enhance the trapping and cementation of sediment.

Eroding organisms can be classified into grazers, etchers and borers (Hutchings, 1986). Grazers are

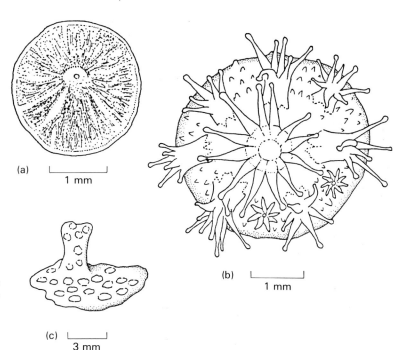

Fig. 11.9 (a) Metamorphosed coral planula, a few days after settlement. (b) Encrusting colony, about 7 weeks after settlement (large ancestral polyp, representing metamorphosed planula, in centre). (c) Vertical growth, underway about 10 months after settlement. From various authors, cited in Hughes (1983).

mainly sea urchins, parrotfish, surgeonfish, filefish and other fish with chisel- or beak-like teeth. Etchers include bacteria, fungi and filamentous endolithic algae, while borers are predominantly clionid sponges, bivalves, such as giant clams and date mussels, spionid polychaetes and sipunculans.

Reef formation reflects the balance between accretion and erosion, sometimes being positive and sometimes negative. In general, potentially rapid accretion, associated with high calcification rates on the windward side of a reef, is prevented by severe erosion, resulting in the transportation of sediments to the leeward side. Windward slopes are thus dominated by erosion, and leeward environments by accretion (Davies, 1983). A detailed calcium carbonate budget, showing net accretion, has been made for a leeward fringing reef in Barbados (Table 11.2).

Reef formation is further modified on a geological time-scale by changes in sea level. Cores have shown that part of the Great Barrier Reef started growing some 8000 years ago at a rate of 1–8 m per 1000 years (Marshall & Davies, 1982). At first, the sea level rose faster than the vertical growth of the reef, but around 6200 years ago, sea

Table 11.2 Calcium carbonate budget for a 10 800 m² fringing reef on the west coast of Barbados. From Scoffin *et al.* (1980)

Accretion (g m⁻² year⁻¹)		Erosion (g m⁻² year⁻¹)	
Scleractinia	129.7	Chiton	97.5
Coralline algae	46.8	Clionid sponges	24.9
Inorganic cements	29.8	Parrot fish	0.4
Total	206.3	Total	122.8
	Balance +83.5		

level stabilized, whereupon the vertical growth decelerated and accretion increased on the leeward side. Three phases in reef development could be recognized: initial vertical growth to sea level, transitional adjustment of coral associations at sea level, and a leeward growth phase involving the lateral spread of reef flats.

Relative sea level sometimes drops, either by withdrawal of water into glacial ice-sheets, or by elevation of the coastline. The old reefs are then exposed above sea level, where they may remain as limestone cliffs, while seaward growth is continued

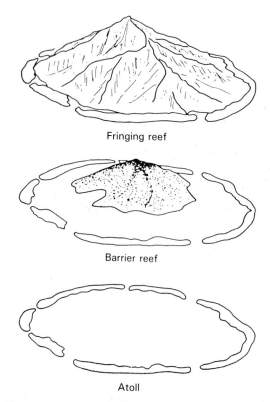

Fig. 11.10 Formation of barrier reefs and atoll by the subsidence of a volcanic island and the compensatory vertical growth of coral. From Barnes & Hughes (1988).

at a new, lower level. This phenomenon is common in the Red Sea and western Indian Ocean. If relative sea level begins to rise more quickly, either due to melting ice-sheets or sinking coastline, fringing reefs become increasingly separated from the coast by a lagoon, gradually evolving into barrier reefs. When this occurs around the shores of a volcanic island, the land itself may eventually erode and sink beneath the sea, leaving a circular barrier reef or atoll (Fig. 11.10).

11.4.3 Reef structure

The physical and biological structure of a reef change in a predictable manner from seaward to landward, and from windward to leeward, sides of a reef. On the most exposed windward reefs, the reef crest is dominated by encrusting calcareous algae, especially *Porolithon*, forming an internally honeycombed limestone with a smooth outer sur-

face that absorbs and deflects the force of persistent swell. On moderately exposed windward reefs, the reef crest is characterized by a few species of stoutly branched corals and calcareous hydrozoans, or 'stinging' corals. These close-growing, robust colonies form ramparts against the heavy seas. Both algal- and coral-dominated reef crests are often dissected by a 'spur-and-groove' system normal to the reef edge (Fig. 11.11). Alternating spurs and grooves are several metres wide and up to several hundred metres long, depending on the slope of the reef. They dissipate the power of waves, estimated to exceed 500 000 kW. Spurs and grooves are themselves the result of wave action. Surging water erodes weak spots, churning up sediment and rolling boulders so that coral growth is prevented. The enlarging grooves accommodate and dissipate much of the surge, creating less turbulent areas on either side where corals survive, grow and act as baffles, further reducing turbulence. Development of the 'spur-and-groove' system is therefore a self-reinforcing process, and is of great importance to the topography and stability of windward reef slopes.

At greater depths on the seaward side of a reef, the force of water movement is less, and so also is the light intensity. Calcification is reduced, resulting in a less firmly cemented framework. Sponges, molluscs and worms that excavate and weaken coral skeletons are abundant, increasing the risk of collapse and slumping of the substratum. Coral growth-forms tend to be foliaceous or plate-like, maximizing the surface area receiving incident light and minimizing any tendency to roll, after dislodgement, down the reef slope into lethal depths. Below about 30–70 m, depending on water clarity, sponges, soft corals, seawhips, gorgonians and azooxanthellate corals replace frame-building corals. Calmer water at depth encourages silting, and the substratum changes from rubble to fine sediments.

On the landward sides of the reef crest, calmer water allows a variety of corals to survive, especially small, branching forms that are able to grow quickly in bright light and to repopulate areas devastated by storms, or by exceptionally low tides. Further landward, the accumulation of sediments, shallow water and frequent exposure to the air

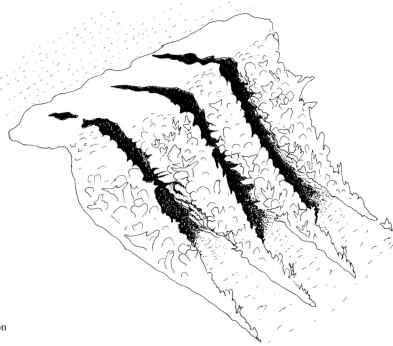

Fig. 11.11 Spur-and-groove formation
on an exposed, windward reef.

prevent the extensive growth of corals. A reef flat is formed, dominated by frondose algae and sea-grasses.

Leeward reefs have similar fore-reef slopes and reef flats, but differ considerably in the region of the reef crest. There is no spur-and-groove system, and coral associations on the upper reef slope merge gradually with those on the outer reef flat. Branching, quick-growing corals are abundant, together with a wide variety of massive forms, including brain corals. Colonies tend to be aggregated, forming patches and large mounds among stretches of sand and rubble.

11.4.4 Biological diversity of reefs

Coral reefs are topographically complex and highly productive; consequently there is an abundance of microhabitats and food, sufficient to support extremely rich biological communities. Immediately obvious are multitudes of fish, echinoderms and sedentary colonial invertebrates, but a significant part of the reef community consists of small animals living within crevices and cavities in the rock,

beneath ledges and within sediments. A coral skeleton taken from the Great Barrier Reef proved to contain 1441 small polychaete worms belonging to 103 species (Grassle, 1973). Most of these were syllids, but several other families were also present. In addition to the polychaetes, which accounted for two-thirds of the animals collected from the coral, there were tanaids, amphipods, isopods and smaller numbers of sipunculans, oligochaetes, decapod crustaceans and ophiuroids. Lumps of reef rock support an abundant 'cryptofauna', rich in harpacticoid copepods, gammarid amphipods, syllid polychaetes and gastropods (Klumpp *et al.*, 1988). On the Great Barrier Reef, the cryptofaunal density within the territories of damselfish can exceed $580\,000\,m^{-2}$, three to four times higher than on adjacent areas. Damselfish feed upon algal mats growing within their territories, which they defend from other grazers. Outside these territories, the cryptofauna is exposed to intense grazing pressure by a variety of animals, whereas inside it suffers only accidental destruction by the damselfish.

Ledges and overhangs are densely colonized by sedentary invertebrates, particularly sponges and

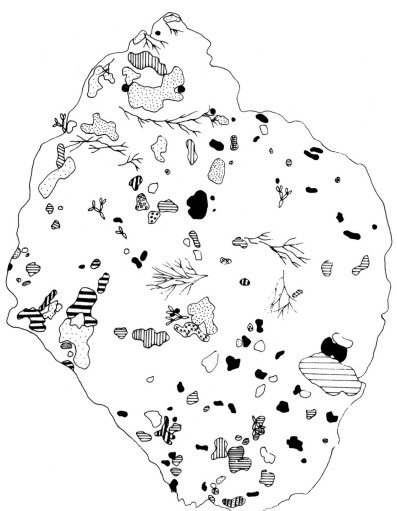

Fig. 11.12 Undersurface of a plate-forming coral, colonized by 32 species of bryozoan. Data from D.J. Hughes (unpublished).

bryozoans (Fig. 11.12). Sediments, ranging from silts to coarse rubble, support rich communities of infaunal species. On the Great Barrier Reef, infaunal densities of 3000–44 000 individuals m^{-2} have been recorded (Riddle, 1988). Community composition varies according to granulometry of the sediment, and particularly to location on the reef.

High macrofaunal diversity, typical of coral reefs, may result partly from topographical complexity and partly from climatic stability (Grassle, 1973). The effect of topographical complexity is illustrated by the number of *Conus* species living in different coral-reef habitats. These carnivorous gastropods achieve ecological separation by utilizing different microhabitats. More microhabitats, and hence more species of *Conus*, are present in subtidal than intertidal habitats, and on rocky rather than sandy substrata (Fig. 11.13).

The climatic stability of coral reefs is testified by their persistence, relatively unchanged, in the tropics for over 50 million years. Because corals are killed by quite minor deviations from optimal conditions, the reef environment itself must have remained virtually unchanged. Of course, in marginal areas, reefs may have been destroyed and recolonized many times. In January 1981, cold air masses over Florida chilled reefs to below 16 °C for 10 days, causing massive mortality of corals and reef fish

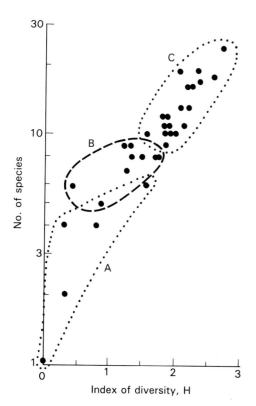

Fig. 11.13 Diversity of *Conus* species in three habitats of different topographical complexity. Habitat A, bare sand (least complex); habitat B, intertidal limestone platform; habitat C, subtidal reef platform (most complex). H is an index of diversity incorporating both the number of species present and their relative abundance. From Kohn (1967).

(Walker *et al.*, 1982). However, this sort of climatic catastrophe is unknown for most of the Caribbean.

The absence of mass extinction, associated with general climatic stability over geological time, has allowed intricate relationships to coevolve among coral-reef organisms, contributing greatly to species diversity. Examples include symbioses, such as that between cleaner fish and their hosts (Fig. 11.14), and mutualisms that enable species to share resources within small areas of habitat. On the Great Barrier Reef, the relatively-small branching coral *Pocillopora damicornis* harbours as many as 16 species of crustacean and fish that use the coral for food and shelter. These animals can potentially exclude one another from the coral, but a system of ritualized signals appeases aggression and facilitates coexistence. Residents attack immigrants, and so the 'team' benefits by the exclusion of competitors (Lassig, 1977).

In contrast to the general effect of climatic stability, occasional environmental catastrophes are instrumental in maintaining the local diversity of corals in shallow reef habitats. Left undisturbed, faster-growing branching corals would begin to overshadow slower growing forms, and would displace them. Hurricanes and severe storms hitting the reef every decade or so break up the corals, however, arresting competitive exclusion (Connell, 1978). Here and elsewhere on the reef, biological disturbances, such as predation and grazing, may be important agents of diversification, retarding the occupation of space by competitively dominant species (Benayahu & Loya, 1977).

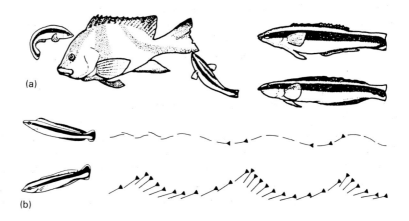

Fig. 11.14 Symbiosis between cleaner fish and its host, which it rids of parasites. A cleaner-mimic exploits the symbiosis. (a) Cleaner wrasse attending a red snapper. The mimic (lower right) sneaks bites out of the snapper's tail. (b) The mimic copies not only the colour, but also the movement pattern of the cleaner fish. From Wickler (1968).

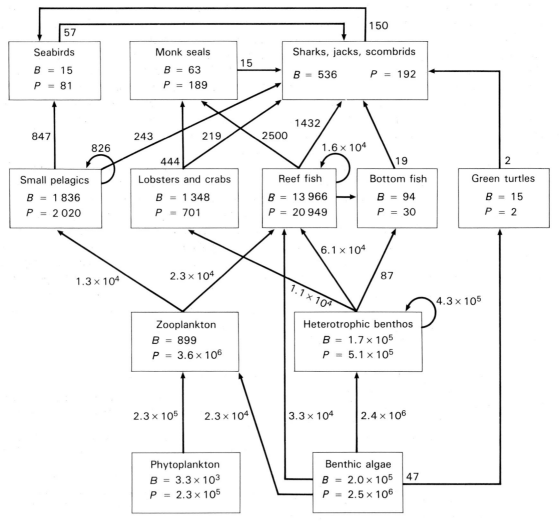

Fig. 11.15 Biomass-budget model for a 1200 km² coral reef. *B*, mean annual biomass (kg km⁻²); *P*, annual production (kg m⁻² year⁻¹). From Grigg *et al.* (1984).

11.4.5 Productivity

Unlike the nutrient-impoverished surrounding oceanic communities, coral reefs have very high rates of gross primary production (Odum & Odum, 1955). This productivity results from the huge surface area of photosynthetic tissues present on the reef, together with favourable light and temperature regimes (Smith, 1981). How can such productivity be sustained in nutrient-poor oceanic waters? Nutrient limitation would seem inevitable. To test

this idea, a lagoon containing a patch reef was fertilized with phosphorus and nitrogen, increasing the community primary productivity by 50% (Kinsey & Domm, 1974; Kinsey & Davies, 1979). In this case, primary productivity did seem to be nutrient-limited.

Given nutrient limitation, it has been reasoned that the high primary productivity of coral reefs must depend on the efficient recycling of nutrients within the community. Highly efficient nutrient recycling might result from the intimate association between zooxanthellae and host tissues (Muscatine & Porter, 1977). More recent work, however, has shown that nutrient limitation may not be so overwhelmingly important as previously thought.

The fertilizer experiment, for example, should be interpreted with caution, on two counts. First, it was conducted in a semi-enclosed lagoon, which received only a limited throughput of oceanic water, so exacerbating local nutrient depletion. Second, it induced only a relatively modest (50%) increase in primary production. Grigg *et al.* (1984) suggest that while nutrient concentrations are low in oceanic water flowing over coral reefs, the flux of nutrients under normal conditions appears sufficient to sustain high primary productivity. Rather than being nutrient-limited systems, coral reefs may be physiologically limited by the speed of the photosynthetic process itself. Only in embayments and lagoons, where circulation is poor, might nutrient depletion be limiting (Smith & Jokiel, 1978).

The high primary productivity of coral reefs might be expected to support high sustainable fishing yields (Marten & Polovina, 1982), yet many coral-reef fisheries are declining (Johannes, 1978). In fact, most of the carbon fixed by photosynthesis is lost before reaching the end of the food chain. Substantial losses of organic carbon occur sporadically by offshore transportation (Qasim & Sankaranarayanan, 1970), but, more importantly, most of the fixed carbon is respired by intermediate predators, resulting in production to respiration ($P : R$) ratios close to unity (Grigg *et al.*, 1984). There simply is not enough fixed carbon left to support high sustainable yields of large carnivores, such as commercially important fish. A model (Fig. 11.15), based on these premises, successfully predicts the production characteristics of a Hawaiian atoll (Table 11.3). Predation therefore, is the major factor controlling production all the way down the food chain. Coral-reef communities are predator-controlled rather than nutrient-limited; in other words they are driven from 'top down' rather than 'bottom up'.

Influence at the bottom of the food chain may be felt by the coral-reef community, however, as a consequence of eutrophication. Cryptofaunal biomass rises markedly in response to increased particulate organic loading (Brock & Smith, 1983), while high nutrient concentrations favour the growth of benthic algae, potentially choking the system (Grigg *et al.*, 1984). Coral reefs appear not to be buffered by their complexity against wholesale perturbations. Sustainable exploitation will require subtle management.

FURTHER READING

Barnes, R.S.K. & Hughes, R.N. (1988) *An Introduction to Marine Ecology.* 2nd edn. Blackwell Scientific Publications, Oxford.

Bayne, B.L. (1969) The gregarious behaviour of the larvae of *Ostrea edulis* L. at settlement. *J. Mar. Biol. Assoc. UK* **49**: 327–56.

Benayahu, Y. & Loya, Y. (1977) Space partitioning by stony corals, soft corals and benthic algae on the coral reefs of the northen Gulf of Eilat (Red Sea). *Helgolander wiss. Meeresunters.* **30**: 362–82.

Bockschoten, G.J. & Bijma, J. (1982) Living bryozoan–stromatolite reefs in the southwestern Netherlands. *Int. Soc. Reef Stud. Second Ann. Meeting*, p.8. Leiden, The Netherlands.

Brock, R.E. & Smith, S.V. (1983) Response of coral reef cryptofaunal communities to food and space. *Coral Reefs* **1**: 179–83.

Caspers, H. (1950) Die Lebensgemeinschaft der Helgolander Austerbank. *Helgolander wiss. Meeresunters.* **3**: 119–69.

Connell, J.H. (1978) Diversity in tropical rainforests and coral reefs. *Science* **199**: 1302–10.

Davies, P.J. (1983) Reef Growth. In: Barnes, D.J. (ed) *Perspectives on Coral Reefs*, pp. 69–106. Australian Institute of Marine Science, Townsville.

Dustan, P. (1975) Growth and form in the reef-building coral *Montastrea annularis. Mar. Biol.* **33**: 101–7.

Fagerstrom, J.A. (1987) *The Evolution of Reef Communities.* John Wiley & Sons, New York.

Table 11.3 Production characteristics of French Frigate Shoals, a 700 km^2 Hawaiian atoll. Observed values are compared with values predicted by the model of Polovina (1984). From Grigg *et al.* (1984)

	Predicted	Observed
Net primary production (10^6 kg wet mass km^{-2} year^{-1})	4.3	6.1
Net community production (10^6 kg wet mass km^{-2} year^{-1})	0.22	0.94
P/R ratio	1.1	—
Number of trophic levels	—	6
Ecotrophic efficiency	—	0.85

Goreau, T.F. (1959) The physiology of skeleton formation in corals. I. A method for measuring the rate of calcium deposition by corals under different conditions. *Biol. Bull.* **116**: 59–75.

Grassle, J.F. (1973) Variety in coral reef communities. In: Jones, O.A. & Endean, R. (eds) *Biology and Geology of Coral Reefs, Vol II. Biology 1*, pp. 247–70. Academic Press, New York.

Grigg, R.W., Polovina, J.J. & Atkinson, M.J. (1984) Model of a coral reef ecosystem. III. Resource limitation, community regulation, fisheries yield and resource management. *Coral Reefs* **3**: 23–7.

Harper, J.L., Rosen, B.R. & White, J. (eds) (1986) The growth and form of modular organisms. *Phil. Trans. R. Soc. Lond.* B **313**: 1–250.

Highsmith, R.C., Riggs, A.C. & D'Attonio, C.M. (1980) Survival of hurricane-generated coral fragments and a disturbance model of reef calcification/growth rates. *Oecologia* **46**: 322–9.

Hughes, R.N. (1979) Coloniality in Vermetidae (Gastropoda). In: Larwood, G. & Rosen, B.R. (eds) *Biology and Systematics of Colonial Organisms*, pp. 243–53. Academic Press, London.

Hughes, R.N. (1983) Evolutionary ecology of colonial reef-organisms, with particular reference to corals. *Biol. J. Linn. Soc.* **20**: 39–58.

Hughes, R.N. (1989) *A Functional Biology of Clonal Animals*. Chapman & Hall, London.

Hughes, T.P. & Jackson, J.B.C. (1980) Do corals lie about their age? Some demographic consequences of partial mortality, fission and fusion. *Science* **209**: 713–5.

Hutchings, P.A. (1986) Biological destruction of coral reefs. A review. *Coral Reefs* **4**: 239–52.

Jackson, J.B.C. & Hughes, T.P. (1985) Adaptive strategies of coral-reef invertebrates. *Am. Sci.* **73**: 265–74.

Johannes, R.E. (1978) Traditional marine conservation methods in Oceania and their demise. *Ann. Rev. Ecol. Syst.* **9**: 349–64.

Joubin, L. (1922) Distribution géographique de quelques coraux abyssaux dans les mers occidentales européennes. *C.R. Acad. Sci. Paris* **175**: 930–3.

Kauffman, E.G. & Sohl, N.F. (1974) Structure and evolution of Antillean Cretaceous rudist frameworks. *Verhandl. Naturf. Ges. Basel* **84**: 399–467.

Kellogg, R.B. & Patton, J.S. (1983) Lipid droplets, medium of energy exchange in the symbiotic anemone *Condylactis gigantea*: a model coral polyp. *Mar. Biol.* **75**: 137–49.

Kinsey, D.W. & Davies, P.J. (1979) Effects of elevated nitrogen and phosphorus on coral reef growth. *Limnol. Oceanogr.* **24**: 935–40.

Kinsey, D.W. & Domm, A. (1974) Effects of fertilization on a coral reef environment — primary production studies. *Proc. 2nd Int. Coral Reef Symp.* **1**: 49–66.

Klumpp, D.W., McKinnon, A.D. & Mundy, C.N. (1988) Motile cryptofauna of a coral reef: abundance, distribution and trophic potential. *Mar. Ecol. Progr. Ser.* **45**: 95–108.

Kohn, A.J. (1967) Environmental complexity and species diversity in the gastropod genus *Conus* on Indo–West Pacific reef platforms. *Am. Nat.* **42**: 85–92.

Lassig, B.R. (1977) Communication and coexistance in a coral community. *Mar. Biol.* **42**: 85–92.

Liddell, W.D., Ohlhorst, S.L. & Coates, A.G. (1984) Modern and ancient carbonate environments in Jamaica. *Sedimenta* **10**: 1–100.

Linke, O. (1951) Neue Beobachtungen über Sandkorallen-Riffe in der Nordsee. *Natur und Volk* **81**: 77–84.

Marshall, J.F. & Davies, P.J. (1982) Internal structure and Holocene evolution of One Tree Reef, southern Great Barrier Reef. *Coral Reefs* **1**: 21–8.

Marten, G.G. & Polovina, J.J. (1982) A comparative study of fish yields from various tropical ecosytems. In: Pauly, D. & Murphy, G.I. (eds) *Theory and Management of Tropical Fisheries*, pp. 255–89. International Center for Living Aquatic Resources Management, Conference Proceedings, 9. Division of Fisheries Research, Commonwealth Scientific and Industrial Research, Cronulla, Australia.

Muscatine, L. & Cernichiari, E. (1969) Assimilation of photosynthetic products of zooxanthellae by a reef coral. *Biol. Bull.* **137**: 506–23.

Muscatine, L. & Porter, J.W. (1977) Reef corals: mutualistic symbioses adapted to nutrient-poor environments. *BioScience* **27**: 454–60.

Odum, H.T. & Odum, E.P. (1955) Trophic structure and productivity of windward coral reef community on Eniwetok Atoll. *Ecol. Monogr.* **25**: 291–320.

Pearse, V.B. & Muscatine, L. (1971) Role of symbiotic algae (zooxanthellae) in coral calcification. *Biol. Bull.* **141**: 350–63.

Polovina, J.J. (1984) Model of a coral reef ecosystem. I. The ECOPATH model and its application to French Frigate Shoals. *Coral Reefs* **3**: 1–11.

Porter, J.W. (1976) Autotrophy, heterotrophy and resource partitioning in Caribbean reef-building corals. *Am. Nat.* **110**: 731–42.

Qasim, S.Z. & Sankaranarayanan, V.N. (1970) Production of particulate organic matter by the reef on Kavaratti Atoll (Laccadives). *Limnol. Oceanogr.* **15**: 574–8.

Riddle, M.J. (1988) Patterns in the distribution of macrofaunal communities in coral reef sediments on the central Great Barrier Reef. *Mar. Ecol. Progr. Ser.* **47**: 281–92.

Safriel, U. (1966) Recent vermetid formation on the Mediterranean coast of Israel. *Proc. Malacological Soc. Lond.* **37**: 27–34.

Schuhmacher, H. & Zibrowius, H. (1985) What is hermatypic? A redefinition of ecological groups in corals and other organisms. *Coral Reefs* **4**: 1–9.

Scoffin, T.P., Stearn, C.W., Boucher, D., Frydl, P., Hawkins, C.M., Hunter, I.G. & MacGeachy, J.K. (1980) Calcium carbonate budget of a fringing reef on the west coast of Barbados. Part II. Erosion, sediments and internal structure. *Bull. Mar. Sci.* **30**: 475–508.

Seed, R. (1969) The ecology of *Mytilus edulis* L. (Lamellibranchiata) on exposed rocky shores. II. Growth and mortality. *Oecologia* **3**: 317–50.

Seed, R. (1976) Ecology. In: Bayne, B.L. (ed) *Marine Mussels: their Ecology and Physiology*, pp. 13–65. Cambridge University Press, Cambridge.

Smith, S.V. (1981) The Houtman Abrolhos Islands: carbon metabolism of coral reefs at high latitude. *Limnol. Oceanogr.* **26**: 612–21.

Smith, S.V. & Jokiel, P.L. (1978) Water composition and biochemical gradients in the Canton Atoll lagoon II. Budgets of phosphorus, nitrogen, carbon dioxide and particulate materials. *Mar. Sci. Comm.* **1**: 165–207.

Stanley, S.M. (1987) *Extinction*. Scientific American Library, New York.

Stenzel, H.B. (1971) Oysters. In: Moore, R.C. (ed) *Treatise on Invertebrate Paleontology, Part N*, pp. 953–1224. University of Kansas Press, Lawrence, Kansas.

Teichert, C. (1958) Cold- and deep-water coral banks. *Am. Assoc. Pet. Geol. Bull.* **42**: 1064–82.

Ten Hove, H.A. (1979) Different causes of mass occurrence in serpulids. In: Larwood, G. & Rosen, B.R. (eds) *Biology and Systematics of Colonial Organisms*, pp. 28–98. Academic Press, London.

Thomassin, B. (1978) Soft-bottom communities. In: Stoddart, D.R. & Johannes, R.E. (eds) *Coral Reefs: Research Methods*, pp. 263–98. UNESCO Monographs on Oceanographic Methodology, no. 5, Paris.

Walker, N.D., Roberts, H.H., Rouse, L.J. & Huh, O.K. (1982) Thermal history of reef-associated environments during a record cold-air outbreak event. *Coral Reefs* **1**: 83–7.

Wickler, W. (1968) *Mimicry in Plants and Animals*. World University Library, McGraw-Hill, New York.

Wulff, J.L. (1984) Sponge-mediated coral reef growth and rejuvenation. *Coral Reefs* **3**: 157–63.

Yonge, C.M. (1963) The biology of coral reefs. In: Russell, F.S. (ed) *Advances in Marine Biology*, Vol. 1, pp. 209–60. Academic Press, New York.

12

Streams and Rivers: One-way Flow Systems

M.J. WINTERBOURN & C.R. TOWNSEND

12.1 INTRODUCTION

Features that characterize streams and rivers and distinguish them from other aquatic environments are: (i) unidirectional flow; (ii) linear form; (iii) fluctuating discharge; and (iv) unstable channel and bed morphology. River channels are self-formed and self-maintained. They evolve in response to climate, geology and weathering, and the forces involved in shaping and maintaining the channel are related to fluid flow. The shearing action of flowing water results in scouring, transport and deposition of bed and bank materials, and a continually changing physical environment (Leopold *et al.*, 1964).

As channels that conduct water, dissolved materials and particulate matter from headwaters to mouth, running waters are peculiarly open ecosystems. Internal biotic cycling is small compared to the throughput of material, and the biotic community gains only a temporary hold on nutrients. A river's narrow, linear form has the further consequence of linking it much more intimately with the surrounding terrestrial ecosystem than is the case for a lake or ocean. This has ramifications for many aspects of running water ecology, as described in Section 12.2 where we consider the catchment as a unit of study.

Terrestrial vegetation can influence trophic structure in rivers in two ways. Firstly, by shading the water it may reduce primary production (of attached algae and macrophytes) and secondly, by shedding leaves, and other parts, it may contribute directly to the food supply of heterotrophs. The flowing nature of rivers, coupled with the short passage time of water, precludes plankton as an important component of the biota, except in downstream reaches of larger rivers. In consequence, the benthos predominates. The relatively extreme hydraulic regime near the bed, and the unstable substratum combine to dislodge many benthic invertebrates into the 'invertebrate drift', to be picked off by fish, or to settle further downstream, unharmed. In the absence of zooplankton, many fish in rivers feed on drift and/or directly on the benthos. We discuss trophic relationships in Section 12.3.

Section 12.4 deals with temporal and spatial patterns in the distribution of the running-water biota. Fluctuations in discharge, and particularly the severe events of spates or floods (when water level rises above bank-full), make the river bed a risky place to live. We pay particular attention to these disturbances, and their consequences for succession, colonization and distribution.

Finally, human settlements are usually associated with rivers that are used variously as convenient gutters for storm water, sewers for waste of all kinds and sources of water for power generation. Some of the impacts that human activities have on the ecology of streams and rivers are discussed in Section 12.5.

12.2 THE CATCHMENT AREA AS A UNIT OF STUDY

Many of the characteristic physical and biological features of running waters are determined by their physical settings; therefore we need to consider them in the context of their catchment area (the river plus its terrestrial drainage basin). Much of the water in a typical stream is derived from groundwater rather than surface runoff and the pattern of flow into a stream is greatly affected by the slope of the valley, the depth and permeability

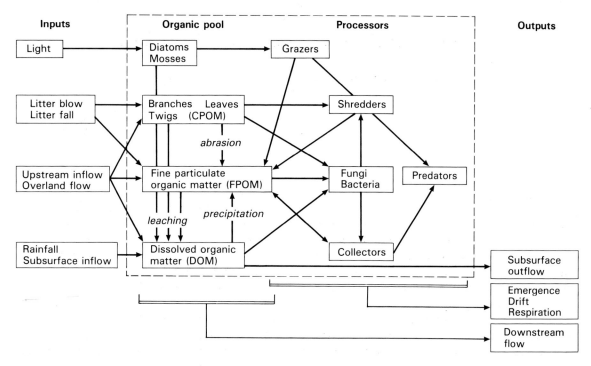

Fig. 12.1 Simplified model of a forest-stream ecosystem showing the principal biological components, energy sources and material pathways. The relative importance of the pathways will differ from stream to stream.

of soil, and patterns of precipitation. The inorganic chemistry of the water is also defined by characteristics of the catchment, in particular its geology. Thus, the major minerals in rocks dominate stream-water chemistry, although the rates at which they are delivered to a stream are controlled largely by terrestrial vegetation that temporarily removes ions from circulation (Hynes, 1975). Except for some streams with unusual sources, for example thermal springs and saline lakes, most unpolluted, running waters are ionically dilute environments.

Land–water linkages that affect the structure and function of running-water ecosystems most strongly, are those involving terrestrial vegetation. In small, forested streams, dead leaves and wood are primary sources of the organic carbon that becomes incorporated into aquatic food webs (Fig. 12.1). In closed-canopy headwater tributaries, stream ecosystems are normally heterotrophic (i.e. depend on a net import of organic matter) with a ratio of internal production to respiration ($P:R$) less than 1. However, where river channels are

wider, or riparian cover is sparse, particulate organic materials of terrestrial origin are less significant as energy sources and $P:R$ may be greater than 1.

Decomposing terrestrial plant material is also the major source of dissolved organic matter (DOM), and this can represent over 70% of annual energy flux in streams. Most DOM enters streams in groundwater, and although chemically complex, two broad, functional categories can be recognized. The first of these consists of refractory, high molecular weight materials, like humic and fulvic acids, most of which are not utilized by the aquatic biota, but are simply transported downstream. The second category is composed of smaller, labile compounds that are taken up rapidly by organisms. Stone surfaces are particularly important sites of DOM uptake by bacteria and fungi, and of subsequent energy transfer to animal consumers higher in the food web.

Although the direct links between riparian and aquatic communities will inevitably be weaker in larger, wider rivers, the unidirectional nature of the flow means that downstream reaches are influenced by upstream ecosystems. The extent of such influence is not well-understood, but it is clear that both particulate and dissolved materials are washed from upstream, often extensively during spates. Furthermore, the physical transformation of organic matter by the feeding activities of river animals results in the downstream displacement of detritus, which may then be incorporated into downstream food webs. The substantial role that benthic invertebrates can play in converting coarse materials to fine particulate detritus was demonstrated by Wallace *et al.* (1982) who observed a marked decrease in the quantity of fine material in transport following elimination of the fauna by an application of insecticide.

The effectiveness of downstream links within running water ecosystems is influenced by channel morphology, gradient, discharge patterns and, of particular interest to many stream ecologists, 'retention devices' formed by fallen trees. The presence of such structures represents another significant influence of the terrestrial environment on the ecology of streams, because fallen trees trap leaves and smaller branches, act as sediment traps and modify flow patterns. In streams with a high density of log jams, downstream transport of particulate material is slowed down and the potential for detritus processing and nutrient cycling is increased. Since there is inevitably a downstream movement of material in running waters, the term 'spiralling' was coined by Webster (see Webster & Patten, 1979) to describe the coupled processes of nutrient cycling and transport. The 'tightness' of the spirals is an index of the degree of retention and re-utilization of any element or material of interest, and as such may provide a useful, comparative measure of stream ecosystem structure and stability (Minshall *et al.*, 1983).

Up to this point we have concentrated on the ways in which running-water ecosystems are influenced by their terrestrial settings. However, interactions also occur in the reverse direction. For example, when rivers flood or (over a longer time-scale) meander across their floodplains, they deposit sand and create gravel bars, swales and natural levees, and these determine, at least in part, the distribution of terrestrial plant species. More specifically, Dahm *et al.* (1987) have shown that river deposits determine the spatial and temporal distribution of anaerobic zones, within waterlogged soils, where organic matter is captured, stored and processed. Trees such as willows, which are adapted to withstand low soil oxygen concentrations, exploit these conditions.

The intimate relationship between land and water is even more apparent on the floodplains of the Amazon, where seasonal floods inundate huge areas of surrounding forest, liberating nutrients from the soil (Fig. 12.2). Although detailed knowledge of biotic interactions on the flooded plains is sparse, Welcomme (1988) has described them as 'giant fish factories' where fish feed and are supported to a large extent by foods of terrestrial origin. On a smaller scale, emerging river insects can provide an energy subsidy to the terrestrial community, mainly as food for spiders and birds. More than 96% of emerging insects ($22.4\,\mathrm{g\,m^{-2}}$ $\mathrm{year^{-1}}$) were exported from a Sonoran Desert stream to the riparian system (Jackson & Fisher, 1986).

12.3 TROPHIC RELATIONSHIPS OF THE RUNNING-WATER BIOTA

Organic energy resources utilized within running-water ecosystems originate within the aquatic environment itself, or are obtained from terrestrial sources. Their relative importance is a matter of continuing interest to stream ecologists. Although many stream-dwelling animals are omnivorous, and to a considerable extent opportunistic feeders, it is possible to classify them more-or-less accurately in terms of the way in which they feed. This functional feeding-group concept (Cummins, 1973) is used in discussing trophic interactions below.

When a dead leaf falls into a woodland stream, soluble substances are rapidly lost by leaching, and a variety of microorganisms, principally bacteria and fungi, begin to colonize its surface. Of particular note are hyphomycete fungi, whose characteristic tetraradiate or sigmoid spores develop into mycelia that penetrate the leaf. Microbial colonization has been termed 'conditioning' by Cummins (1974), who likened the rather inert leaf tissue to a cracker biscuit, and the microbial colonists to more

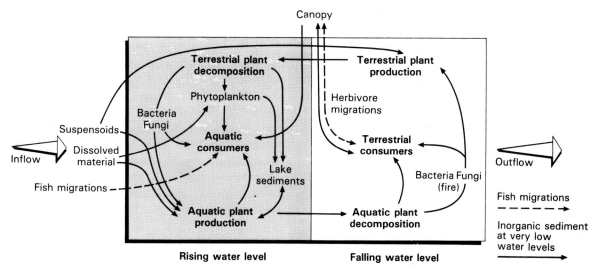

Fig. 12.2 Nutrient pathways in the Amazon River and its floodplain during the period of rising river water and falling water. From Junk (1980).

nutritious peanut butter! Although subsequent research has shown that consumers of dead leaves obtain only a small proportion of their total nutritional requirements from the microflora, hyphomycete fungi appear to be important sources of the lipids needed by some insect larvae prior to pupation (Cargill *et al.*, 1985).

Animals that feed on conditioned leaves include various kinds of insect larvae (craneflies, stoneflies and caddisflies) and crustaceans (gammarid and isopod 'shrimps', prawns and crayfish) and are usually described as shredders, or large-particle detritivores (Fig. 12.3). Representatives of this group may be abundant in retentive streams, especially in pools and organic 'dams' where decomposing plant materials accumulate. However, they tend to be sparsely represented or absent from less retentive streams, as in New Zealand, where frequent spates prevent the build-up of litter and therefore adequate food and habitat for shredders (Rounick & Winterbourn, 1983).

The other characteristic groups of detritivorous animals inhabiting forest streams are generally smaller than shredders. They are collector-filterers, and a more heterogeneous assemblage of animals

variously described as collector-gatherers, browsers and grazers. The former feed on a variety of fine particulate material that is filtered from the water column, whereas the latter sweep or scrape a mixture of loose and attached materials from the stream bed. Again, much of the material ingested is of terrestrial origin, and either enters the water in a finely divided state, or has been fragmented within the stream as a result of shredder feeding and/or microbial decomposition.

The larvae of blackflies (Simuliidae) and hydropsychid caddisflies (Fig. 12.3) are amongst the most characteristic filtering collectors in many parts of the world. The former trap ultra-fine particles, including bacteria, with complex labral fans, whereas the latter use fixed nets, spun from silk and elaborated by their modified salivary glands. The more mobile, surface browsers are a particularly abundant group that includes the larvae of many mayflies, stoneflies, caddisflies, midges and beetles. Most members of this functional feeding group have standard chewing mouth parts, but some mayflies have prominent maxillary and/or labial brushes, which sweep or scrape materials into the mouth.

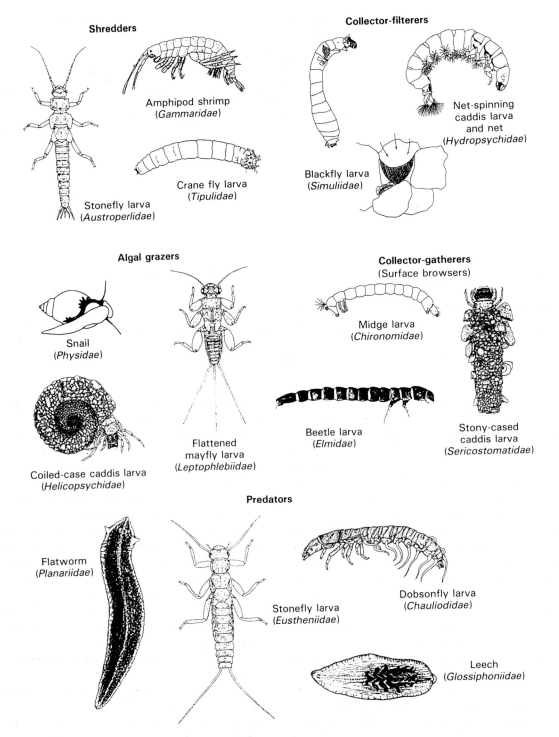

Fig. 12.3 Examples of invertebrates, belonging to different functional feeding groups, in running waters.

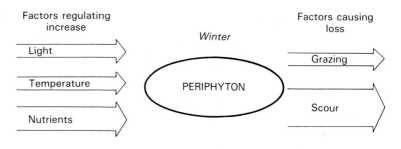

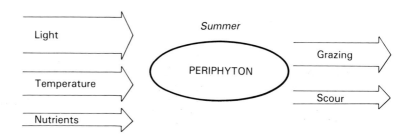

Fig. 12.4 The major physicochemical and biological factors controlling periphyton biomass in Oregon mountain streams during winter and summer. The relative importance of different factors is indicated by the thickness of the arrows. From Rounick & Gregory (1981).

Much of the energy flow through small, forest-stream ecosystems is supported by materials of terrestrial origin, and primary producers within the stream usually play a minor role. However, where the forest canopy opens out, benthic algae are increasingly important foods. Unlike lakes and ponds, little, if any, phytoplankton occurs in running waters, and most aquatic macrophytes are confined to low gradient streams and rivers with fine sediments that enable their roots to become established. Heavily shaded or turbid rivers tend to be devoid of higher plants, although extensive mats of water hyacinths, paspalum, and papyrus line the banks of many tropical rivers.

In headwater tributaries above and below the treeline, carpets of bryophytes (liverworts and mosses) may be prominent on boulders, bedrock and waterfalls. Like living aquatic macrophytes they are eaten by very few animals, but nevertheless a diverse and often dense community of small invertebrates (mites, nematodes, small crustaceans and chironomid larvae) is commonly associated with them. They feed on fine detritus trapped within the mats and on periphytic diatoms that grow on the living fronds. Diatoms are also the dominant autotrophs on stones in forested streams, and along with filamentous algae, unicellular and filamentous bacteria, fungi, detrital particles and slime, may form extensive growths on stones in warm, sunlit reaches.

In running waters, primary production can be limited by light, flow rate, temperature, and the availability of nutrients. Flow-related factors, such as physical abrasion, also affect standing crops and the taxonomic composition of algal assemblages, as do the grazing activities of animals (Fig. 12.4). Where grazing pressure is high, stone-surface communities tend to be dominated by small, closely-adhering diatoms (e.g. *Cocconeis*, *Achnanthes*), whereas erect, stalked and filamentous algae can dominate where grazers are less abundant (Steinman *et al.*, 1987).

Many insect larvae that ingest fine, particulate detritus also eat algae, and the distinction between herbivores and detritivores on the one hand, and detrital collectors and algal grazers on the other is frequently obscure. Nevertheless, algae form a more important part of the diet of some species than others, notably many gastropod snails, helico-psychid caddis larvae and several kinds of mayflies (see Fig. 12.3). The finely-toothed radulae of snails can remove periphyton adhering closely to various surfaces, some caddis larvae have blade-like mandibles well adapted for scraping, and some microcaddises (Hydroptilidae) pierce algal filaments and then suck out the contents.

Midge larvae (Chironomidae) also form close associations with algae as both habitat and food, and a remarkable association is found between the globular cyanophyte, *Nostoc parmelioides*, and two species of *Cricotopus* (Ward *et al.*, 1985). This appears to be a mutualistic relationship as the larvae tunnel inside the algae, where they feed on vegetative cells and heterocysts, while *Nostoc* responds by changing its growth form, thereby increasing its weight-specific photosynthetic rate. In somewhat similar vein, secretions and excretory substances released by other, sessile chironomid larvae stimulate the growth of diatom communities on the outer surfaces of their silken retreats (Pringle, 1985).

Finally, streams and rivers harbour diverse faunas of carnivorous animals (see Fig. 12.3). Flatworms and leeches feed on snails, worms and chironomid larvae (some flatworms may also be scavengers). All 10 insect orders with representatives in running waters include some carnivorous species, and large stonefly and dobson fly larvae are characteristic predators in stony streams throughout the world. They use active hunting or 'sit-and-wait' strategies to capture small invertebrates (mainly insects) that are detected by a combination of mechanical, chemosensory and visual cues (Peckarsky, 1982).

Fish are, of course, the largest and most common vertebrates in most river systems, and although many are carnivorous, others are herbivores, omnivores or detrital feeders. Some feeling for the vast diversity of freshwater fish can be obtained when it is realized that the Amazon system has over 1500 species that inhabit a mosaic of interconnecting rivers, streams, swamps and lakes, and undergo complex migrations between them (Lowe-McConnell, 1986). Although their trophic dependencies are complex, some general feeding patterns can be discerned within the catchment of the Amazon. Thus, algal grazers predominate above the forest, allochthonous materials, both plant and

insect, are used extensively within the forest and a preponderance of zooplankton feeders inhabit the Varzea lakes of the floodplain.

12.4 COLONIZATION, SUCCESSION AND DISTRIBUTION

The distribution of many running-water inhabitants is dependent on the colonizing abilities of specific life-history stages and, in the case of insects with aquatic larvae, a knowledge of the biology of their terrestrial adults is needed in order to better understand their distributions. Unfortunately, these are the most difficult stages to study, and those about which we know least. Nevertheless, recent studies in arctic, desert and forest environments have provided insights into the ways stream communities develop and subsequently change.

Glacier Bay National Park in southeast Alaska, provided Milner (1987) with a unique opportunity to study the colonization and ecological development of streams of known age (0–150 years), in places where glacial ice had recently receded. Pioneer colonists of brand-new streams were chironomids of the genus *Diamesa*, whose larvae were found on stones with the filamentous alga, *Ulothrix*. A more diverse community, comprising additional chironomid species (Orthocladiinae), mayflies and stoneflies was found in 15-year-old streams, while mosses, blackflies, water mites and molluscs had colonized the oldest streams. Fish (Salmonidae) were first observed in a 15-year-old stream and their subsequent population development was associated with the occurrence of increasingly stable flows, pools and a fairly diverse invertebrate fauna. Later development of these arctic-stream ecosystems is probably dependent on successional changes in the surrounding terrestrial ecosystem, particularly the establishment of woody vegetation. By providing cover and inputs of large organic debris, the presence of riparian forest can be expected to increase habitat and trophic diversity and this will be reflected in changes in stream-community structure and function. The likely direction of such changes, in response to forest succession, are suggested by the results of Molles' (1982) study of trichopteran assemblages in the mountains of New Mexico (Fig. 12.5). He found

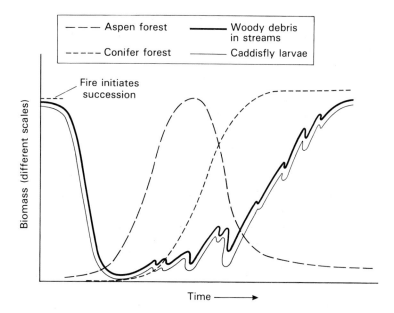

Fig. 12.5 The proposed relationship between forest succession (aspen to conifer) and long-term changes in the biomass of shredding caddisfly larvae (Trichoptera) in streams of the southern Rocky Mountains. From Molles (1982).

that although the species composition of caddisflies was the same in streams draining early-successional aspen forests and climax conifer forests, the biomass of those classified as shredders was 35 times greater in the latter. Molles attributed this difference to the order-of-magnitude greater standing crop of detritus in the conifer-forest streams, itself a consequence of the presence in the streams of more logs, with slower rates of decomposition.

Whereas the successional model shown in Fig. 12.5 considers changes on a time-scale of decades, or even hundreds of years, Fisher *et al.* (1982) were able to demonstrate successional patterns within periods of less than a year. Thus, after an intense summer spate that virtually eliminated algae and reduced the invertebrate standing crop of an Arizona desert stream by 98%, recolonization was rapid. Fast growing diatoms were early dominants, but were overtaken within a few weeks by slower growing, but larger, species of filamentous algae. Insects also recolonized rapidly, mainly by immigration of aerial adults and subsequent oviposition.

Fisher and his co-workers also observed that periodic spates broke otherwise orderly, successional changes in community organization, and acted as 'reset' mechanisms. Their finding that post-disturbance recolonization of almost totally denuded streams was far from deterministic is also

significant, and Fisher (1983) noted that 'there may be several routes to one subsequent community type or several community types generated from one pioneer assemblage'. He also made the observation that a spectrum of successional states can occur in running waters, such that streams which flood infrequently may show marked succession, in contrast to those that are continually disturbed.

Within the limits imposed by colonization dynamics, the distribution and abundance of stream flora and fauna is determined by a complex of physicochemical and biological factors, including stream morphology, current velocity, substrate composition, temperature, food, competition and predation (Townsend, 1980). We concentrate here on the combined roles of flow and substrate composition.

The potential importance of 'stream hydraulics' as a determinant of benthic invertebrate zonation patterns along the lengths of rivers has been emphasized by Statzner and Higler (1986), who drew particular attention to zones of transition of hydraulic stress, with which some changes in species composition appear to be associated (Fig. 12.6). Although an attractive hypothesis, it should be treated cautiously, because adequate descriptions of the velocity and substratum conditions to which organisms are actually exposed are rare. In par-

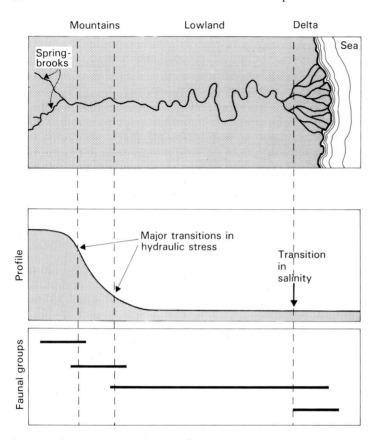

Fig. 12.6 Longitudinal zonation of the benthic invertebrate fauna along a hypothetical, pristine river system. Characteristic faunal assemblages are hypothesized to change at sites of 'hydraulic stress' identified on the river profile. From Statzner & Higler (1986).

ticular, stream flows are incompletely described by mean velocity alone, and measurements of depth, substratum roughness and shear velocity are needed in order to adequately characterize near-bed flows, where plants and animals actually live (Davis & Barmuta, 1989).

Obtaining relevant measurements poses formidable difficulties for the field ecologist, especially in riffles, where much of the fauna is found, and where flows are both chaotic and highly patchy. Furthermore, a great many benthic species spend much of their time occupying dead spaces within the stream bed, a microhabitat to which their small size and often flattened forms are ideally suited. Nevertheless, most benthic algae and many grazing and filter-feeding invertebrates occur on the upper surfaces of stones, and a knowledge of the local hydraulic conditions to which they are exposed must improve our understanding of their biology (Statzner & Muller, 1989).

The physical stability of benthic substrata, and their propensity to be moved or abraided by periodic spates and floods, also have measurable effects on the stream benthos. Stability is hard to define objectively, but the Pfankuch (1975) procedure provides a simple, comparative index for ranking channel stability at the scale of the stream reach. Calculation of the index involves an assessment of 15 features of the stream bed and banks, each of which is assigned a score. Important criteria used for evaluating bank stability are: (i) the degree of vegetative protection; (ii) evidence of undercutting; and (iii) the amount of newly deposited sand or gravel present. Rock angularity and brightness, the size and degree of packing of bed materials, the amount of scouring and deposition and the distribution of mosses and algae are used to assess bed stability. Surveys of New Zealand streams have illustrated the utility of this index, and indicated that the distribution of shredders in forested streams

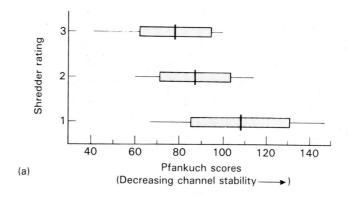

(a)

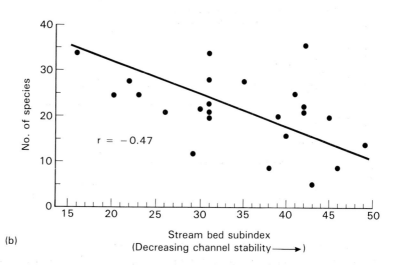

(b)

Fig. 12.7 (a) The relationship between abundance of shredders and the Pfankuch channel stability score in 43 New Zealand forest streams. Shredder abundance ratings shown are: 1, absent; 2, less than 3% of all macroinvertebrates; 3, over 3% of macroinvertebrates. Graphs show means, standard deviations and ranges of scores. (b) The relationship between numbers of invertebrate taxa and the stream-bed component of the Pfankuch stability index in 23 streams in Westland, New Zealand. From Rounick & Winterbourn (1982); Winterbourn & Collier (1987).

could be predicted from the Pfankuch scores (Fig. 12.7a). Furthermore, Winterbourn and Collier (1987) found that the bed component of the index was significantly correlated with invertebrate species richness in streams, and indicated that as stability increased so did the number of species (Fig. 12.7b).

During periods of high discharge, the scouring and deposition of stream-bed sediments can result in destruction of sessile, stone-surface communities, and the downstream displacement of benthic animals. To minimize such losses, members of the benthos are assumed to have evolved refuge-seeking behaviour, and it is generally supposed that many of them burrow deeper into the bed. Others may move to the margins of rivers, where flow conditions are less harsh, and fish are known to take

shelter beneath overhangs, and in the lee of boulders and snags. The ability of stream communities to persist in the face of high discharge events is remarkable (Meffe & Minckley, 1987; Townsend *et al.*, 1987) and testifies to the presence of refuges in which spate-related increases in shear stress are not so great, and which provide centres for recolonization when discharge declines (Townsend, 1989).

Subsequent recolonization of denuded, or partially denuded, substrata is frequently rapid, and field experiments with unidirectional colonization trays indicate that it is brought about by a combination of downstream drift, active upstream migration, upward movements within the bed sediments and, in the case of insects, oviposition (Williams & Hynes, 1976). Of these, drift usually appears to be most important (Townsend & Hildrew, 1976), but as probable source areas of drifting invertebrates may also be affected by floods, it rather begs the question as to the real nature of presumed faunal refuges. In gravel-bed rivers at least, the presence of a permanent hyporheic fauna, comprising 'typical' members of the river fauna in addition to specialized subterranean species, may provide the answer. Although often only tens of centimetres deep, the hyporheos can be much larger than this, and in the valley of the Flathead River, Montana, it extends laterally for up to 2 km, and may support a faunal biomass exceeding that in the river (Stanford & Ward, 1988).

It is worth noting that the phenomenon of invertebrate drift is not simply the result of accidental dislodgement from the substratum, but a well-documented behavioural characteristic of a great many stream-dwelling invertebrates (Brittain & Eikeland, 1988). Indeed, it often has a well-defined diel rhythm, triggered by changes in light intensity that result in many species showing a nocturnal peak in drift activity. Nevertheless, only a small fraction of 1% of the benthos is normally in the drift at any one time, and distances drifted are typically only a few metres (Table 12.1). Just why and how insect larvae launch themselves into the

drift are matters of some debate (Sheldon, 1984). However, some, like the mayfly *Baetis*, enter the drift in response to the close proximity of predatory stoneflies; ironically perhaps, fish including salmonids take advantage of this behaviour by picking them off in the water column.

Lastly, the large- and small-scale distributions and densities of stream-dwelling invertebrates are known to be influenced by competitive and predatory interactions, but the extent to which they help structure benthic communities is a topic of continuing controversy. It is a subject that is beyond the scope of the present chapter, but is considered in Chapter 7 of this volume.

The reader is also referred to the recent appraisal of the topic by Hildrew and Townsend (1987) and an excellent evaluation of the evidence pertaining to the role of predation in streams by Allan (1983).

12.5 THE LAND–WATER INTERFACE REVISITED

The basic characteristics of running-water ecosystems are determined by their setting, in particular the geology and geomorphology of the catchment, latitude, altitude and the nature of the riparian vegetation. Historical factors have helped shape the aquatic communities we see today; rivers have acted as magnets for human colonization and utilization, and it is not surprising that people have profoundly affected many riverine ecosystems.

In the USA today for example, few, if any, rivers remain totally untouched (Dahm *et al.*, 1987). Before the advent of European settlers, beavers were a key geomorphic agent in river systems throughout much of North America, and their removal by early fur-trappers resulted in accelerated gully formation, downcutting of stream beds and flood damage.

Subsequently, other human activities resulted in alteration of the river system: (i) snags, boulders and debris dams were removed systematically from rivers to facilitate their use for transportation; (ii) grazing of riparian ecosystems by domestic stock resulted in flow changes, sedimentation and bank erosion; and (iii) channelization was practiced to provide flow and flood control.

These management practices have fundamentally altered many river ecosystems, which no longer

Table 12.1 Mean distances drifted by the larvae of four insect species in a small stream in the English Lake District. Distances drifted by all species were greater at higher current velocities. Differences between species reflect differences in their morphology and behaviour. Thus, cased caddis larvae sink to the stream bed rapidly, *Baetis* returns actively to the substratum, whereas, *Ecdyonurus* and *Hydropsyche* are more passive drifters. From Elliott (1971)

	Mean distance drifted (m) at current velocity		
	$10\,cm\,s^{-1}$	$30\,cm\,s^{-1}$	$50\,cm\,s^{-1}$
Agapetus fuscipes (cased caddis)	0.13	1.25	1.90
Baetis rhodani (swimming mayfly)	1.00	2.73	4.41
Ecdyonurus venosus (flattened mayfly)	1.59	4.76	8.33
Hydropsyche species (net-spinning caddis)	1.99	6.60	11.49

retain organic matter so efficiently, and have had their riparian/aquatic linkages upset.

The extensive impoundment of rivers by dams, most notably for hydroelectric power generation, has exacerbated the situation further, by altering flow patterns and downstream temperature regimes. As a result, the physical structure of river beds has been changed, upstream–downstream linkages have been cut, and biological communities have been affected by changes to their food supply and the physicochemical environment. For example, below deep-release dams, which modify a river's temperature regime, the longitudinal distributions of various insect taxa have been curtailed or extended so that they correspond with the temperature ranges usually occupied in unregulated rivers.

Riparian zones act as nutrient filters, sediment traps, climatic regulators and wildlife refuges; therefore their disruption can have far-reaching effects on the structure and function of running-water ecosystems. Nowhere is this more apparent than in areas of intensive agriculture or forestry, and in urban areas where the natural vegetation has been removed down to the water's edge. Less obviously, detrimental effects of acid precipitation are mediated via the soils of the riparian zone, from which hydrogen ions and potentially toxic metals are taken into solution. High concentrations of inorganic monomeric aluminium entering streams in groundwater, or recently melted snow, have been responsible for the deaths of fish and other, more subtle, changes at all trophic levels in river ecosystems (Park, 1987).

Although it is apparent that disruption of the riparian zone can have significant consequences for the functioning of riverine ecosystems, it should not be assumed that a particular disruption will apply equally to all groups of organisms, or that the severity of impact will be comparable in all situations. Thus, in a high-rainfall area of New Zealand, where severe spates occur frequently, the removal of riparian forest vegetation has little apparent effect on the structure of stream invertebrate communities (Winterbourn, 1986), whereas in streams draining catchments with contrasting forestry histories, at Coweeta in North Carolina, invertebrate communities differ significantly in density, biomass and representation of invertebrate functional

feeding groups (Wallace, 1987). These examples serve to show that, although individual stream ecosystems share many features with each other, they also possess their own distinctive characteristics that reflect their past histories, and geological and geographical settings. An appreciation of the existence of such differences is necessary for the development of appropriate river management programmes.

FURTHER READING

Allan, J.D. (1983) Predator–prey relationships in streams. In: Barnes, J.R. & Minshall, G.W. (eds) *Stream Ecology: Application and Testing of General Ecological Theory*, pp. 191–229. Plenum Press, New York.

Brittain, J.E. & Eikeland, T.J. (1988) Invertebrate drift: a review. *Hydrobiologia* 166: 77–93.

Cargill, A.S., Cummins, K.W., Hanson, B.J. & Lowry, R.R. (1985) The role of lipids as feeding stimulants for shredding aquatic insects. *Freshwater Biol.* 15: 469–78.

Cummins, K.W. (1973) Trophic relations of aquatic insects. *Ann. Rev. Entomol.* 18: 183–206.

Cummins, K.W. (1974) Structure and function of stream ecosystems. *BioScience* 24: 631–41.

Dahm, C.N., Trotter, E.H. & Sedell, J.R. (1987) Role of anaerobic zones and processes in stream ecosystem productivity. In: Averett, R.C. & McKnight, D.M. (eds) *Chemical Quality of Water and the Hydrologic Cycle*, pp. 157–78. Lewis Publishing, Chelsea, Michigan.

Davis, J.A. & Barmuta, L.A. (1989) An ecologically useful classification of mean and near-bed flows in streams and rivers. *Freshwater Biol.* 21: 271–82.

Elliott, J.M. (1971) The distances travelled by drifting invertebrates in a Lake District stream. *Oecologia* 6: 350–79.

Fisher, S.G. (1983) Succession in streams. In: Barnes J.R. & Minshall, G.W. (eds) *Stream Ecology: Application and Testing of General Ecological Theory*, pp. 7–27. Plenum Press, New York.

Fisher, S.G., Gray, L.J., Grimm, N.B. & Busch, D.E. (1982) Temporal succession in a desert stream ecosystem following flash flooding. *Ecol. Monogr.* 52: 93–110.

Hildrew, A.G. & Townsend, C.R. (1987) Organization in freshwater benthic communities. In: Gee, J.H.R. & Giller, P.S. (eds) *Organization of Communities: Past and Present*, pp. 317–41. Blackwell Scientific Publications, Oxford.

Hynes, H.B.N. (1975) The stream and its valley. *Internationale Vereinigung für Theoretische und Angewandte Limnologie. Verhandlungen* 19: 1–15.

Jackson, J.K. & Fisher, S.G. (1986) Secondary production, emergence, and export of aquatic insects of a Sonoran Desert stream. *Ecology* 67: 629–38.

Junk, W.J. (1980) Die Bedeutung der Wasserstandsschwankungen für die Okolgie von uberschwemmungsgebieten, dargestellt an der Varzea des mittleren Amazonas. *Amazonia* 7: 19–29.

Leopold, L.B., Wolman, M.G. & Miller, J.P. (1964) *Fluvial Processes in Geomorphology*. Freeman, San Francisco.

Lowe-McConnell, R.H. (1986) Fish of the Amazon system. In: Davies, B.R. & Walker, K.F. (eds) *The Ecology of River Systems*, pp. 339–51. Dr W. Junk Publishers, Dordrecht.

Meffe, G.K. & Minckley, W.L. (1987) Persistence and stability of fish and invertebrate assemblages in a repeatedly disturbed Sonoran Desert stream. *Am. Midland Natur.* 117: 177–91.

Milner, A.M. (1987) Colonization and ecological development of new streams in Glacier Bay National Park, Alaska. *Freshwater Biol.* 18: 53–70.

Minshall, G.W., Petersen, R.C., Cummins, K.W., Bott, T.L., Sedell, J.R., Cushing, C.E. & Vannote, R.L. (1983) Interbiome comparison of stream ecosystem dynamics. *Ecol. Monogr.* 53: 1–25.

Molles, M.C. (1982) Trichopteran communities of streams associated with aspen and conifer forests: long-term structural change. *Ecology* 63: 1–6.

Park, C.C. (1987) *Acid Rain*: *Rhetoric and Reality*. Methuen, London.

Peckarsky, B.L. (1982) Aquatic insect predator–prey relations. *BioScience* 32: 261–6.

Pfankuch, D.J. (1975) *Stream Inventory and Channel Stability Evaluation*. United States Department of Agriculture, Forest Service, Region 1, Missoula, Montana.

Pringle, C.M. (1985) Effects of chironomid (Insecta: Diptera) tube-building activities on stream diatom communities. *J. Phycol.* 21: 185–94.

Rounick, J.S. & Gregory, S.V. (1981) Temporal changes in periphyton standing crop during an unusually dry winter in streams of the Western Cascades, Oregon. *Hydrobiologia* 83: 197–205.

Rounick, J.S. & Winterbourn, M.J. (1982) Benthic faunas of forested streams and suggestions for their management. *NZ J. Ecol.* 5: 140–50.

Rounick, J.S. & Winterbourn, M.J. (1983) Leaf processing in two contrasting beech forest streams: effects of physical and biotic factors on litter breakdown. *Arch. Hydrobiol.* 96: 448–74.

Sheldon, A.L. (1984) Colonization dynamics of aquatic insects. In: Resh, V.H. & Rosenberg, D.M. (eds) *The Ecology of Aquatic Insects*, pp. 401–29. Praeger, New York.

Stanford, J.A. & Ward, J.V. (1988) The hyporheic habitat of river ecosystems. *Nature* 335: 64–6.

Statzner, B. & Higler, B. (1986) Stream hydraulics as a major determinant of benthic invertebrate zonation patterns. *Freshwater Biol.* 16: 127–39.

Statzner, B. & Muller, R. (1989) Standard hemispheres as indicators of flow characteristics in lotic benthos research. *Freshwater Biol.* 21: 445–9.

Steinman, A.D., McIntyre, C.D., Gregory, S.V., Lamberti, G.A. & Ashkenas, L.R. (1987) Effects of herbivore type and density on taxonomic structure and physiognomy of algal assemblages in laboratory streams. *J. N. Am. Benth. Soc.* 6: 175–88.

Townsend, C.R. (1980) *The Ecology of Streams and Rivers*. Edward Arnold, London.

Townsend, C.R. (1989) The patch dynamics concept of stream community ecology. *J. N. Am. Benth. Soc.* 8: 36–50.

Townsend, C.R. & Hildrew, A.G. (1976) Field experiments on the drifting, colonization and continuous redistribution of stream benthos. *J. Animal Ecol.* 45: 759–72.

Townsend, C.R., Hildrew, A.G. & Schofield, K. (1987) Persistence of stream invertebrate communities in relation to environmental variability. *J. Animal Ecol.* 56: 597–613.

Wallace, J.B. (1987) Aquatic invertebrate research. In: Swank, W.T. & Crossley, D.A. (eds) *Forest Hydrology and Ecology at Coweeta*, pp. 257–68. Springer-Verlag, New York.

Wallace, J.B., Webster, J.R. & Cuffney, T.E. (1982) Stream detritus dynamics: Regulation by invertebrate consumers. *Oecologia* 53: 197–200.

Ward, A.K., Dahm, C.N. & Cummins, K.W. (1985) *Nostoc* (Cyanophyta) productivity in Oregon stream ecosystems: invertebrate influences and differences between morphological types. *J. Phycol.* 21: 223–7.

Webster, J.R. & Patten, B.C. (1979) Effects of watershed perturbation on stream potassium and calcium dynamics. *Ecol. Monogr.* 19: 51–72.

Welcomme (1988) Concluding remarks I: on the nature of large tropical rivers, floodplains, and future research directions. *J. N. Am. Benth. Soc.* 7: 525–6.

Williams, D.D. & Hynes, H.B.N. (1976) The recolonization mechanisms of stream benthos. *Oikos* 27: 265–72.

Winterbourn, M.J. (1986) Forestry practices and stream communities with particular reference to New Zealand. In: Campbell, I.C. (ed) *Stream Protection*: *The Management of Rivers for Instream Uses*, pp. 57–73. Water Studies Centre, Chisholm Institute of Technology, Victoria, Australia.

Winterbourn, M.J. & Collier, K.J. (1987) Distribution of benthic invertebrates in acid, brown water streams in the South Island of New Zealand. *Hydrobiologia* 153: 277–86.

PART 5
HUMAN EFFECTS

13
Impacts of Man's Activities on Aquatic Systems

B.T. HARGRAVE

13.1 EXPLOITATION

13.1.1 Urban and agricultural water supplies

Human population growth and accompanying growth of agriculture and aquaculture are limited by the amount of fresh water in the hydrosphere. Currently the 5 billion (5×10^9) population of earth is increasing in number by 1.7% year^{-1}. If this rate is maintained, the population will reach 8.5 billion by the year 2025. The rate in non-industrialized countries is twice as high. Globally, water demands are increasing at a rate seven times greater than the population. Requirements for additional supplies of fresh water by human populations have not occurred before on this scale, and fresh water, once considered an abundant commodity, has become a scarce resource in many countries.

Much of the annual renewable freshwater supply potentially available to man is not geographically distributed in a useful way. Regions of low population densities, in northern temperate and subarctic latitudes, often contain large supplies of fresh water, while arid areas are densely populated. For example, 60% of the total annual river discharge in the USSR (4700 km^3) flows through land bordering the Arctic Ocean, where population density is low. Thus, the increasing demand for water use is centred in areas least able to supply the need. Ironically, as discussed below, the high demand coincides with the accumulation of waste products in rivers and lakes, often making the water that is available of poor quality.

Future demands for fresh water, especially in industrialized areas, can best be met by water recycling. Desalination of abundant salt water has high energy costs and will only be economical where conventional water supply costs are abnormally high. Secondary and tertiary waste-water treatment processes can convert industrial waste water to clean water that may be reused. The cost of these treatment systems for providing fresh water will become acceptable only when abundant supplies of fresh water are reduced.

13.1.2 Harvesting food and natural products

In comparison to food taken from the land, human populations generally obtain an insignificant part of their diet from aquatic animals. Exceptions occur in rural coastal communities and in Arctic regions where lipid-rich mammals and fish comprise a major part of the diet of native populations. Fish are also often a major source of protein in heavily populated developing countries, where they are grown in ponds. Some species, such as char, salmon, lobster, shrimp, or sturgeon roe (eggs from ripe female fish) are consumed as luxury foods. Although these species and products are a small part of the total global fisheries harvest, they are threatened because high demand has led to high landed-weight values.

Commercial harvesting of freshwater and marine fish and invertebrates is the primary process by which man obtains food from aquatic systems. While some lakes support commercial fisheries, the total yield for freshwater fish is small (about 10% of the annual global catch). Landed weight of marine fish has increased from 18 million tonnes fresh weight in 1938 to stabilize near 70 million tonnes during the 1970s with only a small annual increase (an average of 1 million tonnes) in the 1980s. There are many examples of changes in species composition that occur as a result of overfishing. As stocks become depleted, fishing pressure is transferred to abundant species that remain. Approximately 30% of the total catch is processed for fish oil and meal, which is used as an additive for farm animal feed.

There is a growing seafood industry based on 'fabricated' seafood (*surimi*), as has been produced in Japan for over a century. Non-commercial species are utilized which would not usually be processed or marketed, and because of this the industry will probably expand in the future. Mechanically deboned fish are washed and pressed to produce a protein-rich compound with a texture and flavour that resembles shellfish such as lobster, shrimp or scallops. Growth of markets for this material has been exponential in recent years with Japan accounting for most of the world production.

Vascular plant species are also a renewable resource used by man. Freshwater reeds (*Phragmites*) have been used historically to make paper, and for roof covering (thatching). The dry stems can be compressed and cemented, to be used in building construction as wall partitions in some eastern European countries. Seagrass such as *Zostera* was used in northern temperate areas for house insulation before synthetic materials became available. Dry seagrass is still used as a foundation windbreak around farm houses in rural coastal areas. Kelp (*Macrocystis*) has been used as an agricultural fertilizer and source of potash for many years in coastal areas of Asia and Europe. It was an important source of potash for explosive production by the USA during the First World War.

Marine macroalgae are harvested to produce emulsifying agents used in dairy industries and in pharmaceutical and cosmetic products. Carrageenan is a gelatinous substance extracted from the red alga *Chondrus* (Irish moss). Algin and agar are similar compounds derived from the subtidal seaweeds *Macrocystis* and *Gracilaria*. They are used as solidifying agents for bacteriological culture media, and as a thickener in ice-cream.

These and other natural products from marine plants have a variety of medical uses. For example, carrageenan is used in treatment of gastric ulcers, since it reacts with the mucoid lining of the stomach to form a layer that decreases the absorption of pepsin across the gut wall. The polysaccharide is also an anticoagulant, although it is less effective than heparin. Sulphated polysaccharides produced by some macroalgae can also increase the growth of connective tissue and collagen. They are therefore applied to bone grafts to stimulate growth of connective tissue.

13.1.3 Energy and mineral supplies

Heat, kinetic and potential energy in aquatic systems are used for human and industrial purposes. Terrestrial hot springs in Iceland and New Zealand are the best known examples of exploitation of superheated groundwater. Domestic heat and hot water are provided to the cities of Reykjavik and Rotorua from natural thermal springs. Hydrothermal vents, formed at continental spreading centres in the ocean floor, are the marine equivalent of terrestrial hot springs, but no method presently exists to utilize the energy contained in heated water released from these vents.

The most common source of aquatic energy harvested by man is the production of hydroelectric power from the kinetic energy of flowing water. Rivers are the usual source of hydraulic power used to rotate electrical turbines. However, tidal power has been generated at two locations by building barrages across coastal embayments in areas with large tidal ranges (La Rance River, France and Kislaya Bay, USSR). Water entering the basins behind the barrages during flood tide is held, and slowly released through turbines in the barriers during low tide.

Large temperature gradients between surface and deep layers of the ocean have been proposed as a source of thermal energy that could be harnessed to produce electricity. The process would use warm surface water to evaporate a working fluid, such as ammonia, which would turn an electric turbine. Cold water pumped from below 1000 m would be used to reliquify the ammonia vapour using heat exchangers. Designs for huge offshore structures have been proposed, but no commercial production of electricity using the method has been attempted.

Large reserves of oil and natural gas exist in sedimentary formations below the sea bed. For example, the Safaniya Field off Saudi Arabia is the largest known offshore structure, with 25 billion barrels of recoverable oil. From its discovery in 1951 until 1976, this single field contained more oil than exists in the combined offshore reserves of Canada and the USA. Subsequent discoveries in the

Caspian Sea, coastal Alaska and the banks of Newfoundland, have increased the potential total oil and gas reserves even further.

Other minerals also lie below the sea bed. Seams of coal, mined from land in eastern Canada, extend outward under the sea for many kilometres. Sulphide-rich mineral deposits, containing iron and manganese, occur around hydrothermal vents. Heated water with high dissolved-metal concentrations is formed when sea water circulates through cracks in volcanic rocks along continental-plate spreading zones. Cold temperatures and high dissolved oxygen in the surrounding sea water cause the minerals in the heated water to precipitate, forming metal-rich sediments. This process accounts for geological formations on land where volcanic activity occurred during a time when the land was submerged.

Ferromanganese nodules are spherical or disc-shaped accretions, up to several centimetres in diameter, that occur at the surface of sediments in some softwater lakes and regions of the deep sea, from the Equator to high latitudes. The nodules are rich in manganese and iron (comprising approximately 30% of their weight) with lesser amounts of silicon and aluminum (10% nickel, 1–2% copper and cobalt by weight). Commercial mining of nodules has been proposed by companies from several countries. The process would involve an extension of the present technology of shallow-water mining of sand and gravel by suction dredges. Methods such as towing a submersible dredge along the bottom are needed at greater depths. Concentrated nodules would be separated from sediment on board ship, with debris released to settle back to the bottom. Cost–profit ratios require processing of large volumes of nodules. This fact, and the potential complication of providing licences for mining in international waters under a proposed International Sea-bed Authority, have delayed development of the deep ocean mining industry.

13.2 DEGRADATION

13.2.1 Waste-water discharge

Treatment of sewage to remove particulate matter, dissolved nutrients and contaminants is usually carried out for aesthetic and human health reasons. Historically, the primary aim has not been to prevent damage to the environment or organisms other than fish. Primary sewage treatment consists of removal of solid material. Screens and settling ponds are used to remove up to 50% of suspended solids. Only small amounts of dissolved nutrients (5–10% of dissolved phosphorous and nitrogen, for example) adsorbed to solids are retained in settled sludge, but up to 95% of faecal coliform bacteria can be removed with this treatment.

Secondary sewage treatment involves the biological process of microbial decomposition. Organic matter is partially digested by bacterial activity through vigorous aeration, followed by sedimentation of solids. Recycling of dried sludge as fertilizer is possible following this treatment. The process can remove from 80 to 90% of the biochemical oxygen demand, and up to 30% of particulate phosphate and inorganic nitrogen, if settling times are sufficiently long. However, because of decomposition of organic matter during the digestion phase, dissolved nutrients may be released from particulate matter and added to effluent water at higher concentrations than occur with primary treatment.

Tertiary treatment results in the most effective removal of nitrogen and phosphorus from sewage. Chemicals such alum and iron are added to separate phosphorous, and occasionally pH is artificially raised with hydroxides to cause nitrogen to be degassed. Flocculated inorganic nutrients and organic matter are allowed to settle. Passage through charcoal (to remove colour and ions), and chlorination (to kill microorganisms) can produce water suitable for drinking. The complete three-treatment system is expensive, but as supplies of clean fresh water are diminished, the added cost may have to be accepted if public health standards and water supply are to be maintained.

Alternatives may exist to these expensive sewage treatment systems. Effluents may be sprayed on farmland, after sterilization by heat and ozone treatment. For nutrient-rich water this serves to simultaneously irrigate and fertilize the soil. However, metal and industrial organic contaminants must be at low levels, or crops will be contaminated. Liquid agricultural waste water, from pig farming for example, can be disposed of in this manner on adjacent land, which is used to grow food for pigs. This is similar to the centuries-old practice in China of collecting human and domestic wastes for placement in shallow ponds used for fish aquaculture.

13.2.2 Eutrophication

Cultural eutrophication of rivers and lakes often involves changes resulting from man's activities, which affect entire drainage basins, not only the water bodies themselves. For example, changes in water flow and water chemical composition in a catchment basin can affect runoff and groundwater over a land surface area much greater than that of the lake that receives the water. Clearing of land for agriculture, or as a result of forest fires, removes forest cover, increases runoff and leaches mineral salts from soil. Schindler *et al.* (1980) showed that increased runoff and loss of nitrogen, phosphorus and potassium occurred in two small Precambrian watersheds, following a severe natural windstorm

and high intensity forest fire. These events accelerated the natural loss of nutrients from the forested watershed and caused nutrient enrichment of the drainage streams.

Vollenweider (1976) used mass-balance models to predict the steady-state phosphorus content of lake water (the ratio of the areal loading rate to the factor of lake mean depth multiplied by the sum of water flushing and phosphorus sedimentation rates). Comparison of various lakes showed that over a range of mean depths and flushing rates ($1-100\,\mathrm{m}$ year^{-1}), the change from oligotrophic to eutrophic status occurred with an increase in areal phosphorus loading from less than $0.1\,\mathrm{g\,P\,m^{-2}}$ year^{-1} to values greater than $0.5\,\mathrm{g\,P\,m^{-2}}$ year^{-1}. Vollenweider suggested that lakes with average whole-lake concentrations of total dissolved phosphorus over $20\,\mathrm{\mu g\,P\,l^{-1}}$ will be eutrophic while those with levels below $10\,\mathrm{\mu g\,P\,l^{-1}}$ will be oligotrophic.

In an earlier study, Schindler (1971) assumed that the quantity of nutrients supplied to a head-

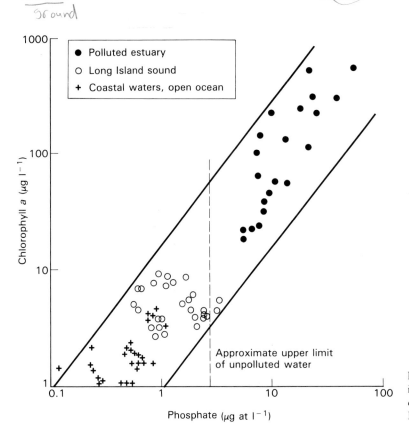

Fig. 13.1 Relationship between inorganic phosphate and chlorophyll *a* content of different water bodies. From Ketchum (1969).

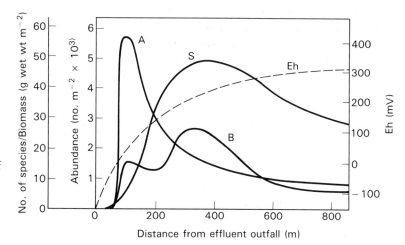

Fig. 13.2 Changes in abundance (A), biomass (B) and species numbers (S) of benthic macrofauna, and surface sediment oxidation–reduction potential (Eh) along a transect from a pulp mill outfall. From Pearson & Stanley (1979).

water lake by weathering of rocks and soil and internal sources, might be directly proportional to the catchment area (A) — the sum of the terrestrial portion of the drainage area and surface area of a lake. On the other hand, dissolved nutrients entering a lake by precipitation and runoff will be diluted in proportion to lake volume (V). The hypothesis that the quantity of nutrients entering a lake per unit time is directly proportional to the ratio A:V was tested for several small Precambrian lakes not influenced by man. Variables such as total suspended phosphorus and nitrogen, midsummer phytoplankton biomass and production, winter oxygen depletion and organic carbon and nitrogen in sediments, were all positively and linearly correlated with this morphometric index.

As discussed in earlier chapters, productivity in aquatic systems is determined by many factors, including those related to basin morphometry. Where light and dissolved oxygen are not limiting, nutrient supplies are critical to sustain biomass production at all trophic levels. However, addition of excess nutrients does not always enhance productivity. In hypereutrophic waters, where hydrogen sulphide and other by-products of microbial metabolism accumulate, the addition of inorganic nutrients will not increase production. Microbial growth through sulphate or methane reduction is usually not limited by inorganic nutrient supply. However, dissolved nutrient enrichment in oligotrophic and moderately eutrophic systems

causes increased growth rates of photosynthetic microalgae and macrophytes when light is not limiting. Thus, there is a positive relationship between dissolved nutrient concentrations, such as phosphate, and biomass of phytoplankton, measured as suspended chlorophyll *a* (Fig. 13.1).

Effects of nutrient enrichment are also evident spatially in sediments of water bodies subject to point source discharge of effluents (Fig. 13.2). Benthic species abundance, biomass and chemical indicators of reducing activity, such as oxidation–reduction potentials (Eh), are all minimal near the effluent source. Organic matter loading creates anoxic conditions close to the discharge, since microbial oxygen consumption exceeds rates of re-aeration. Microbial sulphate reduction also leads to sulphide accumulation in sediments, which creates chemical oxygen demand. In this area, the benthic fauna is dominated by small organisms, such as oligochaetes and annelid worms, that are tolerant of reducing conditions. Large macrofauna that cannot withstand oxygen depletion avoid reduced sediments. Further along the transect, the diversity of the fauna increases, since larger organisms colonize more oxidized sediments.

13.2.3 Accidental releases

Unintentional release of toxic organic and inorganic chemicals into rivers and lakes may occur as a chronic discharge or as an acute input, when the result of an accident. For example, chronic releases of potentially toxic substances occur from pulp and paper production using the kraft bleaching process. The large volume of fresh water required for paper production creates effluents which are enriched with particulate, colloidal and dissolved organic matter.

Chemical spills from industrial sources are sometimes not detected until chronic symptoms of impact are observed. For example, high levels of mercury release from an industrial plant into Minimata Bay, a coastal embayment in Japan, were only detected when people eating fish from the bay, over several months, developed symptoms of mercury poisoning.

Freshwater and marine oil spills are usually visible and well-publicized accidents. The low solubility of many hydrocarbons causes slicks to form on the water surface. This leads to dispersion of spilled oil by wind, waves and, in marine waters, tidal action. It also aids in evaporation of the lighter more volatile hydrocarbons. One- to two-thirds of oil spilled on the sea surface may be released to the atmosphere by evaporation, where hydrocarbons become photochemically oxidized. Biological processes further reduce the concentration of oil in water.

A review of the inputs, fates and effects of oil in the sea was carried out by the US National Academy of Sciences (1985) (Table 13.1). While there are many uncertainties in estimates of annual input, approximately half of the additions result from marine transportation and oil production activities. Pumping of bilge water contributes as much oil as accidental spillage. An almost equal contribution comes from river and municipal runoff and atmospheric input of hydrocarbon combustion products. While shipping accidents are not a dominant source of oil to the oceans on a global basis, they can have a significant local impact on natural populations. The National Academy of Sciences did not find evidence of damage to marine resources on a broad oceanic scale arising from chronic inputs or major oil spills, but impacts were noted in areas directly affected by exposure to high oil concentrations.

A less visible impact of oil on marine organisms occurs at offshore production sites, where drilling muds (slurries of water and cuttings) are circulated

Table 13.1 Estimated inputs of petroleum hydrocarbons to the marine environment (10^6 tonnes year^{-1}). From US National Academy of Sciences (1985)

Source	Best estimate	% of total
Atmosphere	0.3	9.2
Natural sources		
Marine seeps	0.2 ⎱	7.7
Sediment erosion	0.05 ⎰	
Offshore production	0.05	1.5
Transportation		
Tanker operations	0.7 ⎫	
Dry-docking	0.03	
Marine terminals	0.02	45.2
Bilge and fuel oils	0.3	
Tanker accidents	0.4	
Non-tanker accidents	0.02 ⎭	
Municipal and industrial wastes and runoff		
Municipal wastes	0.7 ⎫	
Refineries	0.1	
Non-refining wastes	0.2	36.3
Urban runoff	0.12	
River runoff	0.04	
Ocean dumping	0.02 ⎭	
Total	3.25	

through the borehole to lubricate the cutting heads. The areal extent of impact is visible on the sea bed, where the concentration of aromatic hydrocarbons in sediments decreases exponentially over distances up to several kilometres from drill holes (Fig. 13.3). Benthic monitoring programmes in the North Sea have shown changes in abundance of dominant macrofauna species over areas up to $10 \, km^2$ after 2 years of drilling, production and discharge of ballast water around some offshore sites.

13.2.4 Ocean dumping and disposal

The release of contaminants such as sewage sludge, dredge spoils and industrial–chemical wastes has been carried out at specific sites in the ocean by various dumping and waste disposal methods. Calculations of the 'assimilative capacity' of disposal sites are used to estimate the amount of a given waste material that can be contained within the area, without producing an unacceptable impact on

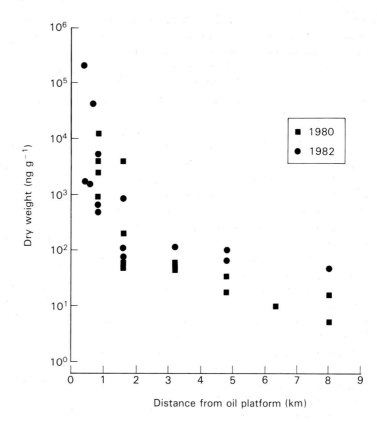

Fig. 13.3 Concentrations of aromatic hydrocarbons (anthracenes and phenanthracenes) in sediments with distance from an offshore oil production platform at Beryl in the northern North Sea, where diesel-based drilling muds have been used for several years. From Davies *et al.* (1984).

living organisms or non-living resources. The argument for managed disposal into the ocean is that dilution by a large water volume, or containment in sediments for long periods of time at sites remote from man, makes this a safe disposal method. In some cases, direct marine disposal offers economic, social and technical advantages over the release of waste onto land or into the air. However, the practice requires an understanding of how added waste materials move after placement, knowledge of how resulting concentrations will affect organisms and how much damage is 'acceptable'. Proponents of oceanic waste disposal point out that mankind has dammed rivers, used lakes for discharge of waste water and constructed ports that have greatly modified estuaries. Therefore, it should be acceptable for man to dispose of some waste in the ocean.

There is evidence that aquatic systems differ in their ability to receive man's wastes. The largest US ocean industrial dumpsite, off the coast of New Jersey, receives hundreds of thousands of cubic metres of waste annually from titanium dioxide production, organic chemical production and water treatment, yet impacts on the planktonic and benthic food web appear to be minimal (Goldberg, 1984).

Puget Sound, an embayment on Washington's coast with restricted exchange to the Pacific, would be a less desirable place for disposal of sewage or industrial sludge. Toxic dinoflagellate blooms and mortality of oyster larvae in recent years may be related to eutrophication by waste-water discharge from Seattle. Complex bottom topography and restricted water exchange with adjacent coastal waters, because of shallow depths within connecting channels, create poor conditions for flushing.

Extremely toxic substances, such as heavy metals, non-degradable organic pesticides, spent nuclear fuel and reprocessing waste derived from nuclear power plants are among materials on a 'blacklist' of substances banned from dumping at sea under the London Dumping Convention (LDC). Low-level radioactive wastes from nuclear industries and scientific research laboratories are included on a 'greylist'. Until 1983, they could be dumped at sea subject to guidelines of the International Atomic Energy Agency (IAEA). A voluntary moratorium imposed by the LDC in 1983 on dumping of these wastes at sea is still in effect, pending evaluation of the scientific, political, legal and economic aspects of the practice.

The burial of high-level radioactive wastes in deep ocean sediments, although presently banned by the LDC, has been considered as an option for nuclear waste disposal by some countries. The waste would be held in water- or air-cooled storage facilities on land for 10 or more years, to allow cooling, before being fused with glass and encased in corrosion-resistant stainless steel cannisters. These would be permanently placed below the sea floor in abyssal plain areas with deep stable clay sediments, using drilled emplacement or free-fall projectiles. The concept is feasible from an engineering perspective, however no abyssal sites deeper than 4500 m have been found with suitable geological characteristics that extend over the required disposal area of 100 km^2.

13.2.5 Airborne contaminants *acid rain*

Observations that CO_2 levels in the atmosphere are increasing were first made in 1938. The changes could alter the earth's climate, affecting both hydrological and carbon cycles in terrestrial and aquatic systems. Systematic measurements of CO_2 at the Mauna Loa Observatory in Hawaii have shown that the concentration has increased from 316 ppm in 1958 to approximately 350 ppm in 1989. When compared to an estimated pre-industrial concentration of 280 ppm, from analysis of gas bubbles trapped in glacial ice, the increment over the first 60 years of this century equals that which has occurred over the last 30 years. Increases since 1959 exceeding 1 ppm year^{-1} (0.4% year^{-1} since 1986) represent a rise of over 8% of the total carbon present in the atmosphere during the past century.

Carbon dioxide, carbon monoxide, methane and water vapour are called 'greenhouse' gases. Although they are present in very small quantities in the earth's atmosphere, they are transparent to visible and short-wave ultraviolet light, but they absorb and reradiate longer-wavelength infrared radiation. These longer wavelengths are emitted from the earth's surface and this causes heat energy to be trapped in the atmosphere. Since air temperatures depend on the balance between incoming radiation, absorption and reradiation, changes in concentration of these trace gases will alter average global temperatures. Although controversy surrounds the rate at which changes will occur, there is agreement that climate changes will occur over the next few decades if the present trend to higher concentration continues.

Climatological models have been used to predict that, if the present trend continues, atmospheric CO_2 will double in concentration over the next century. Only 10% of this increase will be absorbed in the upper, mixed layer of the ocean. Photosynthesis by land plants will increase, but growth of vegetation is rather unresponsive with only a 0.5–2% increment for each 10 ppm increase in CO_2. The most profound impact would be a global average temperature increase of 2–3°C. The greatest increase (three times that close to the Equator) would occur at high latitudes.

This seemingly small increment would warm the surface waters of the ocean, causing some melting of polar ice-packs and alpine glaciers, and increased water volume through thermal expansion. The resulting rise in sea level may be as much as 0.5–1 m over the next century. This exceeds the present rate (15–35 cm rise per century) observed from global tide gauge measurements, corrected for changes due to tectonic activity. Changes in sea level would have drastic consequences for heavily populated coastal regions throughout the world. More than half of the world's population lives in coastal regions that would be affected.

In addition to changes in oceanic circulation and coupled climatic variations, oceanic warming could increase the concentration of methane in the atmosphere. Sediments on oceanic continental slopes between 300 and 600 m contain a 'clathrate' (methane-water ice) as a solid gas hydrate, stable at temperatures of 2°C. The amount of carbon in these solid deposits is estimated to be two to three times that in the atmosphere, terrestrial biosphere and oceans combined. Even a few degrees increase in water temperature may cause melting and release of methane — a greenhouse gas 20 times more efficient in trapping heat than CO_2. If as little as 0.1% of clathrate melts in the next 50 years, atmospheric warming due to this release of methane would equal that predicted for increased levels of CO_2 and global warming would double.

Combustion of fossil fuels adds more than CO_2 to the atmosphere. Besides particulate matter that contributes to atmospheric dust, which absorbs heat, oxides of sulphur and nitrogen are also produced during combustion. In the presence of water vapour, these form acids that lower the pH of rain water and snow. Values are naturally low (pH 5.5–6.0) due to the formation of carbonic acid, but with excess sulphate and nitrate ions, more sulphuric and nitric acids are formed, which can lower the pH even further (pH 4–5). The impact on rivers and lakes in geological areas of granitic rock is greater than in more alkaline watersheds, where the

natural buffering capacity can neutralize excess acidity.

Effect of acid precipitation on water bodies with naturally low pH levels has been observed in the disappearance of fish from rivers and lakes in Sweden, eastern Canada and northeastern USA over the past 30 years. Acidification also causes increased leaching of heavy metals from soils. Metals in ionic form are dissolved and transferred in runoff to lakes and estuaries. Even when bound to particulate matter, the metals can be released into the water by bacterial decomposition of organic matter.

Rainwater, snow and air samples collected in regions such as Bermuda, northern Norway, the Canadian Arctic and Antarctica contain chlorinated hydrocarbons (DDT, toxaphene, lindane, dieldrin and chlordane) — volatile organic pesticides used in agricultural applications in Central America, India, eastern Europe and Asia. Their relatively low vapour pressure causes evaporation from surfaces of

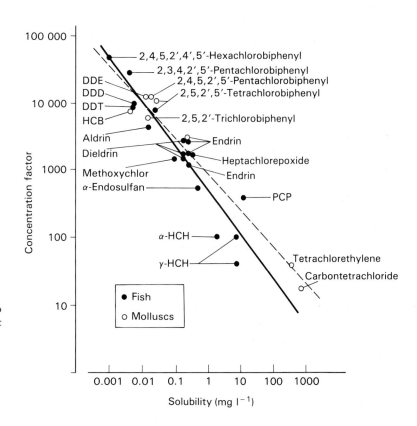

Fig. 13.4 Concentration factor (ratio of concentration in wet tissues to that in sea water on the same weight: weight basis) for various chlorinated hydrocarbons, by various fish and mollusc species, related to the solubility of the compounds in sea water. From Ernst (1980).

vegetation, soil and water. Once in the vapour phase, or adsorbed to airborne particles, they can be transported long distances. The appearance of chlorinated pesticides, such as DDT and toxaphene, in fish and sediments from remote northern lakes can only be explained by atmospheric supply.

Many chlorinated pesticides, such as DDT, are soluble in lipids and since they are slowly metabolized, they accumulate in fatty tissues of aquatic animals (bioaccumulation). The ratio of concentration of a contaminant in organism tissues to that in the surrounding water or sediment is known as the concentration factor. Since a high affinity for fats leads to a low water solubility, there is an inverse logarithmic relationship between the concentration factor of many organochlorines and their water solubility (Fig. 13.4). Ironically, organochlorines which have the lowest dissolved concentrations in water, are the ones most concentrated by aquatic organisms.

Inorganic contaminants, such as heavy metals and radioactive elements, are also transferred to aquatic environments through the atmosphere. Lead measured in Greenland glaciers, formed before 1750, is present at levels of $0.02 \, \mu g \, kg^{-1}$, but ice formed since 1968 contains 10 times this concentration. Similarly, sediments in anoxic basins off California have a pre-industrial sedimentation rate of $2-10 \, mg \, Pb \, m^{-2} \, year^{-1}$ which has increased to $9-21 \, mg \, Pb \, m^{-2} \, year^{-1}$ in recent decades, due to input from industries and gasoline burned by automobiles in nearby Los Angeles.

Radioactive elements are added to aquatic

systems from the atmosphere. Atmospheric testing of nuclear weapons following the Second World War resulted in fallout of nuclear reaction fission products such as ^{90}Sr (strontium-90) and ^{137}Cs (caesium-137) that reached peak concentrations in surface waters of lakes and oceans in 1963. The occurrence of natural radioactive compounds in all water bodies is often not emphasized when concern is expressed about new sources of activity from fallout and nuclear waste disposal (Table 13.2). Globally, however, cumulative fallout from fission products on a per volume basis increased natural levels of radioactivity by about 10% during years of atmospheric testing.

13.2.6 Water regulation   *dams*
disease

Dams are built to regulate river discharge, which prevents flooding and creates reservoirs of water for irrigation in arid areas. They are essential for hydroelectric power production. Increased generating capacity can be achieved by holding back water to build up greater hydraulic pressure. The most immediate negative environmental effects of dam construction are the loss of land by flooding and interference with the migration of fish. A less obvious impact of dams occurs in coastal waters affected by freshwater input.

All dams impede the flow of water to the ocean and hence change the amounts and seasonal timing of freshwater discharge through estuaries and coastal areas. In temperate latitudes, low river discharge occurs during winter, when runoff is stored as snow. Peak flow occurs during melting periods in spring and early summer. This is the season when coastal plankton communities grow rapidly due to the availability of light, thermal and salinity stratification and the availability of dissolved nutrients. For large rivers, the stimulatory effects of enhanced production can extend over thousands of kilometres.

Frontal zones, with relatively large density differences, are created where fresh and salt water meet in estuaries. Aggregations of planktonic organisms used as food by fish larvae can become concentrated in these regions. Alternatively, advection and diffusion may remove fish larvae and their planktonic

Table 13.2 Average concentrations of radioisotopes, occurring naturally, and derived from fallout of nuclear weapon tests in surface water of the North Atlantic Ocean. From Woodhead (1973)

Isotope	Natural radioactivity (pCi l^{-1})	Isotope	Fallout radioactivity (pCi l^{-1})
^{40}K	320	^{3}H	48 (31–74)
^{87}Rb	2.9	^{137}Cs	0.21 (0.03–0.80)
^{234}U	1.3	^{90}Sr	0.13 (0.02–0.50)
^{238}U	1.2	^{14}C	0.02 (0.01–0.04)
^{3}H	0.6–3	^{239}Pu	0.0003–0.0012

$1 \, Ci = 3.7 \times 10^{10}$ disintegrations s^{-1}.

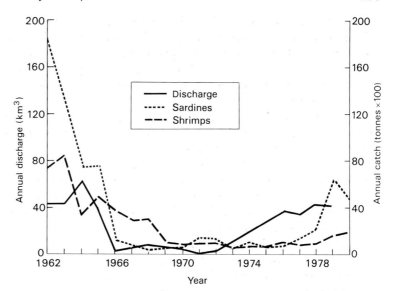

Fig. 13.5 Annual changes in water discharge from the Nile River compared with annual sardine (*Sarinella*) and shrimp (mostly *Penaeus* species) catch in the southeastern part of the Mediterranean Sea before and after 1965, the year the river was dammed by the Aswan High Dam. Data from Wadie (1984).

food from strategic growing areas during periods of high river discharge.

The Aswan High Dam on the Nile River in Egypt provides one example, where a large-scale water regulation project was monitored for its effect on the ocean environment prior to, and following, construction. The historic annual average discharge of the Nile River (1912–1942) into the Mediterranean was 62 km^3. The dam was built between 1960 and 1968, with closure of the Nile for the first time in 1964. After this, the annual flow decreased to only 10% of the previous discharge by the late 1960s (Fig. 13.5).

The dramatic decrease in discharge and shift of Nile peak flow from the autumn rainy season to winter months has altered physical, chemical and biological conditions in the southeastern part of the Mediterranean Sea near the Nile Delta. Coastal erosion has also increased, since siltation during the annual flood period has been reduced. Lower dissolved nutrient concentrations and reduced siltation have decreased primary production in areas of the Mediterranean that were previously very productive. The areal extent of phytoplankton blooms in coastal waters near the mouth of the Nile River has decreased and phytoplankton cell numbers during winter and autumn blooms are now only one-tenth of previous levels.

The impact of the Aswan Dam is most readily seen in the loss of sardine and shrimp harvest during years of minimal discharge (Fig. 13.5). Sardines feed on zooplankton, which would be expected to be immediately affected by decreased phytoplankton production. A trend towards increased yields in 1979 and 1980 reflects higher rates of outflow as Lake Nasser reached its maximum capacity. However, sardine and shrimp catches cannot be expected to recover completely, due to the shift in time of the freshwater pulse as well as the reduced nutrient supply.

13.2.7 Pathogens and marine toxins

Plagues during medieval times were caused by poor sanitary conditions, absence of efficient waste disposal and the lack of potable water in densely populated cities. Cholera was rapidly spread by bacteria in contaminated water, used by large numbers of people who were unaware of the cause of the disease. Infectious diseases are still spread today by means of insanitary water supplies in the less developed, and often over-populated, countries. Water regulation by dam construction also causes the spread of parasitic infections such as malaria and schistosomiasis. Reservoirs often cause water tables in surrounding land to rise. This may allow mosquitos, hosts to amoeboid parasites, to spread into areas previously uninfected because of aridity. Large-scale drainage projects or pesticide applications are then required to reduce mosquito populations.

Infection of aquatic organisms by pathogens may occur in coastal areas subject to eutrophication by sewage and urban water discharge. For example, *Vibrio parahaemolyticus* is a bacterium present in seafood throughout the world. Infection does not usually affect the health of the organisms themselves, but *Vibrio* species were isolated from flounder in the Dutch Wadden Sea, where 50% of the individuals collected were diseased with skin ulcers. Their poor condition (low weight per unit length) may reflect a high mortality rate.

Pathogenic bacteria (for example, *Salmonella* and *Clostridium*) may be present in shellfish, especially in urbanized harbours. Viral diseases such as hepatitis B also occur in these areas, largely as a result of the discharge of untreated sewage. Cooking prior to consumption of shellfish kills the bacteria and usually denatures toxins produced by these pathogens, but cross-contamination can occur through storage with other food.

In addition to diseases, the production of chemical toxins by marine microorganisms can cause human illness, or in some cases death. For example, diarrhetic shellfish poison (DSP) is a class of toxins produced by marine dinoflagellates such as *Dinophysis and Protocentrum*. Symptoms resemble those that follow ingestion of spoiled shellfish. Saxitoxin, the other main class of neurotoxins, produces symptoms of paralytic shellfish poisoning (PSP) such as numbness and loss of sensation. In extreme cases, paralysis of the diaphragm may cause respiratory failure. These toxins are associated with 'red tides' that occur in coastal waters; the description refers to the red or brown discolouration of water, due to concentrations of accessory pigments other than chlorophyll *a*, when cell numbers of dinoflagellates or diatoms increase.

Diatoms (*Nitzschia pungens*) in specific coastal embayments of eastern Canada have recently been found to contain relatively high concentrations (ppm) of the neurotoxin domoic acid. Mussels feeding on blooms of the diatoms are thought to have concentrated domoic acid to levels sufficient to cause illness and death in humans consuming contaminated mussels.

Microalgae may also pose a hazard to fish. *Ptychodiscus brevis*, a dinoflagellate that produces the neurotoxin brevetoxin, is a red-tide organism which is commonly the cause of fish kills in the Gulf of Mexico. Circumstantial evidence has recently linked deaths of dolphins along the eastern coast of the USA to the presence of brevetoxin, perhaps ingested with menhaden on which the dead dolphins had fed. Aerosol or gaseous emission by blooms of the dinoflagellate cause human throat irritation, and dead fish in large numbers are aesthetically unattractive when washed up on beaches.

13.2.8 Biomass removal

Exploitation of aquatic organisms by any means of harvesting necessarily results in the removal of biomass. If harvesting is restricted to individuals of a specific age, growth of individuals in the remaining size classes may increase. Experimental studies with fish populations have shown that maximum sustainable yield is generally reached when about one-third of the population is harvested per reproductive period, and mean biomass is reduced to less than one-half of the unexploited level (Odum, 1971). Any further increase in the rate of biomass removal may result in reduced stock size and a decrease in production. Increased harvesting effort, new efficient vessel and trawl designs and electronic fish-finding equipment have not always produced higher yields, because most stocks are fished at or above the level of maximum sustainable yield.

Removal of a substantial number of one species from a local environment often indirectly affects other species. A change in the competitive equilibrium between species can occur when one or more species is selectively removed by a fishery; remaining species, not subject to harvest, may increase in abundance in response to the removal of predation pressure or the availability of extra food supplies. This may explain cases where sudden shifts in population biomass occur over a few years. The collapse of Pacific sardine, Peruvian anchovey and Namibian pilchard in the past few decades are examples of declines in pelagic stocks that have been attributed to overfishing.

Odum (1982) discussed an energy-hierarchy concept that describes how disturbances to natural food webs may affect the stability of species at different trophic levels. Natural perturbations through environmental change have their greatest effect on energy flow and biomass accumulation at lower trophic levels (for example, plankton). The oscillations tend to be smoothed as energy is transferred to higher trophic levels. Larger body size allows for a wider selection of prey and an increased ability to store energy for periods when food supplies are reduced.

In contrast, man-induced disturbance by unrestrained fishing selectively removes the biomass of highly-valued carnivorous species. They feed high in the food chain (obtaining energy from many species), and the removal of substantial fractions of these populations causes a disturbance that propagates downwards to lower trophic levels. Since there are no natural processes to dampen the perturbations, they may be amplified at lower trophic levels with the result that zooplankton and phytoplankton populations are increased. The suggestion that food web structure is altered and destabilized by the removal of carnivorous species has been supported by experimental studies in lakes and marine intertidal areas, where carnivores have been selectively added or removed.

The development of driftnet fishing for pelagic species in the open ocean has the potential to reduce the biomass of top predators over large areas. Since the early 1980s, vessels, primarily from Japan, Taiwan and South Korea, have used long (15–20 km), lightweight, monofilament nets suspended between the surface and 10 m depth, at night, to capture squid and pelagic fish. Mammals such as dolphins and sea birds are also inadvertently taken. Since the technique is efficient and non-selective, this method of harvesting has been described as the biological equivalent of strip-mining.

An additional destructive feature of driftnets is that discarded or lost nets continue to 'ghost-fish.' Commercial fishing gear is usually made of plastic material that degrades slowly. When discarded or accidentally lost, the nets pose a threat to a wide variety of organisms. They can continue to entangle animals while floating, and even after the weight of trapped biomass causes the net to sink to the bottom. Decomposing carcasses are attractive to predators and scavengers, which themselves can become trapped. The cycle can continue for as long as it takes the synthetic fibres in nets to degrade.

Methods used in harvesting aquatic species may disturb the physical as well as biological environment. Intensive trawling of offshore banks on continental shelves by dragging weighted nets over the bottom can crush and bury organisms not retained in the trawl. Frequent trawling can also cause winnowing of fine sediments, leaving only coarse gravel. Biomass of benthic infauna would be reduced in this coarse substrate, but attached epifauna such as filter feeders would be favoured if disturbance does not prevent colonization.

A more destructive form of biomass harvest is practiced in the Philippines, where dynamite is used on coral reefs. Underwater explosions stun fish, and when they float to the surface they are easily collected. However, the method destroys large areas of coral. Complex marine communities around coral atolls are the basis for sustaining high human populations on islands that have few other natural resources. The harvesting method is effective at providing food over the short term for the dense human populations on the islands, but the long term effects on the marine communities around and on atolls are certain to be negative.

13.3 ENHANCEMENT

13.3.1 Reservoirs

The damming of streams and rivers to create artifical lakes (reservoirs) or flooded ponds has been carried out by human populations for thousands of years. Depending on size, they serve as storage for water supply, for transport or defense purposes, to produce hydroelectric power or as a source of food, either from natural populations or cultured species.

Rice cultivation on flooded soils in Asian countries often requires alteration of water flow through natural drainage basins. The seasonal flooding and drainage of fields is required to ensure the availability of nutrients for intensive rice harvests. Shrimp aquaculture in coastal lagoons is accomplished by a similar practice of diversion and holding of water. At times following spawning, barrier beaches are opened to allow offshore water carry-ing shrimp larvae to enter. The lagoon is then sealed to create a rearing pond. High productivity in the shallow water column supports high growth rates of shrimp, which are harvested after several months.

Dam construction on large rivers accomplishes more than the creation of reservoirs for power production. Water diversion and flow regulation prevent downstream flooding during times of peak runoff. Water is then diverted from wetland areas, which depend on flooding for rejuvenation. Water is also held back in arid regions by construction of barrages on rivers. The Aswan High Dam, dis-cussed earlier, is only one of a series of dams on the Nile River, built after 1898 to provide water for irrigation canals that allowed harvesting of more than one crop per year in the arid climate of Egypt. Lake Nasser, like reservoirs elsewhere, provides a predictable supply of water throughout the year.

13.3.2 Lake fertilization

Freshwater aquaculture, practiced in China and Japan, consists of the addition of fertilizers in the form of sewage wastes to man-made ponds. This recycling of domestic wastes to enhance useable aquatic food resources is comparable to the agri-cultural application of livestock manure to crop lands.

The artificial eutrophication of natural waters, to increase biological production by direct application of fertilizers, has been attempted in a Scottish sea loch and in some lakes. The most complete documentation of stimulated production exists for Great Central Lake, Canada where about 100 tonnes of commercial fertilizer (ammonium nitrate and ammonium phosphate) were added annually over 3 years to determine if increased primary and secondary plankton production would lead to increased growth and survival of juvenile sockeye salmon resident in the lake (LeBrasseur *et al.*, 1974). The study assumed that the production and growth of sockeye smolts and zooplankton was limited by food supply.

During 3 years of fertilization, average summer phytoplankton production in Great Central Lake increased five times, and zooplankton biomass was nine times greater. The mean numbers of adult salmon 3 years after cessation of treatment was seven times greater than abundance over the pre-vious 14 years. However, numbers of sockeye in an adjacent unfertilized lake, which served as a control, also increased during the years of treatment, but increases were about 50% of those in Great Central Lake. Eutrophication, straying of spawning stock from Great Central Lake through a shared tributary and population expansion into previously under-utilized spawning sites could explain the increases in the control lake.

Table 13.3 Lake restoration techniques and their modes of operation. From Welch (1984)

Technique	Mode of operation
Reduction in nutrient inflow	
Waste-water treatment	Modified nutrient input
Waste-water/stormwater diversion	Reduced nutrient input
Treatment of agricultural land	Modified nutrient input
Treatment of inflow	Reduced nutrient input
Product modification (e.g. phosphate-free detergents)	Reduced nutrient input
Disruption of internal nutrient cycles	
Dredging	Nutrient removal, effective with high internal supply
Destratification/aeration	Decreases algal production by increased mixing
Hypolimnetic aeration	Correction of oxygen deficit
Nutrient inactivation	Addition of alum to cause precipitation and complex phosphorus
Bottom sealing	Decreased nutrient release
Lake-level manipulation	Aeration/consolidation of exposed shallow sediment
Biological cycling to higher trophic levels	Regulate fish predators/ increased zooplankton grazing to control algae
Acceleration of nutrient outflow	
Biotic harvesting	Removal of macrophytes
Selective (hypolimnetic) discharge	Removal of nutrient-rich deep water
Dilution/flushing	Nutrient and algal biomass removal

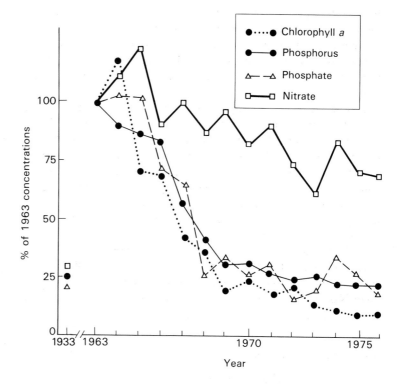

Fig. 13.6 Changes in algal biomass measured as chlorophyll *a* concentration, total dissolved phosphorus, phosphate and nitrate in Lake Washington, expressed as a percentage of values in 1963 following sewage diversion from the lake in that year. Data from Edmondson as summarized by Welch (1984).

13.3.3 Lake restoration

In lakes where impacts of eutrophication are reversible, reduction of dissolved nutrient supplies can lead to decreased phytoplankton production, increased water column transparency and a return of oxygen to previously anoxic deep water, which allows recolonization by benthic fauna and fish. The process is termed 'lake restoration'. Several techniques have been used (Table 13.3) that involve alteration of nutrient cycles, either by reducing input, disruption of internal cycling or accelerating outflow. For example, decreased phosphorus input will not be immediately effective in decreasing algal production in hypereutrophic aquatic systems, because of internal recycling; organically rich sediments in these lakes represent a long-term source of nutrient supply.

Excessive growth of aquatic macrophytes from eutrophication or lack of grazers, often associated with microalgal blooms, requires different treatment. Harvesting is effective on a small scale for clearing water bodies for aesthetic, navigational and recreational purposes. However, the net result on nutrient budgets within a lake or river will be small unless input is low, or if the removed biomass contains a significant fraction of nutrients in the system. If external loading is large relative to the nutrient content of the macrophytes, biomass removal will not reduce this source of nutrients.

One successful example of restoration by diversion of sewage influx was initiated in Lake Washington in 1963. By 1969, following a 75% reduction in phosphorus loading, suspended algal biomass (measured as chlorophyll *a* concentration), total dissolved phosphorus and phosphate decreased below the levels observed in 1933 (Fig. 13.6). Dissolved nitrate increased for 2 years after sewage diversion. However, positive correlations between total phosphorus, phosphate and chlorophyll *a* indicate that phosphorus was responsible for stimulating high algal biomass.

The response of reduced phytoplankton populations in Lake Washington to reduced phosphorus input was attributable to a high flushing rate (30% of the volume per year), its relatively great mean depth (32 m) and an aerobic hypolimnion that caused dissolved phosphate to precipitate with organic matter. Shallow lakes with a seasonal anoxic hypolimnion may also show a dramatic response to reduced loading. Phosphorus, released from sediments when stratification results in oxygen depletion, will stimulate phytoplankton production if it reaches the photic zone. This may be reduced under conditions of stratification, and sedimentation of particulate matter will further lower phosphorus concentrations in the water column. If the external supply of phosphorus is lowered, phytoplankton growth may be correspondingly reduced.

Waste-water treatment is a common practice for inland communities with limited sources of fresh water, and where there is no large body of water to receive sewage effluents. A common goal of this treatment is to reduce total dissolved phosphate to 1 mg P 1^{-1}, 10% of the level usually present in raw sewage. In Lake Shagawa, Minnesota, such treatment removed 85% of the incoming phosphorus (Welch, 1984). Within 2 years, chlorophyll a and total phosphorus in the lake decreased by 50% during the winter and spring, but there was little change in average summer chlorophyll concentrations. Although the period of observation is short, the internal phosphorus supply from anaerobic sediments during summer over 2 years was apparently sufficient to maintain algal biomass, and limit the improvement expected after sewage treatment.

13.3.4 Aquaculture

Aquatic systems can be used for cultivation, just as we use land to grow plant and animal crops. This process, aquaculture, or in marine environments mariculture, is the industrialized production of any aquatic organism. The industry has grown rapidly in recent years, with total global yields now equal to about 10% of the biomass from natural fisheries.

Aquaculture is described as *extensive* when husbandry is practiced for only part of the organism's life cycle, and natural food supplies are utilized. For example, the edible seaweed *Porphyra*, used for Nori production in Japan, is grown in coastal waters on ropes suspended from rafts.

Plants are exposed to natural light and dissolved nutrients. Animal species are also cultivated by allowing individuals to feed on suspended particulate matter, or prey naturally present in their habitats. Mussels and oysters are induced to settle on ropes or suspended cages hung from rafts. This technique is extensively used in Spain to cultivate the mussel *Mytilus edulis*.

Intensive aquaculture provides artificial diets for organisms cultured throughout their entire life cycle. Eggs, from trout in temperate climates and carp species in more tropical climates, are placed in ponds or small lakes that restrict emigration. Species such as trout and salmon can be grown in floating pens or cages. Salmon farming has developed rapidly in Norway, Scotland, on Canada's east and west coasts and in Alaska, using this technique. Food is supplied until these species reach commercial size, before harvesting.

Historically, hatchery programmes have produced juvenile fish for release into lakes and streams to supplement natural stocks. Both freshwater and anadromous species have been grown by rearing fertilized eggs in hatchery and pond systems (farms), constructed for cultivating fish. Sturgeon are propagated in Eurasian countries to replenish natural stocks that have disappeared from river and lake systems. For salmon in Japan, more than 90% of harvested individuals returning from the sea originated in hatcheries. Similarly, Atlantic salmon runs in eastern Canada are supplemented by releasing fry or fingerlings into river systems where small spawning stocks persist.

Fish farms must be operated with care to avoid disease, since bacterial, viral and parasitic infections can spread rapidly. The culture of dense populations of aquatic species in a restricted area can also have an environmental impact which has to be evaluated if high yields are to be maintained. For example, growth will not be maximized if overcrowding of suspension-feeding molluscs in raft cultures reduces food concentration. This can occur when the water residence time around culture sites is long, relative to the filtration rate of the species.

Cage cultures of finfish may also have environmental effects. Food capture by fish fed an artifical diet is not 100% efficient. Unconsumed food and faecal matter accumulate on the bottom under cages, decay and can cause anoxic conditions that kill benthic macrofauna. Antibiotics added to fish food to prevent disease can also accumulate, and could potentially alter the natural microbial flora in sediments and water around the culture site.

13.3.5 Artificial reefs

Anglers know that fish are often concentrated around submerged structures, because of protection from predators and increased food availablity from attached plants and animals. Natural bottom relief (shoals, escarpments and channels) provide refuges where fish congregate. Coral reefs are the best natural example of the development of similar, but far more complex, communities. Symbiosis between algae and coral tissues forms the biological basis which supports complex and diverse food webs.

Reef communities can be enhanced or created by placement of new structures into lake or coastal marine waters to increase local concentrations of fish and invertebrates. Concrete, rubber, plastic or other inert materials are usually used to create an artificial reef. The absence of chemical deterrents to attachment of organisms is desirable. Wood and metal surfaces are less suitable. Fungal decay or boring organisms, such as the ship worm (*Toredo*), cause wood to decay, unless it is preserved by anoxic sediments. Corrosion or oxidation of metal surfaces may slow colonization or favour species tolerant to precipitated metal ions.

Over time, a succession of organisms becomes attached to artificial reef structures. Small individuals capable of rapid growth, such as hydroids, bryozoans and cladocerans, are typically opportunistic species that are the first to colonize. With time, slower-growing larger organisms such as sponges, coelenterates (sea anenomes), amphipods, barnacles and attached molluscs (for example, the blue mussel *Mytilus*) are common marine colonizers. Other, large, slower-growing species such as fish, crustaceans and, in the ocean, echinoderms (starfish) and molluscs, complete the community succession.

13.4 RESOURCE PROTECTION

13.4.1 Monitoring change

Changes in aquatic systems and the organisms that they contain may occur because of natural or man-made factors. Detection of these changes requires repeated observations at scales dependent on the level of organization studied (Table 13.4). The response scales to disturbance differ for different levels of biological organization. For example, biochemical indicators of cellular stress due to a pollutant can be most adequately studied over a short time and on a cellular scale to observe toxic effects. On the other hand, acid rain effects in a lake may impact water, sediments and aquatic organisms, as well as surrounding terrestrial communities, soils and groundwater. Detection of changes at this level of ecosystem organization inevitably requires observations of interrelated physical and biological subsystems for several years.

Different monitoring techniques are used to quantify acute (short term) or chronic (long term)

Table 13.4 Time and space scales required to observe responses of biological systems of different levels of organization to physical or chemical stress. From Capuzzo (1981)

Biological level of organization	Time-scales for study	Space scales for study
Biochemical–cellular	Minutes–hours	nm–μm
Organism	Hours–months	μm–cm
Simulated community	Days–years	cm–m
Population dynamics	Months–decades	m–km
Community/ecosystem dynamics	Years–decades	km–global

effects, which may result in mortality or sublethal impacts on aquatic organisms. Bioassays consist of controlled exposure of organisms to a range of treatments. If effects are acute and cause death in a short time, threshold concentrations can be readily established. The concentration of a pollutant that causes 50% mortality during a specific time period (LC_{50}) provides a guideline for setting water quality standards.

Sublethal effects are usually more difficult to measure, since organisms do not die. These effects may decrease energy intake and reduce growth or reproduction. A weakened physiological condition can also make individuals more susceptible to predation, parasitism or disease. In such cases, sublethal effects may cause death indirectly. Since predators may remove any age class from a population, it may be difficult to determine the additional mortality due to a pollution stress. If death occurs over a short period of time, particularly with large animals, such as fish, the accumulated dead carcasses may serve as a qualitative measure of mortality. Analyses of tissues or the water body may show the cause of death, if a simple toxicant or pollutant is involved. The problem becomes more difficult in industrially-polluted environments, where mixtures of toxic compounds are present.

Morphological anomalies, such as skeletal deformation, and genetic disorders, such as tumours and chromosome damage, may be visible signs that organisms have experienced environmental stress. Pathological symptoms often allow the diagnosis of possible causes — for example, external body tumours on the ventral surface of flatfish may arise from exposure to carcinogenic hydrocarbons in the top layer of sediments. However, responses of organisms to environmental change through altered behaviour, growth or reproduction are less easily observed. Changes in an individual's energy budget may occur when less energy is available for growth and reproduction, because of reduced intake, decreased assimilation or increased respiration due to environmental stress. The excess energy available for growth in an individual, after losses due to respiration and excretion (scope for growth), has been used as an index of pollution stress in laboratory and natural populations.

Assessment of changes in community structure are possible in any aquatic system through observation of species abundance and biomass at different trophic levels. For example, monitoring programmes were established to evaluate trends in eutrophication and contaminant input to the North American Great Lakes, to assess results of legislation which decreased the input of dissolved nutrients and industrial discharges. Biomonitoring studies have also been effective in documenting the long-term impact of lake acidification. Observations of fish and crustaceans in lakes in Norway, Sweden and Canada showed the progressive loss of breeding fish populations due to stress by acidic precipitation.

In contrast to monitoring programmes, undertaken because of recognized or suspected environmental damage, are those with the aim of assessing present-day conditions. These are often baseline studies, against which future long-term changes can be assessed. The US 'Mussel Watch Program' conducted from 1976 to 1979 used the common blue mussel *Mytilus edulis* to monitor trace metals, hydrocarbons and radionuclides at selected coastal sites around the world. Contaminants accumulated in mussel tissues are derived from suspended matter ingested as food, as well as by direct absorption from fractions dissolved in sea water. Comparisons of concentrations of various pollutants between sites and over time were used to indicate trends in contaminant input to the populations studied.

Measurement of contaminants accumulated by one species does not indicate the impact on other species in the community. Although coexisting phytoplankton, zooplankton and fish populations have been used to determine impacts of aquatic contaminants, studies with benthic communities have also proven useful as indicators of pollutant effects. Benthic populations that colonize solid substrates or develop as infauna within sediments are usually sedentary. Although many species have pelagic larvae, they are quantified by standard methods, and susceptible to direct and indirect effects of contaminants, particularly those adsorbed to sediments.

These characteristics make benthic monitoring programmes common elements of many environmental impact assessments. While enumeration of the number of individual species of macrofauna or meiofauna in a specific sampling area may be time consuming, the relative frequency of species allows determination of diversity indices. Although many factors combine to determine biomass, species composition and relative abundance, the technique has proven useful for the detection of pollution gradients (see, for example, Figs 13.2 and 13.3). Changes over time or distance from a known source of contamination can be used to follow improvement or further degradation of environmental conditions. Additional measurements are needed to define the cause of any changes observed.

13.4.2 National regulations, international agencies and global agreements

This chapter has shown how man's activities degrade aquatic systems on small and large scales. Dissolved nutrients, added through local sewage discharge, for example, will be mixed throughout a lake or coastal region to stimulate phytoplankton production over a large area. In a lake or river that separates two or more countries, the impacts cross international boundaries, where eutrophication results from multiple sources of effluent discharge. In these cases, international cooperation is needed to solve problems of environmental degradation.

The International Joint Commission (IJC) between Canada and the USA, formed two decades ago, is an example of such a cooperative body. It was established to address transboundary issues, including water quality of the Great Lakes. These water bodies are used jointly for transportation, domestic and industrial water supplies, discharge of sewage wastes and electricity generation by hydroelectric and nuclear power plants. Monitoring programmes were established by each nation, with an agreement to meet regularly to set criteria for water quality and to report the progress in achieving them.

Similar international issues involve countries that share the common resource of the ocean for food, energy supplies and waste disposal. Agencies within the United Nations (UN) have been established over the past few decades to provide standards for global management of all environments in the world. The United Nations Environment Program (UNEP) was created in the early 1970s to assist all nations

to conserve and develop their resources, while ensuring environmental protection. Projects in many developing countries assess water resources and advise on methods for preserving environmental quality in the face of an urgent need to maximize industrial development. The Fisheries and Agriculture Organization (FAO), also within the UN, has a similar aim with respect to sustaining fisheries yields.

The International Atomic Energy Agency (IAEA) was established by the UN to promote the development of nuclear power for the benefit of man after the Second World War. In response to increasing international concern over fallout of radioactive material from atmospheric weapons testing during the 1950s, an international laboratory for monitoring radioactivity in the sea was established by the IAEA in Monaco. Observations of increasing amounts of global fallout, detected on land and in ocean sampling, confirmed the global distribution of contamination arising from these explosions, carried out by a few nations. The evidence led to a global test-ban treaty that most nations have supported for more than two decades.

The IAEA also implements the London Dumping Convention. Nations met in London in 1971 to sign an accord that banned the disposal of specific substances such as high-level nuclear wastes, carcinogenic hydrocarbons and certain toxic metals in the ocean. Other substances such as low-level radioactive waste from research laboratories and less toxic metals could be disposed of at sea under the guidelines administered by the IAEA. At present, there is a moratorium on dumping of low-level waste, pending international agreement about the environmental and legal aspects of the practice.

The Law of the Sea is a treaty, signed by 113 nations in 1983, to provide a legal framework for solving international disputes involving territorial waters, sea lanes and ocean resources. Under this agreement, territorial waters extend 19 km from a nation's coast, but vessels and aircraft are allowed free passage. Every nation has exclusive rights to fish and exploit oil, gas and other resources within a 360 km economic zone extending from their coasts. Several industrialized countries objected to provisions for sea-bed mining in areas outside of the economic zones of any nation and they did not sign the treaty. Minerals on the sea floor, under the agreement, are considered to be resources for the 'common heritage of mankind' to be shared between all nations, even those without the technology that enables them to harvest it. Fisheries resources beyond continental shelf economic zones of individual countries should also be shared. Disputes between nations under the treaty are settled by the International Court of Justice or by a new Tribunal for the Law of the Sea.

The formation of these, and other, UN agencies is a testimony for the need to draw the nations of the world together to solve common environmental problems. As we have learned more about man's impact on the earth's environments, the importance of interconnections between terrestrial, atmospheric and aquatic ecosystems for distributing materials across the surface of the earth has become clear. We also now realize that some impacts of human activity may extend to global scales. International cooperation is needed if solutions to these global environmental problems are to be found.

FURTHER READING

Capuzzo, J.M. (1981) Predicting pollution effects in the marine environment. *Oceanus* **24**: 25–33.

Davies, J.M., Bell, J.S. & Houghton, C. (1984) A comparison of the levels of hepatic aryl hydrocarbon hydroxylase in fish caught close to and distant from N. Sea oilfields. *Mar. Environ. Res.* **14**: 23–45.

Ernst, W. (1980) Effects of pesticides and related organic compounds in the sea. *Helgolander wiss. Meeresunters.* **33**: 301–12.

Goldberg, E.D. (1984) The oceans as waste space: the argument. *Oceanus* **24**: 4–9.

Ketchum, B.H. (1969) Eutrophication of estuaries. In: *Eutrophication: Causes, Consequences, Correctives*, pp. 197–209. National Academy Press, Washington.

LeBrasseur, R.J., McAllister, C.D., Barraclough, W.E., Kennedy, O.D., Manzer, J., Robinson, D. & Stephens, K. (1974) Enhancement of sockeye salmon (*Oncorhynchus nerka*) by lake fertilization in Great Central Lake: summary report. *J. Fish. Res. Bd. Can.* **35**: 1580–96.

National Academy of Sciences (1985) *Oil in the Sea. Inputs, Fates, and Effects.* National Academy Press, Washington.

Odum, E.P. (1971) *Fundamentals of Ecology.* Saunders, Philadelphia.

Odum, H.T. (1982) Pulsing, power and hierarchy. In: Mitsch, W.J., Bosserman, R.W., Ragade, R.H. & Dillon, J.A. Jr (eds) *Energetics and Systems.* Ann Arbor Science Publishers, Ann Arbor, Michigan.

Pearson, T.H. & Stanley, S.O. (1979) Comparative measurements of the redox potential of marine sediments as a rapid means of assessing the effect of organic pollution. *Mar. Biol.* **53**: 371–9.

Schindler, D.W. (1971) A hypothesis to explain differences and similarities among lakes in the Experimental Lakes Area, Northwestern Ontario. *J. Fish. Res. Bd. Can.* **28**: 295–301.

Schindler, D.W., Newbury, R.W., Beaty, K.G., Prokopowich, J., Ruszcznski, T. & Dalton, J.A. (1980) Effects of a windstorm and forest fire on chemical losses from forested watersheds and on the quality of receiving streams. *Can. J. Fish. Aquat. Sci.* **37**: 328–34.

Vollenweider, R.A. (1976) Advances in defining criticial loading levels for phosphorus in lake eutrophication. *Mem. Ist. Ital. Idrobiol.* **33**: 53–83.

Wadie, W.F. (1984) The effect of regulation of the Nile River discharge on the oceanographic conditions and productivity of the southeastern part of the Mediterranean Sea. *Acta Adriat.* **25**: 29–43.

Welch, E.B. (1984) Lake restoration results. In: Taub, F.B. (ed) *Ecosystems of the World 23*, pp. 557–71. Elsevier, Amsterdam.

Woodhead, D.S. (1973) Levels of radioactivity in the marine environment and the dose commitment to marine organisms. In: *Radioactive Contamination of the Marine Environment.* Proceedings of Symposium in Seattle, 10–14 July 1972, pp. 499–525. International Atomic Energy Agency, Vienna.

Index

Page numbers referring to figures are in *italic*; those which refer to tables are in **bold**.